Studies in Nonlinear Aeroelasticity

Earl H. Dowell   Marat Ilgamov

Studies in
Nonlinear Aeroelasticity

With 152 Illustrations

Springer-Verlag
New York Berlin Heidelberg
London Paris Tokyo

Earl H. Dowell
J.A. Jones Professor and Dean
School of Engineering
Duke University
Durham, NC 27706
USA

Marat Ilgamov
Vice Chairman of the Presidium
Kazan Branch of the USSR
Academy of Sciences
Kazan 420029
USSR

Library of Congress Cataloging-in-Publication Data
Dowell, E. H.
    Studies in nonlinear aeroelasticity / E.H. Dowel and M. Ilgamov.
        p.   cm.
    Includes bibliographical references.
    1. Aeroelasticity.   2. Nonlinear theories.   I. Il'gamov, M. A.
QA930.D66   1988
533'.62--dc19   88-12382

© 1988 by Springer-Verlag New York Inc.
Softcover reprint of the hardcover 1st edition 1988

Camera-ready copy provided by the authors.

9 8 7 6 5 4 3 2 1

ISBN-13:978-1-4612-8397-3   e-ISBN-13:978-1-4612-3908-6
DOI:10.1007/978-1-4612-3908-6

Preface

The great bulk of the literature on aeroelasticity is devoted to linear models. Theoretical work relies heavily on linear mathematical concepts, and experimental results are commonly interpreted by assuming that the physical model behaves in a linear manner. Nevertheless, significant work has been done in nonlinear aeroelasticity, and one may expect this trend to accelerate for several reasons: our ability to compute has increased at an astonishing rate; as linear concepts have been assimilated widely, there is a natural increase in interest in the foundations of nonlinear modeling; and, finally, some phenomena long recognized to be of interest, but beyond the effective range of linear models, are now known to be essentially nonlinear in nature.

In this volume, an exhaustive review of the literature is not attempted. Rather the emphasis is on fundamental ideas and a representative selection of problems.

Despite obvious successes in research on problems of aeroelasticity and the existence of a broad literature, including a number of excellent monographs, up to now little attention has been devoted to a general nonlinear theory of interaction. For the most part nonlinearity has been considered either solely in the description of the behavior of a shell or in the description of the motion of a gas. However, it is sometimes necessary to pose the problem of interaction with consideration of the nonlinear factors associated with boundary conditions on the contact surface of a shell and gas as well. This allows us to pose and solve new classes of problems and, in addition, to propose a basis for simplification in particular problems. It also indicates flaws in certain applied theories and ways of making these theories more precise. Just as in the equations for describing the motion of a shell and gas, consideration of nonlinearities under contact conditions provides not only a refinement of the results of the linear theory, but in particular cases leads to new qualitative results. Such a formulation of the problem is also included in the present work.

Some problems considered here are of a pure methodological character and illustrate the application of the general theory; some problems elucidate physical phenomena; other problems have a direct practical interest.

In Chapter 0, the subject of nonlinear aeroelasticity is defined, and an overview is provided. Various generic topics are discussed qualitatively including static and dynamic nonlinearities, with particular reference to the literature on flutter of an airfoil, flutter of plates and shells, beamlike bluff bodies in a flowing fluid, airfoils oscillating in a separated flow, and aeroelasticity of rotorcraft.

In Chapter I, the fundamentals of continuum mechanics are reviewed. Specifically considered are methods for describing motion and fluid structure interaction, fundamental relations of the nonlinear theory of thin elastic shells, fundamental equations of hydromechanics, and conditions on the surfaces of large discontinuities.

In Chapter II, a general formulation and a classification of interaction problems of a shell and fluid are given. The discussion treats kinematic and dynamic conditions at a contact surface, reduction of contact conditions to the original surface, classification of aeroelasticity problems, interaction for large bending of a shell, interaction for medium bending of a shell, interaction for small bending of a shell, and interaction between a plate and a viscous fluid.

Chapter III, turns to the first class of aeroelastic problems, the behavior of a cylindrical shell or curved plate with axis transverse to an incompressible flow. Discussed here are static bending of a cylindrical shell, dynamic instability of a cylindrical shell, behavior of a curved plate, separated flow, and some results of experiments.

In Chapter IV, the difficult problem of interaction of a fluid with a permeable shell is considered. Such a physical model is of interest with respect to parachute dynamics, for example.

Chapter V introduces the topic of self excited, nonlinear oscillations of airfoils, beams, plates, and shells in incompressible and compressible flow. Typically, bodies in high speed flows are more slender and streamlined with respect to the oncoming flow. In this case their geometry may be treated more simply. However, the added complexities of a compressible fluid provide more than enough challenge to the aeroelastician. The three major topics considered are: effect of structural nonlinearities of plates and shells on dynamic instability (flutter), fluid mechanical nonlinearities and bluff body oscillators, and aerodynamic nonlinearities (in transonic flow) and their effect on airfoil flutter.

The last of these is considered more fully in Chapter VI. The first is explored in a broader context in Chapter VII.

In Chapter VII, several related problems involving chaotic oscillations in mechanical systems are considered that deserve additional attention from research workers. Of these problems, the two discussed in some detail are free and forced response of beams (plates and shells) in the presence of fluid flow and turbulence within the framework of the Lorenz equations as a nonlinear chaotic, pseudo limit cycle oscillation.

In Chapter VIII, the literature on the effects of compliant walls on transition and turbulence is critically reviewed. This is one of the most difficult and challenging subjects in aeroelasticity and is still imperfectly understood. Here an effort is made to provide a conceptual framework for the phenomena of interest and to state as clearly as possible our present understanding of them.

This book has three appendices which contain important background discussions relevant to the book's central themes. Appendix A discusses observation and evolution of chaos for an autonomous system. Appendix B discusses an approximate method for calculating the vortex—induced oscillation of bluff bodies in air and water. Appendix C discusses unsteady separated flow models.

This book combines two major fields of study, fluid and structural dynamics, and is written by co-authors from two distinct cultures. It is hoped that the technical and cultural interactions will have proven useful to the reader as well as to the co-authors, and a challenge to all of us for the future.

Acknowledgments

The first author would like to acknowledge the generous support over many years of several organizations and individuals. These include:

AFOSR: Anthony Amos
ARO: Gary Anderson
NASA Langley Research Center: Herman Bohon, John Mixson,
Clemans A. Powell, Wilmer H. Reed III, and Harry Runyan
NASA Lewis Research Center: Gerald Brown and Krishna R.V. Kaza
NSF: Elbert Marsh
ONR: Michael Reischman

He would also like to acknowledge the stimulation provided by his long time colleagues and mentors, Holt Ashley, John Dugundji, Francis Moon, and Herbert Voss. To those many students and now colleagues from whom and with whom he learned so much, your citations in the references will formally thank you for your essential contributions.

Contents

List of Symbols for Chapters I-IV

x_1, x_2, x_3	Euler variables
$\alpha_1, \alpha_2, \alpha_3$	Lagrange variables
x, θ, r	Cylindrical coordinates
t	Time
R, h	Radius and thickness of shell
$\vec{r}$	Radius vector
$\vec{v}, v_1, v_2, v_3$	Velocity vector and its components
$\vec{u}, u_1, u_2, u_3,$ u, v, w	Displacement vector and its components
ρ, ρ_0	Density of fluid
p, p_0	Pressure
$\rho_\infty, p_\infty, V_\infty$	Parameters of flow at infinity
γ, γ_+	Adiabatic coefficient (Chapters I, II), parameter of hydrostatic load (Chapter III) and parameter of permeability (Chapter IV)
a_0, a_∞	Sound velocity
$\varepsilon, \varepsilon_{ij}$	Small quantity, relative elongations and shear (Chapters I, II), coefficient of perforation (Chapter IV)
χ, χ_{ij}	Change of curvature and twist
$\vec{K}_1, \vec{K}_2, \vec{K}_3$	Unit vectors at $t = 0$

$\vec{K}_1^*, \vec{K}_2^*, \vec{K}_3^*$	Unit vectors at $t > 0$
K_{ij}, K_{ij}^*	Curvatures of coordinate lines
e_{ij}, ω_k, E_i	See (2.3), (2.4), Chapter I
H_j, H_j^*	Lamé coefficients at $t = 0$ and $t > 0$ on coordinates α_1, α_2
$\left.\begin{array}{l} N_{ij}, Q_k, M_{ij} \\ N_{ij}^*, Q_k^*, M_{ij}^* \end{array}\right\}$	Internal forces and moments
B, D	Rigidities of tension and of bending
E, ν, ρ_s	Elastic modulus, Poisson's ratio, and density of shell
X_i, X_i^*	Volume and surface forces on shell
R_m, r_n, R_m^*, r_n^*	Components of external forces of nonaerodynamical character
h_i	Lamé coefficients on Euler coordinates
$\vec{F}, \tilde{U}, \tilde{U}'$	Vector of volume forces and internal energy
$\tilde{\sigma}, \tilde{\sigma}'$	Entropy (Chapters I, IV)
ϕ, ϕ_∞	Velocity potential
ψ	Stream function, disturbance of potential
μ_0	Coefficient of viscosity
μ, μ_+, μ_{++}	Parameter of aeroelasticity (Chapter III)

$\vec{Z}, Z_i, Z_i^*, \tilde{W}$ Body forces and energy in the flow

$L_m, L_m^*, \ell_n, \ell_n^*$ Differential operators of shell theory

H_ν, h_μ Differential operators of hydromechanics

$(M), (M^*)$ Points of space

$\vec{V}, V_i$ Approximate values of velocity (Chapter II), velocity of fluid flow across the shell (Chapter IV)

m, n, k, λ, μ Numbers for estimate of the shell deformation and fluid motion (Chapter II)

M_∞ Mach number

P_ϕ(or p_1), P_ψ(or p_2)   Pressure on a rigid shell and its disturbance due to displacement of the shell

$c, \bar{\lambda}$ Velocity of wave propagation in the shell and the wavelength

W_n, A_n, B_n Amplitudes of shell deflection and of velocity potential

n, s, m Indices for sums (Chapters III, IV)

λ, ν, X See (3.7), (3.11) (Chapter III)

p_i Pressure within the shell

k, σ See (4.15), Chapter III

θ_0 Mean angle of attack of curved plate

θ_s Separation angle of flow

U, U_n Dimensionless curvature of cylindrical shell and its amplitudes

C	Constant of integration [see (6.5), (6.6), Chapter III]
$\bar{p}_0, q$	See (6.9), Chapter III
Re, C_p	Reynolds number and pressure coefficient
V_i, V_3^*	Approximate velocity components at the point M* (Chapter II), components of velocity of fluid flow across the shell (Chapter IV)
$\vec{a}_1, \vec{a}_2, \vec{a}_3$	Unit vectors (see (1.11), Chapter IV)
d, b, α, β	See (2.1), Chapter IV
S_0, S	Hole area and general area of perforated plate
k	See (2.3), Chapter IV
$\varepsilon, \varepsilon_m$	Coefficients [see (2.8), (2.9), Chapter IV]
$\tilde{\mu}$	Coefficient of rigidity
d, ℓ	Hole diameter and distance between holes
p_{02}	See (4.3), Chapter IV
ϕ_1, ϕ_2, p_1, p_2	Potential and pressure from opposite sides of shell (Chapter IV)

Chapter 0

NONLINEAR AEROELASTICITY: AN OVERVIEW

Summary

A conceptual framework is provided for the field of nonlinear aeroelasticity. Various important physical phenomena are identified, and the contents of the present book are placed in context.

0.1 Introduction

For many years, <u>linear</u> models, both theoretical and experimental, have served the aeroelastician extraordinarily well. Most of our understanding of aeroelastic phenomena, such as divergence, flutter, control surface reversal, and gust response, has been obtained by the study of such models. These models are lucidly presented in the now classic texts by Scanlan and Rosenbaum [1], Fung [2], Bisplinghoff [3,4], Ashley [3,4] and Halfman [3], and more recently by Försching [5] and Dowell, Curtiss, Scanlan, and Sisto [6]. Indeed as Ashley [7] has noted, (linear) aeroelasticity is entering the mainstream of aerospace engineering and is now routinely discussed within the framework of broader (or narrower depending on your perspective) fields such as flight mechanics or vehicle dynamics and control. Indeed, perhaps the last research frontier of <u>linear</u> aeroelasticity has been the use of feedback control systems to stabilize or modify the behavior of aeroelastic systems, e.g., increase the flutter speed or reduce the response to gusts [8]. Of course, the ultimate last frontier of any linear system study is optimization. And a good deal of work has been done on this as well, especially within the framework of optimal control of aeroelastic systems and the optimization of structural weight or stiffness distributions [9]. The advent of composite materials has opened up a rich array of possibilities in the latter area.

In this book we consider a different research frontier for aeroelasticity, that is, the study of nonlinear aeroelastic systems. Of course, it has been known for many years that nonlinear effects in aeroelasticity may be important. Indeed "nonlinear" effects are often cited as a possible explanation for any difference between theory and experiment. Many times these nonlinear effects are small, hence the major successes of linear models. However, sometimes nonlinear effects are more important, and, occasionally they are crucial.

For example, it has been known for many years that plate or shell flutter [10] leads to limit cycle motions of small amplitude (typically on the order of the plate or shell thickness); thus, the flutter motion does not usually lead to immediate catastrophic failure. This is unlike wing flutter where the nonlinear effects normally are small, the flutter motions are large (on the order of a fraction of the wing span), and failure is usually immediate. Control surface flutter may also be of limited amplitude due to actuator nonlinearities [11,12,13].

Over the last several decades, the transonic flow regime has been penetrated by aircraft and missile designs and by the aeroelastician with theoretical and experimental models. It was long thought that nonlinear aerodynamic effects are important in the transonic flow regime and, for this reason, theoretical progress was much slower there than in the modeling (by linear theory) of the subsonic and supersonic flow regimes. However, finite difference methods for solving nonlinear partial differential equations and the development of a distinct field of study known as "computational fluid dynamics" have provided an effective means for calculating the aerodynamic forces on airfoils and wings which are oscillating in transonic flow [14]. Indeed, a pleasant surprise has been the degree to which a transonic (not to say classical subsonic or supersonic) linear model still proves useful to the aeroelastician. However, it was only by studying the nonlinear models that this discovery was made [15].

Of course, there are other aeroelastic phenomena that have defied rational modeling, at least until recently, and often this is thought to be due to important nonlinear effects. Among those that are of significant interest are:

- oscillations of bluff bodies in a flowing stream [16]
- large oscillations of streamlined bodies where the flow may separate during at least a portion of the cycle of oscillation [17]
- various structural damping mechanisms including hydraulic and dry friction damping [11,12,13,18,19].

Some progress has been made in understanding each of these.

Several of these nonlinear phenomena and associated models will be discussed in the following.

0.2   Categorization of Generic Nonlinear Effects

It is helpful, before proceeding to the particular discussion of physical phenomena, to distinguish among and define three classes of linear and/or nonlinear models.

- Fully Linear Models

    Here the statically deformed structural shape of the body and the steady flow field about the body may be assumed to have trivial solutions, i.e., the static shape of the body does not significantly influence its dynamic response and the steady flow field deviation from a uniform stream flow due to the presence of the body does not change appreciably the unsteady, time-dependent aerodynamic forces that act upon the body when it is set into oscillatory motion. Moreover, the dynamic motions are sufficiently small that there are no significant nonlinear effects due to body motion.

In this case, both the static equilibrium problem (usually trivially so) and the dynamic motion problem may be treated by linear models.

● Dynamically Linear Models

Here the statically deformed shape of the body and/or the steady flow field about the body do influence the subsequent dynamic response. For example, if a plate is (statically) buckled from its initial flat configuration, this may significantly change its dynamic response from that for the initially flat plate [10]. Another example is the dynamic aeroelastic behavior of a rotor blade which is often significantly modified by its statically deformed shape [20].

An aerodynamic example is that of an airfoil in transonic flow. The (rigid or statically deformed) shape of the airfoil determines the strength and location of steady shock waves. These shock waves, which result from the steady flow field, have a decisive effect on the unsteady aerodynamic forces that act upon the oscillating airfoil [15,21].

By definition, of course, if the behavior of an aeroelastic system is to be modeled as dynamically linear, the dynamic motion must remain sufficiently small or to say it another way, we must be content to study those phenomena that involve small dynamic motions.

For this class of models, the static problem must be treated by a nonlinear model, but the dynamic motion about this static equilibrium may be treated by a linear dynamic model. The solution, in sequence, of first the nonlinear static model and then the dynamically linear model usually has major conceptual and computational advantages over a fully nonlinear model.

- Fully Nonlinear Models

    For this category of models, the dynamic motions themselves are so large that a nonlinear dynamic model must be used. Hence, usually there is no advantage to solving the static and dynamic problems separately and sequentially. Examples in this category are

    - determination of limit cycle oscillations of either the body or fluid

    - study of deliberately large motions to achieve a desired effect, e.g., large control surface oscillations or large wing and body motions (supermaneuvarability)

    - deliberate (or inadvertent) use of dynamically nonlinear devices, e.g., dry friction or hydraulic dampers.

    In what follows, both dynamically linear and fully nonlinear models will be discussed. Fully linear models will not be discussed.

0.3 The Earlier Literature

    There is, of course, a vast array of publications in aeroelasticity on fully linear models. This will not be discussed here; the reader is referred to the cited references. A recent discussion of that literature is contained in Reference 6. A review of much of the linear and nonlinear model literature on plate and shell flutter is contained in Reference 10. See also References 22 and 23. The discussion of plate and shell flutter presented in what follows will be confined for the most part to the more recent work on chaotic oscillations. The subject of unsteady transonic aerodynamics and aeroelasticity has been reviewed in Reference 15. The present discussion of this subject will be focused on the most recent findings on flutter analysis techniques for fully nonlinear models and the significance (or lack thereof) of multiple aerodynamic solutions in the transonic flow regime.

The reader may find it helpful to review References 6, 10, and 15 before reading the present discussion.

References

1. Scanlan, R. H. and Rosenbaum, R., Introduction to the Study of Aircraft Vibration and Flutter, The Macmillan Company, New York, NY, 1951. Also available in Dover Edition.

2. Fung, Y. C., An Introduction to the Theory of Aeroelasticity, John Wiley and Sons, Inc., New York, NY, 1955. Also available in Dover Edition.

3. Bisplinghoff, R. L., Ashley, H. and Halfman, R. L., Aeroelasticity, Addison-Wesley Publishing Company, Cambridge, MA, 1955.

4. Bisplinghoff, R. L. and Ashley, H., Principles of Aeroelasticity, John Wiley and Sons, Inc., New York, NY, 1962. Also available in Dover Edition.

5. Försching, H. W., Fundamentals of Aeroelasticity, In German. Springer-Verlag, Berlin, 1974.

6. Dowell, E. H., Curtiss, H. C., Jr., Scanlan, R. H. and Sisto, F., A Modern Course in Aeroelasticity, Sijthoff and Noordhoff, the Netherlands, 1980.

7. Ashley, H., The Constructive Uses of Aeroelasticity, Polish Academy of Sciences, Engineering Transactions, Vol. 30, No. 3-4, 1982, pp. 369-396.

8. Many Authors, Proceedings of the Aeroservoelastic Specialists Meeting, AFWAL-TR-84-3105, Vol. I, II, October 1984.

9. McIntosh, S. C. and Ashley, H., On the Optimization of Discrete Structures with Aeroelastic Constraints, Computers and Structures, Vol. 8, 1978, pp. 411-419.

10. Dowell, E. H., Aeroelasticity of Plates and Shells, Noordhoff International Publishing, Leyden, 1975.

11. Breitbach, E., Effects of Structural Nonlinearities on Aircraft Vibration and Flutter, AGARD Report R-665, 1978.

12. Laurenson, R. M. and Trn, R. M., "Flutter Analysis of Missle Control Surfaces Containing Structural Nonlinearities," AIAA J., Vol. 18, 1980, pp. 1245-1251.

13. Lee, C. L., "An Iterative Procedure for Nonlinear Flutter Analysis," AIAA J., Vol. 24, 1986, pp. 833-840.

14. Jameson, A., The Evolution of Computational Methods in Aerodynamics, J. Applied Mechanics, Vol. 50, No. 4, Dec. 1983, pp. 1052-1070.

15. Dowell, E. H., Unsteady Transonic Aerodynamics and Aeroelasticity, in Recent Advances in Aerodynamics and Aeroacoustics, Springer-Verlag, New York, 1986.

16. Dowell, E. H., Nonlinear Oscillator Models in Bluff Body Aeroelasticity, J. Sound Vibration, Vol. 75, 1981, pp. 251-264.

17. Chi, R. M. and Srinivasan, A. V., Some Recent Advances in the Understanding and Prediction of Turbomachine Subsonic Stall Flutter, ASME Paper No. 84-GT-151, 1984.

18. Dowell, E. H., Damping in Beams and Plates Due to Slipping at the Support Boundaries, J. Sound Vibration, Vol. 105, 1986, pp. 243-253.

19. Tang, D. -M. and Dowell, E. H., Nonlinear Dynamics of a Helicopter Model in Ground Resonance, J. Am. Helicopter Soc., Vol. 32, No. 1, 1987, pp. 45-53.

20. Hodges, D. H. and Ormiston, R. A., Stability of Hingeless Rotor Blades in Hover and Pitch-Link Flexibility, AIAA J., Vol. 15, No. 4, April 1977, pp. 476-482.

21. Bland, S. R. and Edwards, J. W., Airfoil Shape and Thickness Effects on Transonic Airloads and Flutter, J. Aircraft, Vol. 21, No. 3, March 1984, pp. 209-217.

22. Bolotin, V. V., Nonconservative Problems of the Theory of Elastic Stability, Pergamon Press, 1963.

23. Volmir, A. S., Shells in a Flowing Liquid and Gas, Nauka, Moscow, 1976. (In Russian.)

Chapter I

SOME BACKGROUND FROM CONTINUUM MECHANICS

Introduction

Insofar as aeroelasticity deals with compatible (joint) fluid and deformed solid body motions and also with their interaction, then it is appropriate to begin their study by considering the general principles of continuum mechanics. Moreover, aeroelasticity may be considered as one of the richest domains of continuum mechanics from the point of view of the display of general principles and of the manner for describing the motion of a deformed body.

In continuum mechanics one may choose either the Euler or the Lagrange points of view. However, in solving the problems of deformed solid body mechanics, one commonly uses only the Lagrange variables, while, for fluid mechanics problems, mainly the Euler variables. In aeroelasticity problems one often has to use both methods of motion description, since the formulation of contact conditions between a liquid and an elastic body includes both systems of variables. Thus, a more detailed knowledge of the peculiarities of both methods for the motion description and the transitions between them is required than in the classical study of fluid or solid mechanics.

The reader may be acquainted already with such aspects of mechanics as kinematics and dynamics of a deformed medium, hydromechanics and nonlinear shell theory. As is known there are many excellent references on these questions. For example, one can recommend References 1-4.

However, for the convenience of the reader of this volume, in this chapter we present some basic information on relationships and equations from the above mentioned sources. Subsequently, we shall make use time and again of these references in later chapters.

1.1. Methods for describing motion and fluid-structure interaction

In the Euler method of motion description, coordinates x_1, x_2, x_3 are fixed in space and at these fixed points the medium parameters (displacement, velocity, acceleration, strain, stress, pressure etc.) are determined in the course of time. Coordinates x_1, x_2, x_3 and the time t are the Euler independent variables.

Coordinates of medium points at the initial instant (moment) t = 0, we denote by x_1^0, x_2^0, x_3^0 or α_1, α_2, α_3. At later times (t > 0) these fixed points retain their numerical values. Coordinates α_1, α_2, α_3 and time t are called the Lagrange variables. In this method of motion description, the parameters are determined at individual moving points of the medium.

For example, we consider the bending of a thin bar (rod, pin). In its undeformed state (t = 0) marks are placed at points $\alpha_1^{(0)}$, $\alpha_1^{(1)}$, $\alpha_1^{(2)}$, $\alpha_1^{(3)}$ (Fig. 1).

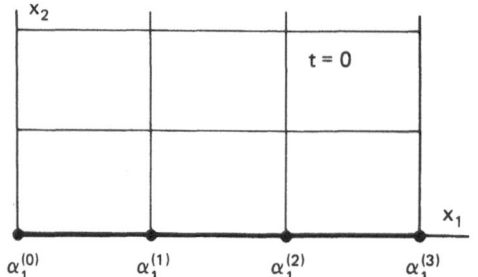 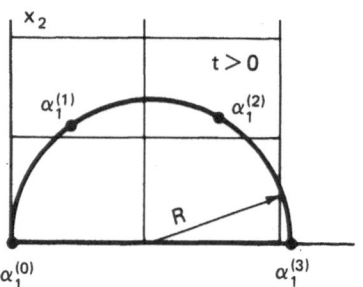

Fig. 1. Lagrange coordinate line along a thin rod.

For t > 0, assume that the bar takes the form of semicircle with a radius R. Numerical values of the Lagrange coordinates $\alpha_1^{(0)}$, $\alpha_1^{(1)}$, $\alpha_1^{(2)}$, $\alpha_1^{(3)}$ are retained even though their positions in space and the distances between them are changed. If at the initial moment the Euler coordinates are equal to $x_1 = \alpha_1$, $x_2 = 0$, then for t > 0, as is seen from Fig. 1,

$$x_1 = R(1 - \cos \frac{\alpha_1}{R}), \qquad x_2 = R \sin \frac{\alpha_1}{R} \qquad (1.1)$$

For further bending of a bar its ends may be superimposed. Thus, one point in space coincides with two body points having the Lagrange coordinates $\alpha_1(0)$ and $\alpha_1(3)$.

Here the most important property of the Lagrange coordinates is to distort together with the medium under its deformation as is shown in Fig. 1. This system is "frozen" to the body, therefore one says that it is the material system with the accompanying coordinate system.

In the general case of the three-dimensional medium motion, the motion law has the form

$$x_i = x_i(\alpha_1, \alpha_2, \alpha_3, t) \qquad (1.2)$$

instead of (1.1). Solving (1.2) for coordinates α_1, α_2, α_3, we may obtain

$$\alpha_i = \alpha_i(x_1, x_2, x_3, t) \qquad (1.3)$$

Velocity and acceleration are calculated for the fixed medium particle with respect to the reference system, the Euler coordinates

$$\vec{v} = \left(\frac{\partial \vec{r}}{\partial t}\right)_{\alpha_i}, \qquad \vec{a} = \left(\frac{\partial \vec{v}}{\partial t}\right)_{\alpha_i}$$

in which $\vec{r}(x_1, x_2, x_3, t)$ is the particle radius-vector and indices α_i indicate that the derivatives are calculated at constant values of the Lagrange coordinates α_1, α_2, α_3 for the particle analyzed.

The individual (fundamental) derivative with respect to t of the scalar function, $f(\alpha_1, \alpha_2, \alpha_3, t)$, which depends on the Lagrange variables, is equal to $(\partial f/\partial t)_\alpha$. If the function is expressed in terms of the Euler variables, $f(x_1, x_2, x_3, t)$, then taking into account that according to (1.2) $f[x_1(\alpha_1, \alpha_2, \alpha_3, t),\ldots]$, we obtain

$$\left(\frac{\partial f}{\partial t}\right)_{\alpha_i} = \left(\frac{\partial f}{\partial t}\right)_{x_i} + \frac{\partial f}{\partial x_1}\left(\frac{\partial x_1}{\partial t}\right)_{\alpha_i} + \ldots$$

$$= \left(\frac{\partial f}{\partial t}\right)_{x_i} + v_1\frac{\partial f}{\partial x_1} + v_2\frac{\partial f}{\partial x_2} + v_3\frac{\partial f}{\partial x_3}$$

(1.4)

It should be noted that $(\partial f/\partial t)_{\alpha_i}$ is considered as the complete (whole) derivative df/dt, while $(\partial f/\partial t)_{x_i}$ is the partial derivative $\partial f/\partial t$.

In order to consider other derivatives, we demonstrate the transition of the continuity and ideal compressible liquid motion one-dimensional equations from the Euler variables $(x_1 = x, \alpha_1 = \alpha)$

$$(\partial\rho/\partial t)_\alpha + \rho\partial v/\partial x = 0,$$

$$\rho(\partial v/\partial t)_\alpha + \partial p/\partial x = 0$$

(1.5)

to the Lagrange variables. Introducing the displacement u of a fixed liquid particle, instead of (1.2) we may write

$$x = \alpha + u(\alpha,t)$$

(1.6)

Furthermore, we have

$$\frac{\partial p}{\partial x} = \frac{\partial p}{\partial \alpha}\frac{\partial \alpha}{\partial x}, \qquad \frac{\partial v}{\partial x} = \frac{\partial v}{\partial \alpha}\frac{\partial \alpha}{\partial x}$$

The derivatives $\partial\alpha/\partial x$, $\partial v/\partial\alpha$ entering here are determined from (1.6)

$$\frac{\partial \alpha}{\partial x} = \frac{\partial(x-u)}{\partial x} = 1 - \frac{\partial u}{\partial \alpha}\frac{\partial \alpha}{\partial x}, \qquad \frac{\partial \alpha}{\partial x} = \left(1 + \frac{\partial u}{\partial \alpha}\right)^{-1},$$

$$\frac{\partial v}{\partial \alpha} = \frac{\partial}{\partial \alpha}\left(\frac{\partial x}{\partial t}\right)_\alpha = \frac{\partial}{\partial \alpha}\left(\frac{\partial u}{\partial t}\right)$$

Therefore,

$$\frac{\partial p}{\partial x} = \frac{\partial p}{\partial \alpha} (1 + \frac{\partial u}{\partial \alpha})^{-1}, \qquad \frac{\partial v}{\partial x} = \frac{\partial^2 u}{\partial \alpha \partial t} (1 + \frac{\partial u}{\partial \alpha})^{-1}$$

Taking into account, moreover, that

$$(\frac{\partial \rho}{\partial t})_\alpha = \frac{\partial \rho}{\partial t}, \qquad (\frac{\partial v}{\partial t})_\alpha = [\frac{\partial}{\partial t} (\frac{\partial x}{\partial t})]_\alpha = \frac{\partial^2 u}{\partial t^2},$$

we reduce equations (1.5) to the form

$$(1 + \frac{\partial u}{\partial \alpha}) \frac{\partial \rho}{\partial t} + \rho \frac{\partial^2 u}{\partial \alpha \partial t} = 0,$$

$$\rho(1 + \frac{\partial u}{\partial \alpha}) \frac{\partial^2 u}{\partial t^2} + \frac{\partial p}{\partial \alpha} = 0$$

Let us divide the first of these equations by $\rho(1 + \partial u/\partial \alpha)$ and then integrate with respect to t. Equating an integrating constant to the value of the undisturbed liquid density ρ_0, we find the continuity equation in Lagrange variables to be

$$(1 + \partial u/\partial \alpha)\rho = \rho_0 \qquad (1.7)$$

Then the motion equation takes the form

$$\rho_0 \partial^2 u/\partial t^2 + \partial p/\partial \alpha = 0 \qquad (1.8)$$

If one uses the adiabatic gas law

$$p/p_0 = (\rho/\rho_0)^\gamma \qquad (1.9)$$

then the following nonlinear equation ($a_0^2 = \gamma p_0/\rho_0$) may be obtained:

$$\frac{\partial^2 u}{\partial t^2} + \frac{a_0^2}{(1 + \partial u/\partial \alpha)^{\gamma + 1}} \frac{\partial^2 u}{\partial \alpha^2} = 0 \qquad (1.10)$$

Now we consider methods for the description of the interaction of a flexible shell and an ideal liquid. In the shell theory one usually applies the Lagrange variables, and, in hydromechanics, mainly the Euler variables; thus, in the interaction problem it is natural to use these traditional methods. The solutions for both the shell and the liquid must be consistent on their contact surface. Such a method of interaction description may be called an Euler-Lagrange one.

Let the shell be placed in a fluid flow (Fig. 2). The shell fixed point with a Lagrange coordinate $\alpha_1^{(1)}$ coincides for $t = 0$ with a point $(x_1^{(1)}, x_2^{(1)})$ in space. As a result of the deformation $(t > 0)$ it moves to a point $(x_1^{(2)}, x_2^{(2)})$ in space.

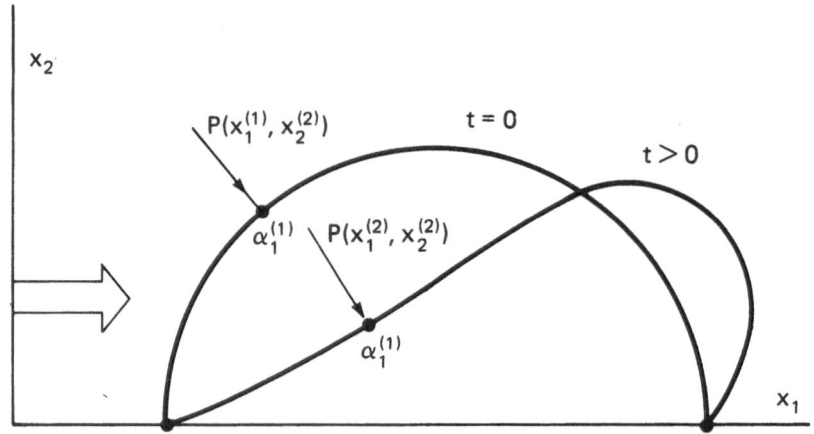

Fig. 2. Bending of the thin shell in a fluid flow.

Equations for the shell motion are constructed for the moving fixed elementary volume of the shell at a point $\alpha_1^{(1)}$ by projection of all the forces, acting on this volume,

$$L_m(\alpha_1^{(1)},t) = \delta_{2m}p(x_1^{(2)}, x_2^{(2)},t) \quad (m = 1,2) \quad (1.11)$$

Here L_m is the nonlinear differential operator of the shell theory in Lagrange variables, while the pressure $p(x_1, x_2, t)$ is a function of the Euler variables, $\delta_{21} = 0$, $\delta_{22} = 1$.

As a point $(x_1^{(2)}, x_2^{(2)})$ is determined from the problem solution, then, consequently, one has to find the pressure at an unknown point in space where the shell point having the coordinate $\alpha_1^{(1)}$ arrives.

The same is true for the determination of the liquid velocity. Its value should be known in order to satisfy kinematic condition on the contact surface. Thus, the boundary domain, in which one has to find the solution of a hydromechanics problem, is not known beforehand. The satisfaction of the contact surface conditions is one of the principal complexities of concrete problems that are solved by an Euler-Lagrange method.

The situation is somewhat simplified, if the shell bending is comparatively small, so that the domain occupied by the liquid changes little after the contour deformation. Then the fluid pressure and velocity at a point $(x_1^{(2)}, x_2^{(2)})$ may be approximately expressed by the parameters determined from the known point $(x_1^{(1)}, x_2^{(1)})$, by using an analytical extension of functions and their decomposition in a Taylor series.

It should be noted that for many cases in writing contact conditions one can equate the deformed and undeformed, initial surfaces. These are mainly the cases for small bending of the shell and for small change of the pressure and velocity fields in a liquid. For example, this is usually done in the classical divergence and flutter studies of lifting surfaces.

The Euler-Lagrange method of interaction description and its application are analyzed in Chapters II, III, and IV. The fact that in many cases the known hydromechanics solutions may be used we consider as this method's advantage.

If the motion of both interacting mediums is described by Lagrange variables, then this method is the united Lagrange method of interaction description. More general is the arbitrary Lagrange-Euler method of interaction description. The latter two methods we do not consider in this book.

1.2.   Fundamental relations of the nonlinear theory of thin elastic shells

We summarize some information [3] from the theory of elastic shells based on the assumption that the strains and the relative thickness of the shell may be neglected in comparison with unity, i.e., assuming that

$$h\chi/2 < \varepsilon_p \ll 1, \; \varepsilon < \varepsilon_p, \; h/R \ll 1,$$

where ε_p is the limit of proportionality of the shell material; h is thickness; R is the smallest radius of curvature of the middle surface of the shell; ε, χ are the largest values of the relative elongations and shear, ε_{ij}, and of the change of curvature and twist, χ_{ij}. In the general case, the values of the displacements are not limited per se.

Also, the Kirchoff-Love hypothesis is used in which the perpendicular to the middle surface σ before deformation remains perpendicular to the deformed surface and at the same time normal stresses perpendicular to σ are small in comparison with the stresses tangential to the surface parallel to σ.

Displacements of points of the middle surface are determined by the vector

$$\vec{u} = u_1\vec{K}_1 + u_2\vec{K}_2 + u_3\vec{K}_3, \tag{2.1}$$

in which $\vec{K}_1$, $\vec{K}_2$, $\vec{K}_3$ are the unit vectors tangent to the orthogonal curvilinear Lagrange lines α_1, α_2, α_3. Unit vectors $\vec{K}_1$, $\vec{K}_2$ lie in the plane tangential to the middle surface of the shell before deformation and $\vec{K}_3$ is directed along the outward normal.

Unit vectors corresponding to coordinates α_1, α_2, α_3 of the deformed middle surface, $\vec{K}_1^*$, $\vec{K}_2^*$, $\vec{K}_3^*$, are expressed in terms of the vectors $\vec{K}_1$, $\vec{K}_2$, $\vec{K}_3$ by the formulae [3]

$$\vec{K}_1^* = (1 + e_{11})\vec{K}_1 + e_{12}\vec{K}_2 + \omega_1\vec{K}_3,$$

$$\vec{K}_2^* = e_{21}\vec{K}_1 + (1 + e_{22})\vec{K}_2 + \omega_2\vec{K}_3, \qquad (2.2)$$

$$\vec{K}_3^* = E_1\vec{K}_1 + E_2\vec{K}_2 + E_3\vec{K}_3$$

Here one has introduced the following definitions:

$$e_{11} = \frac{\partial u_1}{H_1\partial\alpha_1} + \frac{u_2}{H_1H_2}\frac{\partial H_1}{\partial\alpha_2} + K_{11}u_3,$$

$$e_{12} = \frac{\partial u_2}{H_1\partial\alpha_1} - \frac{u_1}{H_1H_2}\frac{\partial H_1}{\partial\alpha_2} + K_{12}u_3, \qquad (2.3)$$

$$\omega_1 = \frac{\partial u_3}{H_1\partial\alpha_1} - K_{11}u_1 - K_{12}u_2$$

The remaining expressions for e_{22}, e_{21}, ω_2 (the quantities which are not specified here) may be obtained by permutation of the indices 1 and 2. In (2.3) K_{ij} are the curvatures of coordinate lines α_1, α_2 and the twists, H_1, H_2 are the Lamé coefficients. Values for E_i have the form

$$E_1 = e_{12}\omega_2 - (1 + e_{22})\omega_1,$$

$$E_2 = e_{21}\omega_1 - (1 + e_{11})\omega_2, \qquad (2.4)$$

$$E_3 = (1 + e_{11})(1 + e_{22}) - e_{12}e_{21}$$

As the relative elongations and shears are small and the wall is thin, the vectors $\vec{K}^*_1$, $\vec{K}^*_2$, $\vec{K}^*_3$ remain orthogonal. The meanings of the quantities e_{ij}, ω_i, E_i are seen from the formulae

$$\vec{K}^*_i \cdot \vec{K}_i = \cos(\vec{K}^*_i \, \vec{K}_i) = 1 + e_{ii}, \quad \cos(\vec{K}^*_i \, \vec{K}_j) = e_{ij}(i \neq j)$$

$$\cos(\vec{K}^*_i \, \vec{K}_3) = \omega_i, \quad \cos(\vec{K}^*_3 \, \vec{K}_i) = E_i, \quad \cos(\vec{K}^*_3 \, \vec{K}_3) = E_3$$

which are obtained from (2.2) by the scalar multiplication on $\vec{K}_i$. Consequently, parameters e_{ij}, ω_i, E_i characterize angles of rotation of the unit vectors during the shell deformation.

Relative elongations and shears equal

$$\varepsilon_{11} = e_{11} + \frac{1}{2}(e^2_{11} + e^2_{12} + \omega^2_1),$$

$$\tag{2.5}$$

$$2\varepsilon_{12} = e_{12} + e_{21} + e_{11}e_{21} + e_{22}e_{12} + \omega_1\omega_2$$

while the curvature changes are as follows:

$$x_{11} = K_{11}e_{22} - K_{12}e_{21}$$

$$- \frac{1}{H_1}\left(E_1 \frac{\partial e_{11}}{\partial \alpha_1} + E_2 \frac{\partial e_{12}}{\partial \alpha_1} + E_3 \frac{\partial \omega_1}{\partial \alpha_1}\right) - \frac{\omega_2}{H_1 H_2}\frac{\partial H_1}{\partial \alpha_2},$$

$$\tag{2.6}$$

$$x_{12} = K_{12}e_{11} - K_{11}e_{21}$$

$$- \frac{1}{H_1}\left(E_1 \frac{\partial e_{21}}{\partial \alpha_1} + E_2 \frac{\partial e_{22}}{\partial \alpha_1} + E_3 \frac{\partial \omega_2}{\partial \alpha_1}\right) + \frac{\omega_1}{H_1 H_2}\frac{\partial H_1}{\partial \alpha_2}$$

Formulae (2.5) characterize the change of dimension shape of the middle surface element, while (2.6) characterize its twisting.

Internal forces and moments are connected with ε_{ij}, χ_{ij} by the relations

$$N_{11} = B(\varepsilon_{11} + \nu\varepsilon_{22}), \quad N_{12} = B(1-\nu)\varepsilon_{12} \qquad (B = \frac{Eh}{1-\nu^2})$$

$$\text{(2.7)}$$

$$M_{11} = D(\chi_{11} + \nu\chi_{22}), \quad M_{12} = D(1-\nu)\chi_{12} \qquad (D = \frac{Eh^3}{12(1-\nu^2)})$$

in which B, D are the rigidities of tension-compression and of bending. E, ν are the elastic modulus and Poisson's ratio.

The quantities not specified here may be obtained from (2.5), (2.6), (2.7) by permutation of the indices 1 and 2.

Vector equations of motion may be presented in a scalar form by projections on the axes α_1, α_2, α_3 of the shell deformed state (unit vectors $\vec{K}^*_1$, $\vec{K}^*_2$, $\vec{K}^*_3$) and for its initial state (unit vectors $\vec{K}_1$, $\vec{K}_2$, $\vec{K}_3$). The forces and moments components for the first case are denoted by $(N^*_{11},\ldots)$. Thus, we have [3]

$$\frac{\partial(H^*_2 N^*_{11})}{\partial\alpha_1} + \frac{\partial(H^*_1 N^*_{21})}{\partial\alpha_2} + N^*_{12}\frac{\partial H^*_1}{\partial\alpha_2} - N^*_{22}\frac{\partial H^*_2}{\partial\alpha_1}$$

$$+ H_1 H_2(Q^*_1 K^*_{11} + Q^*_2 K^*_{12} + X^*_1) = 0,$$

$$\frac{\partial(H^*_1 N^*_{22})}{\partial\alpha_2} + \frac{\partial(H^*_2 N^*_{12})}{\partial\alpha_1} + N^*_{21}\frac{\partial H^*_2}{\partial\alpha_1} - N^*_{11}\frac{\partial H^*_1}{\partial\alpha_2} \qquad \text{(2.8)}$$

$$+ H_1 H_2(Q^*_2 K^*_{22} + Q^*_1 K^*_{12} + X^*_2) = 0,$$

$$\frac{\partial(H_2 Q^*_1)}{\partial\alpha_1} + \frac{\partial(H_1 Q^*_2)}{\partial\alpha_2}$$

$$- H_1 H_2(N^*_{11} K^*_{11} + N^*_{22} K^*_{22} + 2N^*_{12} K^*_{12} - X^*_3) = 0$$

Here X_i^* are the projections of both the volume and surface forces,

$$K_{ij}^* = K_{ij} + X_{ij}, \qquad H_i^* = H_i(1 + \varepsilon_{ii}) \qquad (2.9)$$

In some terms of (2.8) one retains the coefficients H_i^*, although $\varepsilon_{ii} \ll 1$. This is done because the substitution for H_i^* by the coefficients H_i may in some cases lead to a loss of accuracy [3].

The moment equations have the form

$$\frac{\partial(H_2 M_{11}^*)}{\partial\alpha_1} + \frac{\partial(H_1 M_{21}^*)}{\partial\alpha_2} + M_{12}^* \frac{\partial H_1}{\partial\alpha_2} - M_{22}^* \frac{\partial H_2}{\partial\alpha_1} - H_1 H_2 Q_1^* = 0,$$

$$(2.10)$$

$$\frac{\partial(H_1 M_{22}^*)}{\partial\alpha_2} + \frac{\partial(H_2 M_{12}^*)}{\partial\alpha_1} + M_{21}^* \frac{\partial H_2}{\partial\alpha_1} - M_{11}^* \frac{\partial H_1}{\partial\alpha_2} - H_1 H_2 Q_2^* = 0$$

Projections of the main vectors and main moments of the outward forces on the axis before deformation (unit vectors $\vec{K}_1$, $\vec{K}_2$, $\vec{K}_3$), are denoted by N_{ij}, Q_i, M_{ij}, X_i. The corresponding equations of motion have the same form (2.8), (2.10), if instead of N_{ij}^*, M_{ij}^*, Q_i^*, H_i^*, K_{ij}^*, X_i^*, these values are taken without stars.

Between the projections mentioned, there is the obvious connection

$$\vec{K}_1 N_{11} + \vec{K}_2 N_{12} + \vec{K}_3 Q_1 = \vec{K}_1^* N_{11}^* + \vec{K}_2^* N_{12}^* + \vec{K}_3^* Q_1^*,$$

$$\vec{K}_1 X_1 + \vec{K}_2 X_2 + \vec{K}_3 X_3 = \vec{K}_1^* X_1^* + \vec{K}_2^* X_2^* + \vec{K}_3^* X_3^*, \dots$$

Therefore, according to (2.2) the following equality, for example, holds:

$$X_1 = (1 + e_{11})X_1^* + e_{12}X_2^* + E_1 X_3^*$$

1.3. Fundamental equations of hydromechanics

We consider these equations using the Euler variables, as we shall consider the joint Euler-Lagrange method for description of the liquid and the shell interaction. First of all we write the equations of conservation of mass, of momentum, and of energy in an integral form [1,2].

We denote the density, the internal energy, the vectors of velocity, and the volume and surface forces by symbols ρ, $\tilde{U}$, $\vec{v}$, $\vec{F}$, $\vec{p}_n$. The arbitrary movable deformed volume containing the same particles of liquid is denoted by V, while the surface which surrounds this volume is denoted by $\sum$. Then a formula for differentiation of the integral taken over the movable volume has the form

$$\frac{d}{dt} \int_V f(x_1, x_2, x_3, t)dV = \int_V [\frac{\partial f}{\partial t} + \vec{\nabla}(\vec{v}f)]dV,$$

$$(3.1)$$

$$\vec{\nabla} = \frac{\vec{K_1}\partial}{h_1 \partial x_1} + \frac{\vec{K_2}\partial}{h_2 \partial x_2} + \frac{\vec{K_3}\partial}{h_3 \partial x_3},$$

in which h_1, h_2, h_3 are the Lamé coefficients.

The laws of conservation of mass, momentum, and energy for the individual volume have the form

$$\frac{d}{dt} \int_V \rho dV = 0,$$

$$\frac{d}{dt} \int_V \rho \vec{v} dV = \int_V \rho \vec{F} dV + \oint_\Sigma \vec{P}_n d\Sigma, \qquad (3.2)$$

$$\frac{d}{dt} \int_V \rho(\tfrac{1}{2} v^2 + \tilde{U}) dV = \int_V \rho \vec{F} \cdot \vec{v} dV + \oint_\Sigma \vec{P}_n \cdot \vec{v} d\Sigma$$

In the case of an ideal liquid $\vec{P}_n = -n\vec{p}$, where $p > 0$ is the pressure. We usually take the volume forces to be zero $(\vec{F} = 0)$.

As a result of the application of formulae (3.1), (3.2), we obtain the differential equations describing the motion of an ideal liquid:

$$\frac{\partial \rho}{\partial t} + \vec{\nabla} \cdot (\rho \vec{v}) = 0, \qquad (3.3)$$

$$\frac{\partial \vec{v}}{\partial t} + \vec{v} \cdot \vec{\nabla} \vec{v} + \frac{1}{\rho} \vec{\nabla} p = 0 \qquad (3.4)$$

$$\frac{\partial \tilde{U}}{\partial t} + \vec{v} \cdot \vec{\nabla} \tilde{U} + \frac{p}{\rho} \vec{\nabla} \cdot \vec{v} = 0 \qquad (3.5)$$

The last of these is equivalent to the equation

$$\partial \tilde{\sigma}/\partial t + \vec{v} \cdot \vec{\nabla} \tilde{\sigma} = 0 \qquad (3.6)$$

which follows from the first law of thermodynamics and (3.5). Here $\tilde{\sigma}$ is the entropy. From (3.6) it follows that the entropy is conserved for the moving individual particle.

The equation of state

$$p = f(\rho, \tilde{\sigma}) \qquad (3.7)$$

completes the system (3.3) - (3.5). For a barotropic
process, (3.7) has the form p = f(ρ), in particular,

$$p/p_\infty = (\rho/\rho_\infty)^\gamma \qquad (3.8)$$

in which γ is the adiabatic curve coefficient and p_∞,
ρ_∞ are the values of the pressure and the density at
infinity.

The second term in (3.4) may be represented as

$$\vec{v}\cdot\nabla\vec{v} = \vec{\nabla}(\tfrac{1}{2}v^2) - \vec{v} \times rot\ \vec{v}$$

For irrotational motions ($\nabla\times\vec{v} = 0$), one introduces the
velocity potential φ

$$\vec{v} = \nabla\phi \quad (\vec{v}_\infty = \nabla\phi_\infty), \qquad (3.9)$$

and the system of equations (3.3), (3.4), (3.8) may be
reduced to a form ($a_\infty^2 = \gamma p_\infty/\rho_\infty$)

$$a^2\nabla^2\phi = \frac{\partial^2\phi}{\partial t^2} + [\frac{\partial}{\partial t} + \frac{1}{2}(\vec{\nabla}\phi\cdot\vec{\nabla})](\vec{\nabla}\phi)^2,$$

$$\qquad (3.10)$$

$$a^2 = a_\infty^2 - (\gamma-1)[\frac{\partial\phi}{\partial t} + \frac{1}{2}(\vec{\nabla}\phi)^2 - \frac{1}{2}(\vec{\nabla}\phi_\infty)^2]$$

$$\frac{p}{p_\infty} = \{1 - \frac{\gamma-1}{a_\infty^2}[\frac{\partial\psi}{\partial t} + \frac{1}{2}(\vec{\nabla}\psi)^2 - \frac{1}{2}(\vec{\nabla}\psi_\infty)^2]\}^{\frac{\gamma}{\gamma-1}} \qquad (3.11)$$

In the case of an ideal incompressible liquid
($\rho = \rho_\infty$ = const, $a_\infty = \infty$), instead of (3.10) and (3.11)
the following equations are obtained:

$$\nabla^2\psi = 0, \quad p = p_\infty - \rho_\infty[\frac{\partial\psi}{\partial t} + \frac{1}{2}(\vec{\nabla}\psi)^2 - \frac{1}{2}(\nabla\psi_\infty)^2] \qquad (3.12)$$

Motion of a viscous incompressible liquid is described by Navier-Stokes equations

$$\vec{\nabla}\vec{v} = 0,$$

$$\frac{\partial \vec{v}}{\partial t} + \vec{v}\cdot\vec{\nabla}\vec{v} = -\frac{1}{\rho}\vec{\nabla}p + \frac{\mu_0}{\rho}\nabla^2\vec{v}, \qquad (3.13)$$

in which μ_0 is the coefficient of viscosity. These usually will not be employed here.

1.4. Conditions on the surfaces of large discontinuities

If on the surface σ the velocity vector $\vec{v}$, or the pressure, density, or internal energy p, ρ, $\tilde{U}$ undergo a discontinuity, while outside the σ they are continuous, then such a fluid flow is called a motion with a large discontinuity. The surface σ is to be understood as a movable and, in general, permeable shell surface. In view of the small thickness h of its wall, σ is identified with a shell middle surface. The discontinuities of the flow parameters cannot be arbitrary. They must satisfy certain conditions.

For their construction one proceeds from the integral conservation law (3.2). We suppose that the movable volume V includes the elementary section $\Delta\sigma$ of the surface σ and a small thickness $h_0/2$ of the layer from each side of σ. Proceeding to the limit as $h_0 \to 0$, $\Delta\sigma \to 0$ the kine-matic equality [1]

$$\lim_{\substack{\Delta\sigma\to0 \\ h_0\to0}} \frac{1}{\Delta\sigma}\frac{d}{dt}\int_V f\,dV = f\vec{v}\cdot\vec{K}_3^* - f'\vec{v}'\cdot\vec{K}_3^*, \qquad (4.1)$$

holds in which $\vec{K}_3^*(\alpha_1, \alpha_2, t)$ is the outward normal to deformed surface $\sigma(t)$ connected with a principal basis in accordance with (2.2) and f, $\vec{v}$ and f', $\vec{v}'$ are the values of functions from different sides of this surface.

All the integrand functions in the surface integrals on $\sum$ have finite values, but they are different on the different sides of σ. Furthermore, we accept that for $h_0 \to 0$ the limit equalities

$$\lim_{h_0 \to 0} \int_V \rho \vec{F} dV = \int_\sigma \vec{Z} d\sigma,$$

$$(4.2)$$

$$\lim_{h_0 \to 0} \int_V \rho \vec{F} \cdot \vec{v} dV = \int_\sigma \tilde{W} d\sigma$$

hold, in which $-\vec{Z}$, $-\tilde{W}$ are the outward oriented medium forces and the energy in the flow. It is obvious that if $\rho\vec{F}$, $\rho\vec{F} \cdot \vec{v}$ are finite in volume V, then $\vec{Z} = 0$, $\tilde{W} = 0$. For any continuous field of volume forces (for example, gravity), the case mentioned is realized.

On the basis of (4.1) and (4.2), from conservation laws (3.2) we obtain

$$\rho(\vec{v}\cdot\vec{K_3^*}) = \rho'(\vec{v}'\cdot\vec{K_3^*}),$$

$$\vec{Z} + \vec{p_{K_3^*}} - \rho\vec{v}(\vec{v}\cdot\vec{K_3^*}) = \vec{p_{k_3^*}'} - \rho'\vec{v}'(\vec{v}'\cdot\vec{K_3^*}),$$

$$\vec{Z}\cdot\vec{v} + \tilde{W} + \vec{p_{K_3^*}}\cdot\vec{v} - \rho(\frac{1}{2}\,v^2 + \tilde{U})(\vec{v}\cdot\vec{K_3^*})$$

(4.3)

$$= \vec{p_{K_3^*}'}\cdot\vec{v}' - \rho(\frac{1}{2}\,v'^2 + U')(\vec{v}'\cdot\vec{K_3^*})$$

Conditions on the discontinuities (4.3) for the actual values of the parameters from two sides of the discontinuity surface σ allow one to determine $\vec{Z}$ and $\tilde{W}$ of the flow about the shell.

Conditions (4.3) are given in a coordinate system connected with the surface σ. Sometimes it is more convenient to deal with relations written for the fixed Euler coordinate system. Velocity of motion of the shell fixed point in this system is equal to $\partial\vec{u}/\partial t$, where $\vec{u}(\alpha_1, \alpha_2, t)$ is the vector of displacement (2.1). Its normal component equals $(\partial\vec{u}/\partial t)\cdot\vec{k}^*$, therefore, in (4.3) instead of the values $(\vec{v}\cdot\vec{K_3^*})$, $(\vec{v}'\cdot\vec{K_3^*})$, the differences $(\vec{v}-\partial\vec{u}/\partial t)\cdot\vec{K_3^*},...$will appear.

For the case of an ideal liquid $(\vec{p_{K_3^*}} = -p\vec{K_3^*})$ and the fixed coordinate system, conditions (4.3) take the form

$$\rho(\vec{v} - \partial\vec{u}/\partial t)\cdot\vec{K_3^*} = \rho'(\vec{v}' - \partial\vec{u}/\partial t)\cdot\vec{K_3^*}, \qquad (4.4)$$

$$p\vec{K}_3^* + \rho v[(\vec{v} - \partial \vec{u}/\partial t)\cdot\vec{K}_3^*] - \vec{Z}$$

$$= p'\vec{K}_3^* + \rho'v'[(\vec{v}' - \partial \vec{u}/\partial t)\cdot\vec{K}_3^*], \tag{4.5}$$

$$\rho\vec{v}\cdot\vec{K}_3^* + \rho[(\vec{v} - \partial \vec{u}/\partial t)\cdot\vec{K}_3^*](\tfrac{1}{2}v^2 + \tilde{U})$$

$$- \tilde{W} - \vec{Z}\cdot(\vec{v} - \partial \vec{u}/\partial t) \tag{4.6}$$

$$= p'\vec{v}'\cdot\vec{K}_3^* + \rho'[(\vec{v}' - \partial \vec{u}/\partial t)\cdot\vec{K}_3^*](\tfrac{1}{2}v'^2 + \tilde{U}')$$

Here the hydrodynamics parameters from both sides of the surface (ρ, $\vec{v}$, ρ',..) are functions of Euler coordinates, while the parameters $\vec{K}_3^*$, $\vec{u}$ concerning the shell are functions of Lagrange coordinates. These relations have numerous applications in mechanics.

System (4.4) - (4.6) is not complete for problems of interaction with the shell. One also has to consider the physical law for the surface σ and its carrying capacity. This law depends on its porosity, pressure differential, and many other factors. These questions are considered in Chapter IV.

We mention only one limiting case here. When

$$\vec{v}\cdot\vec{K}_3^* = (\partial \vec{u}/\partial t)\cdot\vec{K}_3^* \tag{4.7}$$

the fluid does not flow through the shell. From (4.4) it follows that the densities from two sides of the shell undergo an arbitrary discontinuity ($\rho \neq \rho'$). Condition (4.5) reduces to an equality

$$\vec{Z} = (p - p')\vec{K}_3^* \tag{4.8}$$

References

1. Sedov, L. I.   Continuum mechanics, Fizmatgiz, Moscow, 1970, p. 492.

2. Hodge, P. G.   Continuum mechanics, McGraw-Hill Book Co., 1970.

3. Mushtari, Kh.M., and Galimov, K. Z., Non-linear theory of thin elastic shells, NSF-NASA, Washington, 1961, p. 374.

4. Dowell, E. H., Curtiss, H. C., Scanlan, R. H. and Sisto, F., A modern course in aeroelasticity, Sitjhoff-Noordhoff, Leyden, The Netherlands, 1979.

CHAPTER II

GENERAL FORMULATION AND CLASSIFICATION OF INTERACTION PROBLEMS OF A SHELL WITH A FLUID FLOW

Introduction

Using previously known information we can formulate a general nonlinear problem of aeroelasticity. First of all we establish the contact conditions for the case of an impermeable surface and an ideal fluid. Interaction of an impermeable shell with a liquid or gas can be described relatively simply and this represents an advantage in practice over the more complicated problem of a permeable shell. The theory of permeable shells interacting with a fluid will be considered in Chapter IV. Furthermore, a procedure for introducing the conditions at the deformed surface to the original surface is applied.

Our classification of the problems of aeroelasticity is based on estimates of the magnitudes of the normal component of shell displacement, rotations of its linear elements, and rates of change of the shell displacement. From these parameters general expressions are proposed for the case of large bending of shells, when the normal displacement has an order of its linear size and also in the case of medium bending (in particular, for a shallow shell) and of small bending (linear theory). The behavior of the fluid is taken into account near the shell.

Classification is in terms of orders of magnitudes and qualitative estimates. Therefore supplementary simplifications of the general expressions may be appropriate in specific problems.

Note that the classification introduced is not comprehensive. Further considerations involve the types of boundary conditions that may occur. Experience with solution of particular problems will allow one to penetrate into the essence of the phenomena considered.

Thus, in this Chapter the discussion is about contact conditions, whose formulation bears on problems of structural-fluid interaction. As far as boundary support conditions along its edges conditions for the fluid at infinity, symmetry, etc., they are established in the context of specific problems.

At the end of the Chapter the conditions at the contact surface are formulated for the case of a viscous fluid and a simple example is considered, showing the most important nonlinear conditions.

2.1. Kinematic and dynamic conditions at a contact surface

The shell motion is described by the Lagrange method. The corresponding orthogonal curvilinear coordinate system (α_1, α_2) coincides with the shell middle surface, and α_1, α_2 are directed along the lines of principal curvature of the surface. These coordinate lines α_1, α_2 are deformed with the shell middle surface; fixed points of this surface have the same numerical values for the coordinates α_1, α_2 when the surface moves and deforms. Unit vectors along these directions are $\vec{K}_1$, $\vec{K}_2$ and the external normal to the surface is $\vec{K}_3$ (Fig. 1). Unit vectors $\vec{K}_1^*$, $\vec{K}_2^*$, $\vec{K}_3^*$ of the deformed surface are expressed in terms of unit vectors $\vec{K}_i$ of the undeformed surface by the formulae (2.2) of the preceding chapter. In particular,

$$\vec{K}_3^* = E_1 \vec{K}_1 + E_2 \vec{K}_2 + E_3 \vec{K}_3 \qquad (1.1)$$

Shell motion equations obtained by projecting all the forces in the directions $\vec{K}_1$, $\vec{K}_2$, $\vec{K}_3$ are introduced in Chapter I, Section 1.4. Let us introduce these in symbolic form

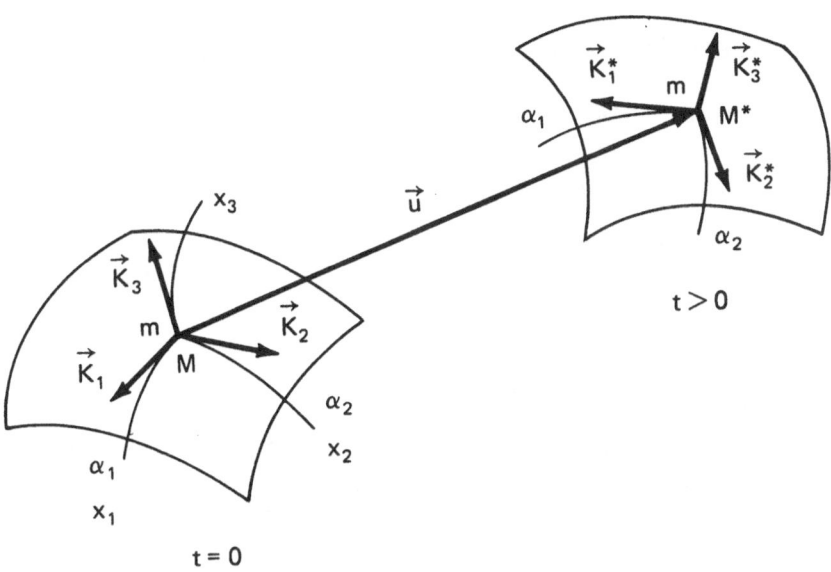

Fig. 1. Two positions of the shell element in a
        space.

$$L_m(u_1, u_2, u_3) = R_m + Z_m \qquad (m = 1,2,3)$$
$$\ell_n(u_1, u_2, u_3) = r_n \qquad (n = 1,2,3,4)$$

(1.2)

Here u_1, u_2, u_3 are projections of the displace-
ment vector: $u_i = \vec{K}_i \cdot \vec{u}$, L_m are nonlinear differential
operators of shell theory; R_m are projections of ex-
ternal forces of nonaerodynamic character (support
forces, own weight, etc.); Z_m are forces from the
fluid flow; ℓ_n are differential operators, defined on
the shell boundary edges; and r_n are displacements,
strains, or forces, applied to these edges. In the
variant of shell theory based on the Kirkhoff-Love hypothesis, numbers m, n are varied from 1 to 3 and from
1 to 4. If other models of shell theory are used,
then only these numbers m, n are changed in (1.2).

The equations of shell theory (2.8), (2.10)
[Chapter I], obtained by projection of all forces in

the directions of $\vec{K}^*_1$, $\vec{K}^*_2$, $\vec{K}^*_3$, may also be introduced
in symbolic form as

$$L^*_m (u_1, u_2, u_3) = R^*_m + Z^*_m \qquad (m = 1,2,3)$$

$$\ell^*_n (u_1, u_2, u_3) = r^*_n \qquad (n = 1,2,3,4) \tag{1.3}$$

Here R^*_m, Z^*_m are the projections indicated above of the
forces in the direction $\vec{K}^*_m$. In this variant, projections of the displacement vector u on $\vec{K}_m$ (t = 0) also
are used.

All variables, included in (1.2), (1.3) besides
Z_m and Z^*_m, are functions of the coordinates α_1, α_2.

We describe the fluid motion by the Euler method.
Fixed in space the Euler coordinates are x_1, x_2, x_3.
Accordingly, the equations for an ideal compressible
fluid [Chapter I] are introduced in the form

$$H_\nu(v_1, v_2, v_3, p, \rho) = Q_\nu \qquad (\nu = 1,\ldots,5),$$

$$h_\mu(v_1, v_2, v_3, p, \rho) = q_\mu \qquad (\mu = 1,\ldots,5) \tag{1.4}$$

Here H_ν are the differential operators of hydromechanics and Q_ν are given quantities (e.g., volume
forces). Operators h_μ and quantities q_r are defined
on the boundaries of volume occupied by the fluid,
which are not on the contact surface with the shell.

Now let us write the contact conditions on the
deformed shell surface. Since the shell wall thickness is small, we identify the contact surface with
the middle surface of the shell.

We assume that the curvilinear Euler coordinates
x_1, x_2 coincide with the Lagrange coordinates α_1, α_2
in the undeformed state of the shell (at t = 0) and

direct a coordinate x_3 along the external normal of the shell (Fig. 1). At the initial point of time ($t = 0$), a fixed point m of the shell with Lagrange coordinates α_1, α_2 coincides with point M of space with Euler coordinates $x_1 = \alpha_1$, $x_2 = \alpha_2$, $x_3 = 0$. At an arbitrary time $t > 0$ of the shell motion the point m coincides with point M* of space.

The Euler coordinates of point M* will have other values, while the Lagrange coordinates of point m will have the same numerical values at the moments $t = 0$ and $t > 0$. Hence, for $t > 0$ the coordinates α_1, α_2 of the fixed point of the shell and the coordinates x_1, x_2 do not coincide. However, we formulate the contact conditions at point M* of space, since the shell motion velocity $\partial \vec{u}/\partial t$ and the force on it from the fluid, $\vec{Z}$, are determined for the shell by the fixed point m coinciding with M*.

For the impermeable shell from (4.7) of Chapter I a kinematic condition is written as

$$\vec{v} \cdot \vec{K}^*_3 = (\partial \vec{u}/\partial t) \cdot \vec{K}^*_3 \qquad (M^*) \qquad (1.5)$$

which means an equality of normal components of the fluid and the shell motion velocities. Here and below the symbol (M*) indicates that the condition is imposed at point M* of space with a radius vector $\vec{r} + \vec{u}$, where $\vec{r}$ is a radius vector of the point M.

To reduce (1.5) to scalar form we use (1.1) and introduce into the analysis the components of the shell displacement and the fluid velocity in the local bases $\vec{K}_1$, $\vec{K}_2$, $\vec{K}_3$:

$$\vec{u} = u_1\vec{K}_1 + u_2\vec{K}_2 + u_3\vec{K}_3,$$

$$\vec{v} = v_1\vec{K}_1 + v_2\vec{K}_2 + v_3\vec{K}_3$$

(1.6)

Substituting (1.6) into (1.5) gives

$$(\frac{\partial u_1}{\partial t} - v_1) E_1 + (\frac{\partial u_2}{\partial t} - v_2) E_2 \qquad (1.7)$$

$$+ (\frac{\partial u_3}{\partial t} - v_3) E_3 = 0 \quad (M*)$$

This condition may also be given in the form

$$v_3^* = \frac{\partial u_1}{\partial t} E_1 + \frac{\partial u_2}{\partial t} E_2 + \frac{\partial u_3}{\partial t} E_3 \quad (M*) \qquad (1.8)$$

From (4.8) of Chapter I the dynamic conditions in the case of an impermeable shell acquire the form

$$\vec{Z} = (p - p')\vec{K_3^*} \quad (M*) \qquad (1.9)$$

where p, p' are pressures from two sides of the shell. Hence,

$$Z_1^* = Z_2^* = 0, \ Z_3^* = p - p' \quad (M*) \qquad (1.10)$$

From (1.9), (1.1) we also obtain

$$Z_1 = (p - p') E_1, \ Z_2 = (p - p') E_2, \qquad (1.11)$$

$$Z_3 = (p - p') E_3 \quad (M*)$$

Consequently, application of the variant of equations (1.3), (1.10) is more advantageous because of the greater simplicity of the right-hand sides of (1.3) than application of the variant of (1.2), (1.11), where none of the components Z_m are equal to zero.

The system of equations and contact conditions (1.2), (1.4), (1.7), (1.11) or the system (1.3), (1.4), (1.7), (1.10), is a mathematical formulation of the problem of aeroelasticity.

We note once again that in these systems, the parameters u_i, E_j are functions of Lagrange coordinates α_1, α_2 and v_n, p, ρ, are functions of Euler coordinates x_1, x_2, x_3. By the physical formulation of the problem, the quantities R_m, r_m are most often functions of α_1, α_2 and the quantities Q_ν, q_μ are functions of x_1, x_2, x_3.

With the application of numerical methods of integration, when the solution is found at each step in time, the formulation of the problem of interaction given above is complete. Here the position of the shell and the point M* may be identified with their position, which was found at previous steps in time, because of the smallness of the shell displacement during this time.

If the solution is carried out with the aid of analytical methods, then conditions (1.5), (1.7), (1.10), (1.11) must be introduced in such a form where the implicit functions $\vec{v}(\vec{r} + \vec{u})$, $p(\vec{r} + \vec{u})$ would be absent, but the functions $\vec{v}(\vec{r})$, $p(\vec{r})$ at point M with known Euler coordinates would be present. This question is covered in the following section.

2.2. Reduction of contact conditions to the original surface

We assume that the velocity vector $\vec{v}$ and the pressure p in the fluid are functions that may be analytically continued (extended) in the neighborhood of the point M of space, with which the fixed point m of the shell coincides at the initial moment of its motion. Then the implicit functions $\vec{v}(\vec{r} + \vec{u})$, $p(\vec{r} + \vec{u})$ may be approximately expressed by $\vec{v}(\vec{r})$, $p(\vec{r})$ in the neighborhood of the point M on nondeformed original surface by means of expansion into a Taylor series

$$\vec{v}(\vec{r} + \vec{u}) = \vec{v}(\vec{r}) + (\vec{u}\cdot\vec{\nabla})\vec{v} + \frac{1}{2}(\vec{u}\cdot\vec{\nabla})^2\vec{v} + \dots \quad (2.1)$$

$$p(\vec{r} + \vec{u}) = p(\vec{r}) + (\vec{u}\cdot\vec{\nabla})p + \frac{1}{2}(\vec{u}\cdot\vec{\nabla})^2 p + \dots \quad (2.2)$$

Here $\vec{v}$ and p on the right-hand side and the Hamilton operator

$$\vec{\nabla} = \frac{\vec{K}_1}{h_1}\frac{\partial}{\partial x_1} + \frac{\vec{K}_2}{h_2}\frac{\partial}{\partial x_2} + \frac{\vec{K}_3}{h_3}\frac{\partial}{\partial x_3} \quad (2.3)$$

are calculated at the point M with known Euler coordinates x_1, x_2, x_3.

Kinematic condition (1.5), taking into account (2.1), has the form

$$[\frac{\partial\vec{u}}{\partial t} - \vec{v} - (\vec{u}\cdot\vec{\nabla})\vec{v} - \frac{1}{2}(\vec{u}\cdot\vec{\nabla})^2\vec{v} - \dots]\cdot\vec{K}_3^* = 0 \ (M) \ (2.4)$$

For its reduction to scalar form we use (1.1) and (1.6). In the expression

$$(\vec{u}\cdot\vec{\nabla})\,\vec{v}$$

$$= (\frac{u_1}{h_1}\frac{\partial}{\partial x_1} + \frac{u_2}{h_2}\frac{\partial}{\partial x_2} + \frac{u_3}{h_3}\frac{\partial}{\partial x_3})(v_1\vec{K}_1 + v_2\vec{K}_2 + v_3\vec{K}_3)$$

the derivatives of unit vectors do not equal zero because of the curvilinearity of the coordinates x_i. We use the following formulae for differentiation [1]

$$\frac{\partial \vec{K}_1}{\partial x_1} = - \frac{\partial h_1}{h_2 \partial x_2} \vec{K}_2 - \frac{\partial h_1}{h_3 \partial x_3} \vec{K}_3 ,$$

$$\frac{\partial \vec{K}_1}{\partial x_2} = \frac{\partial h_2}{h_1 \partial x_1} \vec{K}_2 , \quad \frac{\partial \vec{K}_1}{\partial x_3} = \frac{\partial h_3}{h_1 \partial x_1} \vec{K}_3 ,$$

$$\frac{\partial \vec{K}_2}{\partial x_1} = \frac{\partial h_1}{h_2 \partial x_2} \vec{K}_1 , \quad \frac{\partial \vec{K}_2}{\partial x_3} = \frac{\partial h_3}{h_2 \partial x_2} \vec{K}_3 , \qquad (2.5)$$

$$\frac{\partial \vec{K}_2}{\partial x_2} = - \frac{\partial h_2}{h_1 \partial x_1} \vec{K}_1 - \frac{\partial h_2}{h_3 \partial x_3} \vec{K}_3$$

$$\frac{\partial \vec{K}_3}{\partial x_1} = \frac{\partial h_1}{h_3 \partial x_3} \vec{K}_1 , \quad \frac{\partial \vec{K}_3}{\partial x_2} = \frac{\partial h_2}{h_3 \partial x_3} \vec{K}_2 ,$$

$$\frac{\partial \vec{K}_3}{\partial x_3} = - \frac{\partial h_3}{h_1 \partial x_1} \vec{K}_1 - \frac{\partial h_3}{h_2 \partial x_2} \vec{K}_2$$

With these results (2.4) may be given in the following form:

$$(\frac{\partial u_1}{\partial t} - V_1) E_1 + (\frac{\partial u_2}{\partial t} - V_2) E_2 \qquad (2.6)$$

$$+ (\frac{\partial u_3}{\partial t} - V_3) E_3 = 0 \quad (M)$$

Expressions for E_1, E_2, E_3 are given in (2.4) of Chapter I. Quantities V_1, V_2, V_3 are velocity components

at the point M* expressed approximately by their
values at the point M:

$$V_1 = v_1 + \frac{u_1}{h_1} \left[\frac{\partial v_1}{\partial x_1} + \frac{\partial h_1}{h_2 \partial x_2} v_2 + \frac{\partial h_1}{h_3 \partial x_3} v_3\right]$$

$$+ \frac{u_2}{h_2} \left[\frac{\partial v_1}{\partial x_2} - \frac{\partial h_2}{h_1 \partial x_1} v_2\right] \tag{2.7}$$

$$+ \frac{u_3}{h_3} \left[\frac{\partial v_1}{\partial x_3} - \frac{\partial h_3}{h_1 \partial x_1} v_3\right] + V_1' \left(\frac{u_1^2}{h_1^2} \frac{\partial v_1}{\partial x_1}, \ldots\right)$$

The expressions for V_2 and V_3 are obtained from
(2.7) by the circular rearrangement of the indices 1,
2, 3. The functions V_1', V_2', V_3' corresponding to the
third term on the right-hand side of (2.1) have a
rather cumbersome form. Let us write their expressions for the two-dimensional case (x_2, x_3):

$$V_2' = V_2'' + \frac{u_2 u_3}{h_2 h_3} \frac{\partial}{\partial x_3} \left(\frac{\partial h_2}{h_3 \partial x_3}\right) v_3 ,$$

$$\tag{2.8}$$

$$V_3' = V_3'' - \frac{u_2 u_3}{h_2 h_3} \frac{\partial}{\partial x_3} \left(\frac{\partial h_2}{h_3 \partial x_3}\right) v_2$$

Here

$$V_2'' = \frac{u_2^2}{2h_2^2} \left[\frac{\partial}{\partial x_2} \left(\frac{\partial v_2}{\partial x_2} - \frac{\partial h_2}{h_3 \partial x_3} v_3 \right) \right.$$

$$- \frac{\partial h_2}{h_3 \partial x_3} \left(\frac{\partial v_3}{\partial x_2} - \frac{\partial h_2}{h_3 \partial x_3} v_2 \right) \left. \right]$$

$$+ \frac{u_3^2}{2h_3^2} \left[\frac{\partial}{\partial x_3} \left(\frac{\partial v_2}{\partial x_3} - \frac{\partial h_3}{h_2 \partial x_2} v_3 \right) \right.$$

$$- \frac{\partial h_3}{h_2 \partial x_2} \left(\frac{\partial v_3}{\partial x_3} + \frac{\partial h_3}{h_2 \partial x_2} v_2 \right) \left. \right] \qquad (2.9)$$

$$+ \frac{u_2 u_3}{h_2 h_3} \left[\frac{\partial^2 v_2}{\partial x_2 \partial x_3} + \frac{\partial h_3}{h_2 \partial x_2} \frac{\partial h_2}{h_3 \partial x_3} v_2 \right.$$

$$+ \frac{\partial h_2}{h_3 \partial x_3} \frac{\partial v_3}{\partial x_3} - \frac{\partial h_3}{h_2 \partial x_2} \frac{\partial v_3}{\partial x_2} \left. \right]$$

The function V_3'' is obtained from (2.9) by rearranging indices 2 and 3.

It should be mentioned that in the condition (2.6) the expressions for E_i contain derivatives on α_1, α_2 and the expressions for V_i contain derivatives on x_1, x_2, x_3. Although the Euler and the Lagrange coordinates in (2.6) may be identified because of the special choice of the coordinate systems coinciding at the initial moment of time, nevertheless, the derivatives on α_j and x_j must differ because they are not equal to each other ($\partial/\partial x_j \neq \partial/\partial \alpha_j$). The variables x_j and α_j may be identified only after the completion of the differentiation operation. The same comment can be made with respect to the Lamé coefficients $H_i(\alpha_1, \alpha_2)$ and $h_j(x_1, x_2, x_3)$ entering into the motion equations of the shell and the fluid.

Since now the contact conditions are satisfied on the original surface (t = 0), after the completion of the differentiation operation one may note that

$$h_1(x_1,x_2,0) = H_1(\alpha_1,\alpha_2), \quad h_2(x_1,x_2,0) = H_2(\alpha_1,\alpha_2)$$

The dynamic conditions (1.10), taking into account (2.2), acquire a form (without loss of generality p' may be omitted)

$$Z_1^* = Z_2^* = 0,$$

$$Z_3^* = p + \frac{u_1}{h_1}\frac{\partial p}{\partial x_1} + \frac{u_2}{h_2}\frac{\partial p}{\partial x_2}$$

$$+ \frac{u_3}{h_3}\frac{\partial p}{\partial x_3} + \frac{u_1^2}{2h_1^2}\frac{\partial^2 p}{\partial x_1^2} + \dots \quad \text{(M)}$$

(2.10)

In (2.2) and (2.10) the differentiation operation is done on the scalar quantity p; therefore, the series may be continued easily.

The conditions (1.11) are written analogously

$$Z_1 = E_1 \left(P + \frac{u_1}{h_1}\frac{\partial p}{\partial x_1} + \frac{u_2}{h_2}\frac{\partial p}{\partial x_2} + \frac{u_3}{h_3}\frac{\partial p}{\partial x_3} + \dots \right),$$

$$Z_2 = E_2 \left(p + \frac{u_1}{h_1}\frac{\partial p}{\partial x_1} + \frac{u_2}{h_2}\frac{\partial p}{\partial x_2} + \frac{u_3}{h_3}\frac{\partial p}{\partial x_3} + \dots \right), \quad \text{(M)}$$

(2.11)

$$Z_3 = E_3 \left(p + \frac{u_1}{h_1}\frac{\partial p}{\partial x_1} + \frac{u_2}{h_2}\frac{\partial p}{\partial x_2} + \frac{u_3}{h_3}\frac{\partial p}{\partial x_3} + \dots \right).$$

Thus, the mathematical formulation of the problem is the system of equations and conditions (1.2), (1.4), (2.6), (2.11) or the system (1.3), (1.4), (2.6), (2.10).

2.3. Classification of aeroelasticity problems

Naturally in the formulation of particular physical problems, the general equations of the shell motion, the fluid motion, and the contact conditions are simplified. In this book we shall not consider any class of shell theory or hydromechanics problems where the well known simplifications are made (as, for example, plane problems, nonmoment shells, incompressibility, supersonic and subsonic flow, etc.). These questions are considered in appropriate courses. They will only be discussed in this book with respect to solutions of specific problems.

Here we limit ourselves to analysis of the contact conditions. It is precisely these that give birth to the problem of interaction. In this analysis we shall consider such general factors as the shell displacement magnitude in the normal direction, the shell element rotation, the rate of change of the shell displacement field, and the fluid velocity and pressure fields.

We shall make the following assumptions. Let H be the characteristic linear size of the shell. For α_i and x_i we take such nondimensional coordinates that all Lamé coefficients h_i (and also H_1, H_2 which coincide with h_1, h_2 on the shell surface $\alpha_3 = 0$) are of the same order both with respect to one another and to H.

We shall assume that the geometrical parameters of the shell change smoothly, i.e., the relations $\partial H_i / \partial \alpha_j \sim H_i$ are correct. The corresponding derivatives on x_j possess this same property as well. By the assumptions indicated above, it may be shown that the shell surface curvatures k_{ij} have the order

$$K_{ij} \sim H_i^{-1} \sim h_i^{-1} \sim H^{-1} \tag{3.1}$$

For a qualitative estimate of the shell deformation we introduce numbers m, n, k so that

$$\frac{u_3}{H} \sim \varepsilon^m, \quad \omega_i \sim \varepsilon^n, \quad \frac{\partial u_i}{\partial \alpha_j} \sim \varepsilon^k u_i \tag{3.2}$$

where ε is a quantity that is small in comparison with unity or is the error admitted in the calculations, a symbol $\sim$ denotes that the compared quantities are of the same order, and ω_j characterizes the rotation of the shell element under its bending.

For a qualitative estimate of fluid motion we introduce numbers λ and μ so that

$$\frac{\partial v_i}{\partial x_j} \sim \varepsilon^\lambda v_i, \quad \frac{\partial p}{\partial x_j} \sim \varepsilon^\mu p \qquad (3.3)$$

The numbers k, λ, μ characterize the rate of change of the shell displacement and fluid flow field near the shell surface.

Estimates must be based on the maximum values of ω_i, $\partial u_i/\partial \alpha_j$, $\partial v_i/\partial x_j$, $\partial p/\partial x_j$.

The following cases can occur:

1) The fields in the shell and in the fluid change smoothly, i.e., the derivatives on the coordinates have the same order as the functions themselves (or even strongly decrease). In this case, as one can see from (3.2) and (3.3),

$$K > 0, \quad \lambda > 0, \quad \mu > 0 \qquad (3.4)$$

This occurs, for example, with a transverse unseparated stationary flow of an ideal incompressible liquid around a cylindrical shell.

2) The shell field changes rapidly, and the derivatives are greater than the functions themselves; hence,

$$K < 0, \quad \lambda < 0, \quad \mu < 0 \qquad (3.5)$$

Such may occur in the cases of impulse problems, of high-frequency oscillations, etc.

We assume that ε has the order of relation of the wall thickness of the shell to its characteristic linear dimension h/H. In such a case the degree of accuracy of the thin shell equations, which are based on the Kirchoff-Love hypothesis and have an error h/H in comparison with 1, is the same as for the boundary conditions on the contact surface of the shell with a fluid.

Now we show how the estimates can be done. Consider, for example, in Z_3^* (2.10) the following terms:

$$p + \frac{u_3}{h_3} \frac{\partial p}{\partial x_3} + \frac{u_3^2}{2h_3^2} \frac{\partial^2 p}{\partial x_3^2} \tag{3.6}$$

According to (3.1), $h_i \sim H$. Consequently, $u_3/h_3 \sim \varepsilon^m$ as it follows from (3.2). Therefore, taking into account (3.3), we have the following order of magnitudes:

$$\sim p \ (1 + \varepsilon(m + \mu) + \varepsilon^2(m + \mu))$$

If $m + \mu > 1$, then in (3.6) only the first term p remains. If $m + \mu < 1$, $2(m + \mu) > 1$, then the term that contains u_3/h_3 in the first degree must remain as well. The concept of both classifications of the interaction problems and corresponding simplifications of the contact conditions consists in such estimates.

In specific interaction problems the nonlinearities in all three systems of equations (hydromechanics, shell theory, and contact conditions) may play an equally important part, but not always. In the cases when the nonlinearities in the contact conditions are not the principal ones, a neglect of only the values $\varepsilon^1 \sim h/H$ as compared with 1 can result in unnecessarily rigorous (exact) relations. Then one can neglect the somewhat greater quantities, for example, $\varepsilon^{2/3}(m + \mu > 2/3)$. Note that if $\varepsilon = h/H = 10^{-2}$, then $\varepsilon^{2/3} \approx 0.05$. While for the criterion ε^1 the terms of only order 0.01 are rejected. However, in our further general estimations we shall adhere to the above indicated rule concerning the neglect of values ε as compared with unity.

In conclusion we note that one can deduce some simple relations between the numbers m, n, k and also between λ and μ for some classes of problems. In these cases the analysis of orders of magnitude is simplified as well.

For example, consider the relation between the values λ and μ in the case of potential steady flows of a compressible liquid which has the parameters V_∞, ρ_∞, p_∞, at a large distance from the obstacle.

Then from [3] we have the following expression for the hydrodynamical pressure:

$$p = p_\infty [1 - \frac{\gamma - 1}{2a_\infty^2} (v^2 - v_\infty^2)]^{\frac{\gamma}{\gamma-1}} \qquad (3.7)$$

Using (3.3), $\partial v/\partial x_i \sim \epsilon\lambda v$, by the direct differentiation of (3.7) we obtain

$$\frac{\partial p}{\partial x_i} \sim \epsilon\lambda v^2 \rho_\infty [1 - \frac{\gamma-1}{2a_\infty^2} (v^2 - v_\infty^2)]^{\frac{1}{\gamma-1}}$$

Solving for v^2 with the aid of (3.7),

$$v^2 = v_\infty^2 + \frac{2a_\infty^2}{\gamma-1} [1 - (\frac{p}{p_\infty})^{\frac{\gamma-1}{\gamma}}]$$

we find

$$\partial p/\partial x_i \approx C\epsilon\lambda p \qquad (3.8)$$

where it is noted that

$$C = \frac{2\gamma}{\gamma-1} [1 - \frac{2+(\gamma-1)M_\infty^2}{2} (\frac{p}{p_\infty})^{\frac{1-\gamma}{\gamma}}], \quad M_\infty = \frac{V_\infty}{a_\infty} \qquad (3.9)$$

45

For $M_\infty^2 \ll 2/(\gamma-1)$ the diagram of function $C(p/p_\infty)$ is shown in Fig. 2 for the following values of γ: $\gamma=1.4$ (air); $\gamma=7$ (water); $\gamma=\infty$ (incompressible fluid). In the most interesting case, $p/p_\infty > 1/3$, this function is of order of unity. Since in accordance with (3.3) $\partial p/\partial x_i \sim \varepsilon^\mu p$, we conclude that $\mu = \lambda$. Analogously, other cases when there is an equality of the values μ and λ or a simple relation between them can be deduced. However, for the general case we shall retain both values of λ and μ in our estimates.

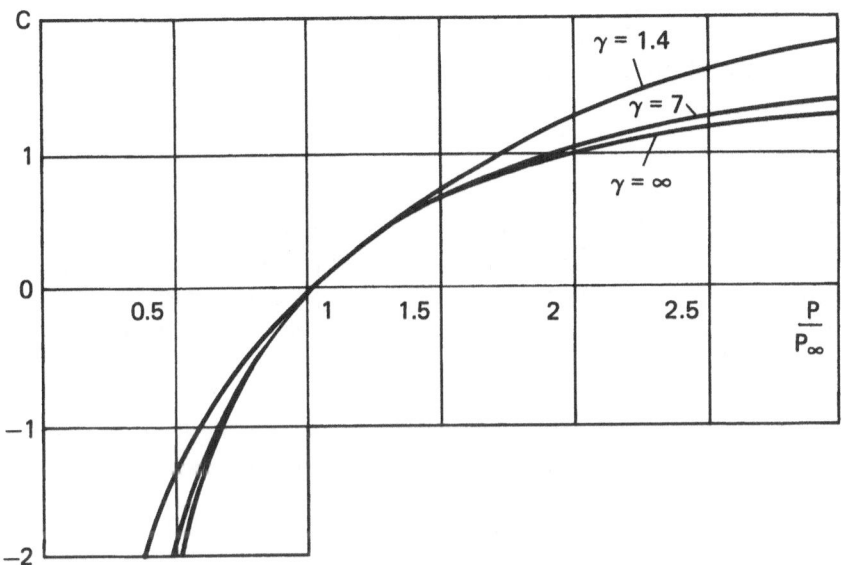

Fig. 2. The coefficient of a pressure change.

2.4. Interaction for large bending of the shell

Large bending of a shell is that case when the displacements of a shell are comparable with its characteristic linear dimension, while the section rotation is of order of unity [2]. Furthermore, large bending is characterized by an identity of the orders of maximum values of the tangent components and normal components of displacements

$$u_1 \sim u_2 \sim u_3 \qquad (4.1)$$

For such statical bending the displacements change smoothly. For dynamic problems, the displacements may quickly change. However, the displacements in dynamic problems, which are connected with the generation of short waves, cannot be accompanied by the large displacements. Therefore, the basic motion of a shell may be considered as smoothly changing along the coordinates as well.

Thus, for the interaction that is accompanied by large bending, the values m, n, k are sometimes less than 1. For concreteness we assume that

$$m = n = 1/5, \quad k = 0 \qquad (4.2)$$

In this case, for example, the deflection of cylindrical shell that has a ratio of thickness to the radius $h/R = 10^{-2}$, can reach the value equal $1/3$ of the radius and more, as $u_3 \sim \varepsilon^m H \sim (h/R)^m R \sim 10^{-2/5} R \sim R/2.5$. Thus, large bending of the shell takes place.

Now we consider the case (3.4) of a smooth change of the liquid field. Let $\lambda = \mu = 1/5$.

In the kinematic condition (2.6) we at first determine the orders of the terms that enter in V_i. In accordance with (3.1), (3.2), (4.1) $h_i \sim H$, and also $u_1/H_1 \sim u_2/H_2 \sim u_3/H_3 \sim \varepsilon^m$. If we admit further that the order of the values that are put in square brackets in (2.7) is defined by the order of their first terms

$$\left[\frac{\partial v_1}{\partial x_2} - \frac{\partial h_2}{h_1 \partial x_1} v_2 \right] \sim \varepsilon^\lambda v_1, \dots,$$

then it may be written that

$$V_i \sim v_i \left(1 + \varepsilon^{m+\lambda} + \varepsilon^{2(m+\lambda)} + \dots \right) \qquad (4.3)$$

According to (2.4) of Chapter I and to (3.2), (4.1), the values E_i have an order of magnitude

$$E_{1,2} \sim \varepsilon^n + \varepsilon^{n+m} + \varepsilon^{n+m+k}$$

$$E_3 \sim 1 + \varepsilon^m + \varepsilon^{m+k} + \varepsilon^{2m} + \varepsilon^{2(m+k)} \qquad (4.4)$$

From formulae for e_{ij}, ω_i from (2.3) of Chapter I, it follows that

$$e_{ij} \sim \varepsilon^m + \varepsilon^{m+k}, \quad \omega_i \sim \varepsilon^n \qquad (4.5)$$

From (4.3) and (4.4) we conclude that for the accepted values of parameters m, n, k, λ in V_i, besides the main terms, the terms of orders $\varepsilon^{m+\lambda}$ and $\varepsilon^{2(m+\lambda)}$ must remain, while in E_i all the terms must remain. This means that in the case of the large bending relations (2.7), (2.8), (2.9) do not change.

Now in (2.6) it is necessary to compare the terms $E_i \partial u_i / \partial t$ and also $E_i V_i$ with each other. From (4.1) and (4.4) we conclude that all the terms $E_i \partial u_i / \partial t$ are of identical order, while for $E_i V_i$ the estimates are more complicated. From (4.3) and (4.4) it follows that $E_3 V_3 \sim v_3(1 + \varepsilon^m + ...)$, where the unity in brackets is a main term with which all other terms should be compared. This comparison shows that term $E_3 V_3$ must be replaced by the sum $E_3 V_3 + V_3'$ in which V_3 is taken according to (2.7), but without terms V_3'.

Analogous comparison of orders of the values $E_1 V_1$ and $E_2 V_2$ with v_3 in $E_3 V_3$ shows (v_1, v_2, $\sim v_3$) that in these terms V_1, V_2 must be taken according to (2.7) without terms V_1', V_2'.

Thus, for dynamical interaction that is accompanied by large bending of the shell and by a smooth change of the velocity field in the fluid having the above values of parameters m, n, k, λ, the kinematic condition (3.6) takes the form

$$\left(\frac{\partial u_1}{\partial t} - V_1\right) E_1 + \left(\frac{\partial u_2}{\partial t} - V_2\right) E_2 \qquad (4.6)$$

$$+ \left(\frac{\partial u_3}{\partial t} - V_3\right) E_3 - V_3' = 0$$

where V_1, V_2 are given by expressions (2.7) without terms V_1^f, V_2^f, and V_3' - by expressions (2.8). The E_i are given in (2.4) of Chapter I. Here and subsequently the sign M in the contact conditions is omitted.

In relations (2.10), (2.11) the terms of type

$$\frac{u_1}{h_1}\frac{\partial p}{\partial x_i} \ , \ \frac{u_1^2}{2h_1^2}\frac{\partial^2 p}{\partial x_1^2}$$

having the order $\varepsilon^{m+\mu}p$, $\varepsilon^{2(m+\mu)}p$, should remain because of (4.2). So, dynamic conditions (2.10) have the form

$$Z_1^* = 0, \ Z_2^* = 0,$$

$$Z_3^* = p + \frac{u_1}{h_1}\frac{\partial p}{\partial x_i} + \frac{u_2}{h_2}\frac{\partial p}{\partial x_2} + \frac{u_3}{h_3}\frac{\partial p}{\partial x_3}$$

$$+ \frac{u_1^2}{2h_1^2}\frac{\partial^2 p}{\partial x_1^2} + \frac{u_2^2}{2h_2^2}\frac{\partial^2 p}{\partial x_2^2} + \frac{u_3^2}{2h_3^2}\frac{\partial^2 p}{\partial x_3^2}$$

$$(4.7)$$

If the velocity field change is such that λ has a negative value, then the series (4.3) does not converge or converges badly, and the contact conditions considered are inapplicable.

When the shell deflection along the normal to its middle surface exceeds the wall thickness, but at the same time it is much less than the characteristic linear dimensions of a shell, one can assume for the estimates

$$m = 2/3, \ n = 1/3, \ k = -1/3 \qquad (4.8)$$

Here we admit simplifications for the large change of the pressure and velocity fields (λ, $\mu < 0$) as well. The corresponding interaction theory, which is inter-

mediate between the large and medium bending, was constructed in [5]. Here we shall not consider this
theory.

2.5. Interaction for medium bending of the shell

The shell bending is said to be medium, if the
maximum deflection is of the same order of magnitude
as the shell wall thickness or greater, but this
deflection is small in comparison with the other
linear dimensions of a shell. The medium bending is
characteristic of shallow shells, that is for shells
having the curvature k_{ij} and the foundation dimension
L (or the halfwave length L) that satisfy the condi-
tion $k_{ij}L \sim \varepsilon^{1/2}$. It should be noted that the above
mentioned intermediate case between the large and
medium bending takes place for a shell with parameters
$k_{ij}L \sim \varepsilon^{1/3}$.

The shallow shell theory is constructed on the
assumption that

$$m = 1, n = 1/2, k = -1/2 \qquad (5.1)$$

In this case instead of (4.1) we have

$$u_1 \sim \omega_1 u_3, u_2 \sim \omega_2 u_3 \qquad (5.2)$$

On the basis of (3.1), (3.2) and (5.1), (5.2) we
determine

$$\frac{u_3}{h_3} \sim \varepsilon, \frac{u_1}{h_1} \sim \varepsilon^{3/2}, \frac{u_2}{h_2} \sim \varepsilon^{3/2},$$

$$(5.3)$$

$$e_{ij} \sim \varepsilon, \omega_k \sim \varepsilon^{1/2}$$

For the parameters E_i from (2.4) of Chapter I, we
obtain

$$E_1 = -\omega_1, E_2 = -\omega_2, E_3 = 1 \qquad (5.4)$$

The angles of rotation with accuracy up to ε as compared to unity are

$$\omega_i = \partial u_3 / H_i \partial \alpha_i \qquad (5.5)$$

In the case analyzed without loss of accuracy one can equate the derivatives with respect to α_1, α_2 and x_1, x_2, respectively.

Furthermore, we compare with each other the terms $E_1 \cdot \partial u_1 / \partial t + E_3 \cdot \partial u_3 / \partial t$ in equation (2.6). From (5.2), (5.4) we have following estimates:

$$\sim \omega_1 \partial (\omega_1 u_3) / \partial t + \partial u_3 / \partial t \sim (1 + \varepsilon) \partial u_3 / \partial t$$

Hence we can conclude that the terms $E_1 \cdot \partial u_1 / \partial t$, $E_2 \cdot \partial u_2 / \partial t$ should be rejected since they are small as compared to $E_3 \, \partial u_3 / \partial t = \partial u_3 / \partial t$.

That is why the kinematic condition (2.6) in the case of interaction of the shallow shell with a rapidly changing fluid field $(-1/2 < \lambda < 0)$ has the form

$$v_3 = \frac{\partial u_3}{\partial t} + \frac{\partial u_3}{h_1 \partial x_1} v_1 + \frac{\partial u_3}{h_2 \partial x_2} v_2 \qquad (5.6)$$

$$- \frac{u_3}{h_3} \left(\frac{\partial v_3}{\partial x_3} + \frac{\partial h_3}{h_1 \partial x_1} v_1 + \frac{\partial h_3}{h_2 \partial x_2} v_2 \right)$$

If the flow field in a liquid is slowly changing $(\lambda > 0)$, then terms with the multiplier u_3/h_3 in (5.6) are rejected. In the case of rapidly changing field of a fluid having the index $-1/2 > \lambda > -2/3$, one must take into account in (5.6) the terms with multipliers $(u_3/h_3)^2$, u_1/h_1, u_2/h_2 as well. These terms are contained in (2.7), (2.8), (2.9).

In order to record the dynamic conditions we shall estimate the following three terms in (4.7):

$$p + \frac{u_1}{h_1} \frac{\partial p}{\partial x_1} + \frac{u_3}{h_3} \frac{\partial p}{\partial x_3} \sim p(1 + \varepsilon^{m+n+\mu} + \varepsilon^{m+\mu}) \qquad (5.7)$$

Here relations (5.2) are taken into account. For the rapidly changing fluid fields ($-1/2 < \mu < 0$), in accordance with (5.1) the term $(u_3/h_3)\partial p/\partial x_3$ has to be retained, while the other terms should be rejected.

Consequently, dynamic conditions for the interacting of a shallow shell with the rapidly changing field in a fluid have the form

$$Z_1^* = 0, \quad Z_2^* = 0, \quad Z_3^* = p + \frac{u_3}{h_3}\frac{\partial p}{\partial x_3} \qquad (5.8)$$

For the slowly changing field in the fluid ($\lambda > 0$), in accordance with (5.7), in the last of (5.8) it remains $Z_3^* = p$.

Dynamic conditions (2.11) for medium bending and rapidly changing field in the fluid ($-1/2 < \mu < 0$) are

$$Z_1 = -\frac{\partial u_3}{h_1 \partial x_1} p, \quad Z_2 = -\frac{\partial u_3}{h_2 \partial x_2} p,$$

$$(5.9)$$

$$Z_3 = p + \frac{u_3}{h_3}\frac{\partial p}{\partial x_3}$$

Here (5.4) and (5.5) are taken into account. In right-hand sides of Z_1 and Z_2, the terms $(u_3/h_3)\partial p/\partial x_3$, having the order $\varepsilon^{m+\mu}$ in comparison with the first terms p, are rejected, since the shallow shell bending weakly depends on loads in the tangential directions.

In the case of a plane plate that undergoes bending with parameters (5.1), the kinematic and dynamic conditions may be written in the form ($u_3 = w$, $h_1 = h_2 = h_3 = 1$)

$$v_3 = \frac{\partial w}{\partial t} + \frac{\partial w}{\partial x_1} v_1 + \frac{\partial w}{\partial x_2} v_2 - w\frac{\partial v_3}{\partial x_3},$$

$$(5.10)$$

$$Z_3^* = p + w\partial p/\partial x_3$$

Hence, one easily obtains, for example, the well known linear flutter theory [3,4] contact conditions about the disturbances of velocity components v_1, v_2, v_3 and the pressure p. Taking into account that for the longitudinal flow around the plate of the uniform fluid with a velocity V_∞ directed along the line x_1, $v_1 = V_\infty + \tilde{v}_1$, $v_2 = \tilde{v}_2$, $v_3 = \tilde{v}_3$, $\partial p/\partial x_3 = 0$ and making a linearization, we have

$$\tilde{v}_3 = \frac{\partial w}{\partial t} + V_\infty \frac{\partial w}{\partial x} \; , \quad Z_3^* = \tilde{p} \tag{5.11}$$

where $\tilde{p}$ is determined from the linearized Bernoulli equation.

Let us make a remark concerning the conditions (5.6), (5.10). If one assumes that the velocity components v_1, v_2, v_3 are of the same order, then the terms of type $v_1\,\partial w/\partial x_1$ are small as compared to term v_3. Such cases can occur, for example, in problems of fluctuation of the shell and the fluid in which a mean flow is absent. However, if the mean flow is present, as in the above mentioned example, then the values v_1 and v_2 can exceed very substantially the value v_3 near the shell surface. Therefore, in the general case the kinematic conditions should have the form (5.6), (5.10). The effect of terms $w\partial v_3/\partial x_3$, $w\partial p/\partial x_3$ in (5.10) will be analyzed further for the secondary flow of a viscous liquid near a plate (section 2.7, this chapter). As is known from section 2.2 of this chapter, these terms occur, since we distinguish between deformed and undeformed contact surfaces.

Now we return to conditions (5.6), (5.8). In the case of a potential fluid flow it is possible to introduce the velocity potential $\phi*$ ($v_i = \partial\phi*/h_i\partial x_i$). As p* we denote the hydrodynamical pressure which corresponds to that potential. Let us represent $\phi*$ and p* as sums

$$\phi* = \phi + \psi \; , \quad p* = p_\phi + p_\psi \tag{5.12}$$

in which the first terms on the right-hand sides describe the field of velocity and pressure in the fluid and invoke the assumption of absolute stiffness of a body surface ($v_3 = \partial\phi/h_3\partial x_3 = 0$), while the second terms represent the effect of disturbances due to displacements of the contact surface. In such a representation the arbitrary constants which are the accuracy of potential determination, are not taken into account. Linearized kinematic and dynamic conditions (5.6), (5.8) take the form

$$\frac{\partial\psi}{h_3\partial x_3} = \frac{\partial u_3}{\partial t} + \frac{\partial u_3}{h_1\partial x_1}\frac{\partial\phi}{h_1\partial x_1} + \frac{\partial u_3}{h_2\partial x_2}\frac{\partial\phi}{h_2\partial x_2}$$

$$- \frac{u_3}{h_3}\left[\frac{\partial}{\partial x_3}\left(\frac{\partial\phi}{h_3\partial x_3}\right) + \frac{\partial h_3}{h_1\partial x_1}\frac{\partial\phi}{h_1\partial x_1}\right. \qquad (5.13)$$

$$\left. + \frac{\partial h_3}{h_2\partial x_2}\frac{\partial\phi}{h_2\partial x_2}\right],$$

$$Z_1^* = 0, \qquad Z_2^* = 0,$$

$$Z_3^* = P_\phi + P_\psi + \frac{u_3}{h_3}\frac{\partial P_\phi}{\partial x_3}$$

The coefficients h_1, h_2, h_3 coincide with H_1, H_2, H_3 on the contact surface $x_3 = 0$.

2.6. Interaction for small bending of the shell

The bending is small if the smoothly changing shell displacement is of the order of the wall thickness and the rotations of the linear elements of the shell are small compared to unity. In this case a linearization of all the relations of thin shell theory takes place. In correspondence with this, the displacement components are of the same order of magnitude

$$u_1 \sim u_2 \sim u_3 \qquad (6.1)$$

while the parameters m, n, k have the values

$$m = 1, \quad n = 1, \quad k = 0 \qquad (6.2)$$

and u_1/H_1, $u_2/H_2 \sim u_3/H_3 \sim \varepsilon$, $\omega_k \sim \varepsilon$. From formulae for e_{ij}, E_i and (6.2) it follows that

$$e_{ij} \sim \varepsilon, \quad E_1 = -\omega_1, \quad E_2 = -\omega_2, \quad E_3 = 1 \qquad (6.3)$$

Here expressions for ω_1, ω_2 are taken without any simplifications to (2.3) of Chapter I. The values $E_1 \partial u_1/\partial t$, $E_2 \partial u_2/\partial t$ in accordance with (6.1)-(6.3) are ε times less than $E_3 \partial u_3/\partial t = \partial u_3/\partial t$.

For the case of a rapidly changing field of velocity and pressure in the fluid ($-1/2 < \lambda, \mu < 0$) near the shell surface we have the following kinematic condition:

$$v_3 = \frac{\partial u_3}{\partial t} + \omega_1 v_1 + \omega_2 v_2 - \frac{u_1}{h_1} \left(\frac{\partial v_3}{\partial x_1} - \frac{\partial h_1}{h_3 \partial x_3} v_1 \right)$$

$$(6.4)$$

$$- \frac{u_2}{h_2} \left(\frac{\partial v_3}{\partial x_2} - \frac{\partial h_2}{h_3 \partial x_3} v_2 \right) - \frac{u_3}{h_3} \left(\frac{\partial v_3}{\partial x_3} + \frac{\partial h_3}{h_1 \partial x_1} v_1 + \frac{\partial h_3}{h_2 \partial x_2} v_2 \right)$$

The corresponding dynamic conditions have the form

$$Z_1^* = 0, \quad Z_2^* = 0,$$

$$(6.5)$$

$$Z_3^* = p + \frac{u_1}{h_1} \frac{\partial p}{\partial x_1} + \frac{u_2}{h_2} \frac{\partial p}{\partial x_2} + \frac{u_3}{h_3} \frac{\partial p}{\partial x_3}$$

In linear theory one can equate the coordinate systems α_1, α_2 and x_1, x_2. The components of external loads Z_i^* in the unit vector $\vec{K}^*$ direction and the components Z_i on $\vec{K}_i$ coincide with each other within a permissible accuracy. Therefore,

$$Z_1 = 0, \quad Z_2 = 0,$$

$$Z_3 = p + \frac{u_1}{h_1}\frac{\partial p}{\partial x_1} + \frac{u_2}{h_2}\frac{\partial p}{\partial x_2} + \frac{u_3}{h_3}\frac{\partial p}{\partial x_3}$$

(6.6)

If the field in the fluid is changed smoothly (λ, $\mu > 0$), then the terms with a multiplier (u_i/h_i) in kinematic and dynamic conditions (6.4), (6.5) are rejected:

$$v_3 = \partial u_3/\partial t + \omega_1 v_1 + \omega_2 v_2,$$

(6.7)

$$Z^*_1 = 0, \quad Z^*_2 = 0, \quad Z^*_3 = p$$

In the case of a potential fluid flow, having represented the potential and hydrodynamic pressure in the form of sums (5.12), we obtain from (6.7) the linearized contact conditions

$$\frac{\partial \psi}{h_3 \partial x_3} = \frac{\partial u_3}{\partial t} + (\frac{\partial u_3}{H_1 \partial x_1} - K_{11}u_1 - K_{12}u_2)\frac{\partial \phi}{h_1 \partial x_1}$$

$$+ (\frac{\partial u_1}{H_2 \partial x_2} - K_{22}u_2 - K_{21}u_1)\frac{\partial \phi}{h_2 \partial x_2},$$

(6.8)

$$Z^*_1 = 0, \quad Z^*_2 = 0, \quad Z^*_3 = p\phi + p\psi$$

Let us consider, for example, the expression for the hydrodynamic pressure in the case of steady flow of an incompressible fluid:

$$p = p_\infty + \frac{p_\infty}{2}[V^2_\infty - (\nabla\phi)^2]$$

(6.9)

The value of p_ϕ coincides with (6.9) and $p_\psi = -\rho_\infty \nabla\phi\nabla\psi$. As $\partial\phi/\partial x_3 = 0$ by force of the assumption in (5.12), then

$$Z_3^* = p_\infty$$

$$(6.10)$$

$$+ \frac{\rho_\infty}{2} [V_\infty^2 - (\nabla\phi)^2] - \rho_\infty \left(\frac{1}{h_1^2} \frac{\partial\phi}{\partial x_1} \frac{\partial\psi}{\partial x_1} + \frac{1}{h_2^2} \frac{\partial\phi}{\partial x_2} \frac{\partial\psi}{\partial x_2}\right)$$

The contact conditions (6.4)-(6.8) are accurate when the shell deformation and its interaction with a fluid happens smoothly (k = 0). If there is a high-frequency wave-generation in the shell, when a half-wave occupies a shallow region of the shell surface, the linearization of shell relations occurs for deflection values substantially smaller than the wall thickness, and for small rotations ω_i as compared to unity. In this case

$$m = 3/2, \quad n = 1, \quad k = -1/2 \qquad (6.11)$$

Let us consider the order of tangential components of the displacement u_1, u_2. If, for instance, along the cylindrical shell circumference with radius R, 2n half-waves occur, then $n = \pi R/L$, where L is the half-wave length. As it was pointed out earlier, for the shallow shell $L/R \sim \epsilon^{1/2}$. On the other hand, for the main bending strain one can assume $e_{22} = \partial u_2/R\partial x_2 -u_3/R$, whence $u_2 \sim u_3/n$ or $u_2 \sim u_3(L/\pi R) \sim u_3\epsilon^{1/2} \sim \omega_2^{1/2}u_3$. Therefore, we shall have the following estimates:

$$u_1 \sim \omega^{1/2}u_3, \quad u_2 \sim \omega_2^{1/2}u_3 \qquad (6.12)$$

With the help of (6.12) we find that in considering the case of (6.11), even for the rapid change of the field in the fluid ($-1/2 < \lambda, \mu < 0$), the kinematic and dynamic conditions have the form (6.7), but the values ω_i in them are determined as for the shallow shell (5.5). Thus,

$$v_3 = \frac{\partial u_3}{\partial t} + \frac{\partial u_3}{H_1 \partial x_1} v_1 + \frac{\partial u_3}{H_2 \partial x_2} v_2,$$

$$Z_1^* = 0, \quad Z_2^* = 0, \quad Z_3^* = p$$

$$(6.13)$$

In conclusion we shall again return to the question of the possible relations among the values m, n, k. This question was considered in section 2.3.

From the above considered cases of large, medium, and small bending it follows that values m, n, k obey the rule

$$n = m + k \qquad (\text{at } k < 0)$$

As a matter of fact, for large bending m = 1/5, n = 1/5, k = 0; for medium bending m = 1, n = 1/2, k = -1/2; in the intermediate case m = 2/3, n = 1/3, k = -1/3; for small bending m = 1, n = 1, k = 0 and also m = 3/2, n = 1, k = -1/2.

2.7. Interaction between a plate and a viscous fluid

We have considered the formulation and classification of the interaction problems between a shell and an ideal fluid. In this section we assume the fluid to be viscous. Here one could repeat the general analysis discussed above. But we confine ourselves to consideration of the boundary conditions in the simple case of a plane contact surface and of Cartesian coordinates. Assumptions concerning the initial state of the plate and the fluid, the coordinate systems for the two mediums are the same as for the ideal fluid problem.

We shall consider deformations of a plate such that the displacement magnitudes are within the limits of the shallow shell assumptions. Then, having written the contact conditions, as are known from section 2.5, one can not make any distinctions between the Lagrange coordinates α_1, α_2 and the Euler coordinates x_1, x_2.

The contact conditions may be obtained from the conservation equations at a shock (jump) in the case of a viscous fluid. However, in the case of an impermeable plate they may be written simply as they are the condition of adherence of a viscous fluid to the plate and of the stress vectors equality.

A plate fixed point m with a radius vector $\vec{r}$ at an initial moment of time (t = 0) coincides with point M of a space (Fig. 1). In an arbitrary moment of time (t > 0) that point m coincides with point M* of a space with a radius vector $\vec{r} + \vec{u}$, where $\vec{u}$ is displacement vector of the plate fixed point m. If one denotes the vector of a motion velocity of a fluid as $\vec{v}$, then the adherence condition reduces to an equality

$$\vec{v} = \partial \vec{u}/\partial t \qquad (M*) \qquad (7.1)$$

The dynamic conditions are

$$\vec{Z} = \vec{p}_3 - \vec{p}_3' \qquad (M*) \qquad (7.2)$$

where $\vec{p}_3$, $\vec{p}_3'$ are the vectors of a surface tension in a viscous fluid at point M* of a space.

Using expansion (2.1), let us express approximately the velocity vector value $\vec{v}$, which is determined at point M* of a space, as the velocity vector value in the neighborhood of point M of a space. In the case of a Cartesian coordinate system ($h_1 x_1 = x$, $h_2 x_2 = y$, $h_3 x_3 = z$) the velocity vector components in direction of unit vectors k_i have the form

$$V_i = v_i + u_1 \frac{\partial v_i}{\partial x} + u_2 \frac{\partial v_i}{\partial y} + u_3 \frac{\partial v_i}{\partial z} \qquad (7.3)$$

As in the operator $(\vec{u} \cdot \vec{\nabla})\vec{v}$, derivatives of $\vec{k}_i$ with respect to x, y, z are equal to zero. Taking into account all the above, the vector equality (7.1) is reduced to the following kinematic conditions:

$$\frac{\partial u_1}{\partial t} - V_1 = 0, \quad \frac{\partial u_2}{\partial t} - V_2 = 0, \quad \frac{\partial u_3}{\partial t} - V_3 = 0 \quad (M) \quad (7.4)$$

in which the values V_1, V_2, V_3 from (7.3) are calculated at point M of a space.

Unlike the case of an ideal fluid, when reducing the dynamic conditions (7.2) to the undeformed surface, it is necessary to use the formulae of the vector function expansion (2.1)

$$\vec{Z} = \vec{p_3} + (\vec{u} \cdot \vec{\nabla})\vec{p_3} + \ldots \quad (7.5)$$

Here without any loss of accuracy the term $\vec{p_3}'$ is rejected.

It should be noted that now in motion equations (1.2), (1.3) there are nonzero right hand sides. The motion equations (1.2), obtained by projecting all the forces on axes directions before the deformation and taking into account (7.5) have the form

$$L_m(u_1, u_2, u_3) = p_{3m}$$

$$+ (u_1 \frac{\partial}{\partial x} + u_2 \frac{\partial}{\partial y} + u_3 \frac{\partial}{\partial z})p_{3m} \quad (7.6)$$

in which p_{31}, p_{32}, p_{33} are the stress components in a viscous fluid.

The Navier-Stokes equations of a viscous fluid motion, and the continuity and the state equations must be connected to equations (7.4), (7.6). It is natural that the corresponding conditions at infinity or on the surfaces that limit the fluid flow are used.

As the simplest example of interaction of a deforming plate with a viscous fluid we consider the G. I. Taylor problem [6]. An infinitely long and wide plate is in a viscous fluid and it has an oscillatory motion in the form of traveling waves. The plate has oscillations due to some internal power source, whose specific nature is not of interest to us. Now we consider the problem in a plane formulation and the interaction on only one side of the plate.

Unlike the case considered in [6] let the plate be deformed so that its fixed points make oscillations only along the normal to an initial surface

$$u_1 = 0, \ u_3 = b \sin(\bar{\sigma}x - \bar{\omega}t) \qquad (7.7)$$

Here $\sigma = 2\pi/\lambda$, $\omega = c\sigma$; λ is the wavelength, c is the velocity of wave propagation.

Following [6], we assume the fluid to be incompressible, and the Reynolds number to be small as compared with unity. The last assumption allows us to neglect the inertial terms in the motion equations in comparison with those taking into account viscosity. Then the Navier-Stokes and continuity equations (3.13) of Chapter I take the form

$$\partial p/\partial x = \mu_0 \nabla^2 v_1, \ \partial p/\partial z = \mu_0 \nabla^2 v_3,$$

$$\frac{\partial v_1}{\partial x} + \frac{\partial v_3}{\partial z} = 0 \qquad (7.8)$$

By introducing the flow stream function with the help of formulae $v_1 = -\partial\psi/\partial z$, $v_3 = \partial\psi/\partial x$, we satisfy the continuity equation. Furthermore, by eliminating the pressure p from equations (7.8), we obtain

$$\nabla^4 \psi = 0 \qquad (7.9)$$

The function

$$\psi = (Az + B)e^{-\sigma z}\sin(\sigma x - \omega t) - V_\infty z \qquad (7.10)$$

satisfies equation (7.9). Here V_∞ is the sought for velocity of fluid at infinity about the immovable coordinate system x, z. Hence, the plate mean longitudinal velocity relative to the fluid equals $-V_\infty$. It does not depend on time.

So far as the plate oscillation is given by (7.7), then here the dynamic conditions (7.6) are not applied. The kinematic conditions (7.3), (7.4) have the form

$$\frac{\partial u_1}{\partial t} = v_1 + u_1 \frac{\partial v_1}{\partial x} + u_3 \frac{\partial v_1}{\partial z} \,,$$

$$(z = 0)$$

$$\frac{\partial u_3}{\partial t} = v_3 + u_1 \frac{\partial v_3}{\partial x} + u_3 \frac{\partial v_3}{\partial z}$$

or in terms of ψ (taking into account that $u_1 = 0$)

$$\frac{\partial \psi}{\partial z} + u_3 \frac{\partial^2 \psi}{\partial z^2} = 0,$$

$$(z = 0) \qquad (7.11)$$

$$\frac{\partial \psi}{\partial x} + u_3 \frac{\partial^2 \psi}{\partial x \partial z} = \frac{\partial u_3}{\partial t}$$

Substituting (7.7) and (7.10) into (7.11), we obtain

$$A = -b\omega, \quad B = -b\omega/\sigma, \quad V_\infty = b^2 \sigma \omega / 2 \qquad (7.12)$$

The expression for V_ω may be written also as

$$V_\infty = 2c(\pi b / \bar{\lambda})^2 \qquad (7.13)$$

If, for example, $\bar{\lambda}/b = 10$, then $V_\infty \approx 0.2c$. Usually, for the assumptions taken above, the value of V_∞ depends only on the fluid parameters. Hence, the velocity of viscous fluid motion at infinity for the above mentioned assumptions is directly proportional to the velocity and amplitude square of the traveling wave on the plate and inversely proportional to the square of the wavelength. As is seen from (7.10) and (7.12), the velocity oscillations in the fluid are directly proportional to the first degree of amplitude. Thus, an appearance of flows that do not depend on time when the traveling waves are generated on a fluid boundary is caused by the nonlinear effects. Actually, if one neglects the terms having multipliers u_3 in kinematic conditions (7.11), which corresponds to an identification of the deformed and undeformed surfaces of a plate, then $V_\infty \equiv 0$.

It should be noted that after the above mentioned work of G. I. Taylor [6] and M. J. Lighthill [7] very many researches on body motion in a fluid with wavy deformation of the surface were performed.

This simple example was discussed in order to show the principal significance of more general non-linear contact conditions.

63

References

1. Novozhilov, V. V., Foundations of the nonlinear theory of elasticity. Graylock Press, Rochester, N.Y., 1953.

2. Mushtari, Kh.M., and Galimov, K. Z., Non-linear theory of thin elastic shells, NSF-NASA, Washington, 1961, p. 374.

3. Bisplinghoff, R. L., and Ashley, H., Principles of aeroelasticity, John Wiley and Sons, New York-London, 1962, p. 527.

4. Dowell, E. H., Aeroelasticity of plates and shells, Leyden, Noordhoff Int. Publ., 1975, p. 139.

5. Ilgamov, M. A., Boundary conditions on the contact surface between a shell and a fluid in Euler-Lagrange form, Transactions of Tenth All-Union Conf. on the Theory of Shells and Plates, Tbilisi, Metzniereba, 1975, pp. 170-179.

6. Taylor, G. I., Analysis of the swimming of microscopic organisms. Proceedings of the Royal Society, Series A, No. 1099, v. 209, 1951, pp. 447-461.

7. Lighthill, M. J., Note on swimming of slender fish, J. Fluid Mech., 1960, v. 9, pp. 305-317.

CHAPTER III

BENDING OF A CYLINDRICAL SHELL WITH A
TRANSVERSE FLOW AROUND IT

Introduction

In this chapter is shown an application of the general formulation of the interaction problem presented in Chapter II and, in particular, of the condition on the contact surface. For this purpose the behavior of a cylindrical shell and curved plate with a transverse potential flow around them of an ideal incompressible liquid is considered.

In Chapter II, for the simplest example of the interaction of a viscous liquid with a plate, it was shown that taking into account the difference between deformed and undeformed boundaries of a shell may lead to qualitatively different results as compared to the linear case. In this chapter for problems of shell bending and stability the role of nonlinear terms in the contact conditions is also considered. For the solution of such problems the principal assumption is that the fluid flow on the shell surface is unseparated.

Also the data from experimental investigations and their comparison with theoretical results are discussed.

The plane static problem of an ideal incompressible fluid flow around a gas bubble and surfaces consisting completely or partially of membranes has been examined in [1,8,13,14,19,20,22,23-26,32,33]. The analogous problem for flow around cylindrical shells and curved plates was considered in [2-6,10-12,15-17, 21,27-31]. In the formulation mentioned the dynamical behavior of a shell in a flow was considered in [7,10, 34].

3.1. <u>Small bending of a cylindrical shell with an</u>
      <u>unseparated transverse flow around it</u>

The plane strain of an infinite cylindrical thin-walled shell with an unseparated transverse potential flow of an ideal incompressible liquid is considered (Fig. 1). The shell is free of any supporting or other forces of a non-aerodynamical nature. At a great distance away the shell ($r \to \infty$), the pressure, density, and velocity of the homogeneous stream are p_∞, ρ_∞, V_∞, respectively. The pressure within the shell equals p_i and it does not change during the deformation of the shell.

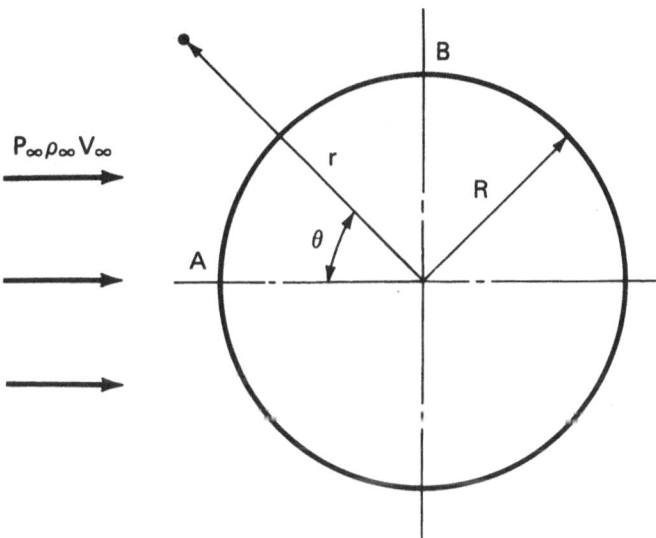

Fig. 1. Cylindrical shell in a fluid flow.

For the plane deformation of a thin-walled cylindrical shell, the assumption of inextensibility of its middle surface is made. It is used here for some simplification of computations. The shell displacement along the normal to its middle surface is considered to be small as compared to its thickness. In this connection the shell bending equation and the equations describing the change in the flow field because of deformation of the contour are linearized. On the other hand, the kinematic and dynamic conditions for small bending are applied.

It is assumed that the difference in pressure outside and inside the shell is small as compared to the value of the pressure at which buckling occurs. Most essential is the assumption that the fluid flow on the shell surface is unseparated. In the next sections of this chapter the influence of the above mentioned factors and also of the shell nonlinear bending will be discussed.

In the formulation mentioned, the problem relative to the velocity potential ϕ of the incompressible liquid motion reduces to solution of the equations

$$\nabla^2\phi = 0, \quad p = p_\infty + \frac{\rho_\infty}{2}\left[V_\infty^2 - (\nabla\phi)^2 - 2\frac{\partial\phi}{\partial t}\right] \qquad (1.1)$$

which satisfy the following condition at infinity:

$$\phi = -V_\infty \, r \cos\theta \quad (r \to \infty) \qquad (1.2)$$

and also the kinematic and dynamic conditions on the shell surface.

For small shell bending, these conditions in the cylindrical coordinate system, θ, r, according to (6.4), (6.5) of Chapter II have the form [$u_2 = v$, $u_3 = w$, $\alpha_2 = x_2 = \theta$, $\alpha_3 = x_3 = r$, $h_2 = r$, $h_3 = 1$, $H_2 = R$, $K_{22} = R^{-1}$, $K_{21} = 0$, $\omega_2 = (\partial w/\partial\theta - v)R^{-1}$, $v_2 = \partial\phi/r\partial\theta$, $v_3 = \partial\phi/\partial r$]

$$\frac{\partial w}{\partial t} = \frac{\partial\phi}{\partial r} - \frac{1}{r^2}\frac{\partial w}{\partial\theta}\frac{\partial\phi}{\partial\theta} + w\frac{\partial^2\phi}{\partial r^2} + \frac{v}{r}\frac{\partial^2\phi}{\partial r\partial\theta} \quad (r=R) \qquad (1.3)$$

$$Z_\theta = 0, \quad Z_r = p_i - p - w\frac{\partial p}{\partial r} - \frac{v}{r}\frac{\partial p}{\partial\theta} \quad (r=R) \qquad (1.4)$$

Here Z_r is the pressure differential.

Linear relations of the shell theory for the plane problem according to (2.8), (2.9), (2.10) of Chapter I are the following ($N_{22} = N_{\theta\theta}$, $Q_2 = Q_\theta$, $X_2 = Z_\theta - \rho_s h\partial^2 v/\partial t^2$, $X_3 = Z_r - \rho_s h\partial^2 w/\partial t^2$):

$$\frac{\partial N_{\theta\theta}}{R\partial\theta} + \frac{Q_\theta}{R} = \rho_s h \frac{\partial^2 v}{\partial t^2} - Z_\theta,$$

$$\frac{\partial Q_\theta}{R\partial\theta} - \frac{N_{\theta\theta}}{R} = \rho_s h \frac{\partial^2 w}{\partial t^2} - Z_r,$$

$$\frac{\partial M_{\theta\theta}}{R\partial\theta} - Q_\theta = 0,$$

(1.5)

$$\varepsilon_{\theta\theta} = \frac{\partial v}{R\partial\theta} + \frac{w}{R} , \quad M_{\theta\theta} = \frac{D}{R^2} \left(\frac{\partial v}{\partial\theta} - \frac{\partial^2 w}{\partial\theta^2}\right)$$

In this system we in succession delete terms Q_θ, $N_{\theta\theta}$, $M_{\theta\theta}$. Moreover, we use the condition of inextensibility of a shell middle surface $\varepsilon_{\theta\theta} = 0$, whence

$$\partial v/\partial\theta = - w , \quad v = - \int w \, d\theta + v_0 \qquad (1.6)$$

For shell displacements induced by its deformation only, $v_0 = 0$. For the ideal liquid $Z_\theta = 0$ according to (1.4). As a result we obtain

$$\frac{D}{R^4} \left(\frac{\partial^6 w}{\partial\theta^6} + 2 \frac{\partial^4 w}{\partial\theta^4} + \frac{\partial^2 w}{\partial\theta^2}\right) + \rho_s h \frac{\partial^2}{\partial t^2} \left(\frac{\partial^2 w}{\partial\theta^2} - w\right) = \frac{\partial^2 Z_r}{\partial\theta^2}$$

(1.7)

Taking into account (1.4), (1.6), condition (1.3) and equation (1.7) take the form

$$\frac{\partial w}{\partial t} = \frac{\partial\phi}{\partial r} - \frac{1}{r^2} \frac{\partial w}{\partial\theta} \frac{\partial\phi}{\partial\theta} + w \frac{\partial^2\phi}{\partial r^2} - \left(\int w d\theta\right) \frac{\partial^2\phi}{r\partial\theta\partial r} \quad (r=R), \quad (1.8)$$

$$\frac{\partial^6 w}{\partial\theta^6} + 2 \frac{\partial^4 w}{\partial\theta^4} + \frac{\partial^2 w}{\partial\theta^2} + \frac{\rho_s h R^4}{D} \frac{\partial^2}{\partial t^2} \left(\frac{\partial^2 w}{\partial\theta^2} - w\right)$$

(1.9)

$$= \frac{R^4}{D} \frac{\partial^2}{\partial\theta^2} \left[p_i - p - w \frac{\partial p}{\partial r} + \left(\int w \, d\theta\right) \frac{\partial p}{r\partial\theta} \right] \quad (r=R)$$

For the static interaction problem being considered we introduce the potential ϕ_1 and the pressure p_1 in the liquid for the rigid shell. We introduce also the potential ϕ_2 and the pressure p_2 due to shell bending ($w_2 = w \neq 0$), so that

$$\phi = \phi_1 + \phi_2 , \qquad p = p_1 + p_2 \qquad (1.10)$$

Substituting (1.10) in (1.1), (1.2), (1.8), (1.9) and taking into account the fact that for the small bending of the bluff body, $p_2 \ll p_1$, $\partial\phi_2/\partial\theta \ll \partial\phi_1/\partial\theta$, etc., we obtain the following boundary-value problem for ϕ_1, p_1:

$$\nabla^2\phi_1 = 0, \quad p_1 = p_\infty + \frac{\rho_\infty}{2}[V_\infty^2 - (\nabla\phi_1)^2], \qquad (1.11)$$

$$\partial\phi_1/\partial r = 0 \ (r=R), \quad \phi_1 = -V_\infty r\cos\theta \quad (r\to\infty) \qquad (1.12)$$

and the problem for ϕ_2, p_2, w is described by

$$\nabla^2\phi_2 = 0, \quad p_2 = -\rho_\infty \left(\frac{\partial\phi_1}{\partial r}\frac{\partial\phi_2}{\partial r} + \frac{1}{r^2}\frac{\partial\phi_1}{\partial\theta}\frac{\partial\phi_2}{\partial\theta}\right), \qquad (1.13)$$

$$\phi_2 = 0 \qquad (r\to\infty), \qquad (1.14)$$

$$\frac{\partial\phi_2}{\partial r} = \frac{1}{r^2}\frac{\partial w}{\partial\theta}\frac{\partial\phi_1}{\partial\theta} - w\frac{\partial^2\phi_1}{\partial r^2} \qquad (r=R), \qquad (1.15)$$

$$\frac{\partial^6 w}{\partial\theta^6} + 2\frac{\partial^4 w}{\partial\theta^4} + \frac{\partial^2 w}{\partial\theta^2}$$

$$(1.16)$$

$$= \frac{R^4}{D}\frac{\partial^2}{\partial\theta^2}[p_i - p_1 - p_2 - w\frac{\partial p_1}{\partial r} + (\int wd\theta)\frac{\partial p_1}{r\partial\theta}] \quad (r=R)$$

Condition (1.15) is obtained by taking into account that according to the first expression of (1.12) $\partial^2\phi_1/\partial r\partial\theta = 0$.

Now we find the solution of the first problem. For ϕ_1 we assume an even function in θ

$$\phi_1 = \sum_{n=o}^{N} \Phi_n(r)\cos n\theta, \qquad (1.17)$$

as the cylinder flowing is symmetrical about line $0;\pi$ (Fig. 1).

Therefore, from equation (1.11) it follows that

$$\frac{d^2\Phi_n}{dr^2} + \frac{1}{r}\frac{d\Phi_n}{dr} - \frac{n^2}{r^2}\Phi_n = 0$$

Substituting the solution of this equation in (1.17) one obtains

$$\phi_1 = A_0\ell nr + B_0 + \sum_{n=1}^{N}(A_nr^{-n} + B_nr^n)\cos n\theta, \qquad (1.18)$$

where according to conditions (1.12) the constants equal

$$A_0 = B_0 = 0, \; A_1 = B_1R^2, \; B_1 = -V_\infty, \; A_n = B_n = 0$$

Therefore, expressions for the potential ϕ_1 and the pressure p_1 for the transverse unseparated fluid flow around a rigid cylinder according to (1.11), (1.18) have the form

$$\phi_1 = -V_\infty(r + R^2/r)\cos\theta, \qquad (1.19)$$

$$P_1 = P_\infty - \frac{\rho_\infty V_\infty^2}{2}(\frac{R}{r})^2(\frac{R^2}{r^2} - 2\cos2\theta) \qquad (1.20)$$

The diagram of the pressure p_1 on cylinder surface (r=R) is shown in Fig. 2 (without taking into account p_∞). The pressure symmetry exists not only about the line $0;\pi$, but the line $\pi/2; 3\pi/2$, as well. The latter is the consequence of the assumption of

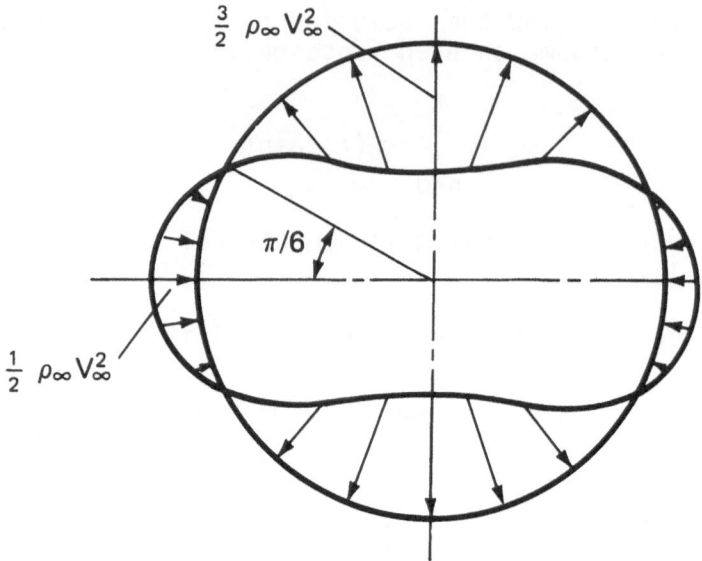

Fig. 2. Distribution of the variable part of pressure on a rigid cylinder under an unseparated flow.

unseparated ideal incompressible flow and it is well known in hydromechanics as D'Alembert's paradox, i.e., the absence of a resultant force acting on the body.

Consider the second part of the problem. The solution ϕ_2 of equation (1.13) has the same form as ϕ_1 in (1.18). According to condition (1.14) $B_0 = B_n = 0$.

Let us represent the deflection function also by a series of even functions of angle θ:

$$w = \sum_{n=0}^{N} W_n \cos n\theta \qquad (1.21)$$

Note that in view of the assumption of inextensibility of the shell middle surface, $W_0 = 0$.

Substituting (1.21) and expressions for ϕ_1 and ϕ_2 in condition (1.15) and equating coefficients of terms like $\cos K\theta$, we find

$$A_0 = 0, \quad A_n = V_\infty R^n(W_{n+1} - W_{n-1}) \quad (n>1)$$

Consequently,

$$\phi_2 = V_\infty \sum_{n=1}^{N} (\frac{R}{r})^n (W_{n+1} - W_{n-1}) \cos n\theta \qquad (1.22)$$

As there is no derivative of p_2 with respect to the radial coordinate on the right-hand side of (1.16) (as distinct from p_1), for computation of p_2 using (1.13) the first term there may be omitted ($\partial\phi_1/\partial r = 0$ for $r = R$). Then

$$P_2 = \rho_\infty V_\infty^2 R^{-1} \sum_{n=1}^{N} n(W_{n+1}-W_{n-1})[\cos(n-1)\theta-\cos(n+1)\theta]$$

$$(1.23)$$

Thus, all the terms on the right-hand side of the bending equation (1.16) are expressed in terms of parameters of the flow and are functions of the shell deflection. Substituting expressions (1.20), (1.21), (1.23) in (1.16) and using the formulae

$$2\cos\alpha\cos\beta = \cos(\alpha - \beta) + \cos(\alpha + \beta),$$

$$2\sin\alpha\sin\beta = \cos(\alpha - \beta) - \cos(\alpha + \beta)$$

one obtains

$$- \frac{1}{2\mu} \sum_{n=1}^{} (n^2-1)^2 n^2 W_n \cos n\theta = 4R\cos 2\theta$$

$$+ \sum_{n=1}^{} n(W_{n+1}-W_{n-1})[(n^2-1)^2\cos(n-1)\theta-(n+1)^2\cos(n+1)\theta]$$

$$+ 2 \sum_{n=1}^{} n^2 W_n \cos n\theta - \sum_{n=1}^{} W_n[(n-2)^2\cos(n-2)\theta+(n+2)^2\cos(n+2)\theta]$$

$$+ \sum_{n=1}^{} \frac{1}{n} W_n[(n-2)^2\cos(n-2)\theta-(n+2)^2\cos(n+2)\theta]$$

Here the relation of dynamic pressure (velocity head) $\rho_\infty V_\infty^2/2$ to the value D/R^3, characterizing the shell stiffness, is denoted by μ

$$\mu = \frac{\rho_\infty V_\infty^2 R^3}{2D} \qquad (1.24)$$

which we shall call the aeroelastic parameter.

In this trigonometric sum we equate all the coefficients of $\cos k\theta$. Therefore, we obtain the system of algebraic equations for W_n:

$$W_3 = 0,$$

$$(\frac{9}{2\mu} - 2)W_2 + \frac{9}{4} W_4 = -R,$$

$$\qquad (1.25)$$

$$(n-3)W_{n-2} + [\frac{(n^2-1)^2}{2\mu} - 2(n-1)]W_n + (n + \frac{1}{n+2})W_{n+2} = 0 \quad (n>3)$$

As follows from equations (1.25) the deflection amplitudes with the odd indices are identically zero. For N=4 from (1.25) the system of two equations follows:

$$(\frac{9}{2\mu} - 2)W_2 + \frac{9}{4} W_4 = -R,$$

$$W_2 + (\frac{225}{2\mu} - 6)W_4 = 0$$

hence

$$W_2 = - \frac{2R\mu}{9-4\mu(1 + \frac{9\mu}{900-48\mu})} , \qquad (1.26)$$

$$W_4 = \frac{4R\mu^2}{[9-4\mu(1 + \frac{9\mu}{900-48\mu})](225 - 12\mu)}$$

Since the external normal to the shell is taken as the positive direction of w, it follows from (1.21) and (1.26) that the cylindrical shell is stretched symmetrically across the stream (Fig. 3). When the velocity increases or the stiffness characteristics of

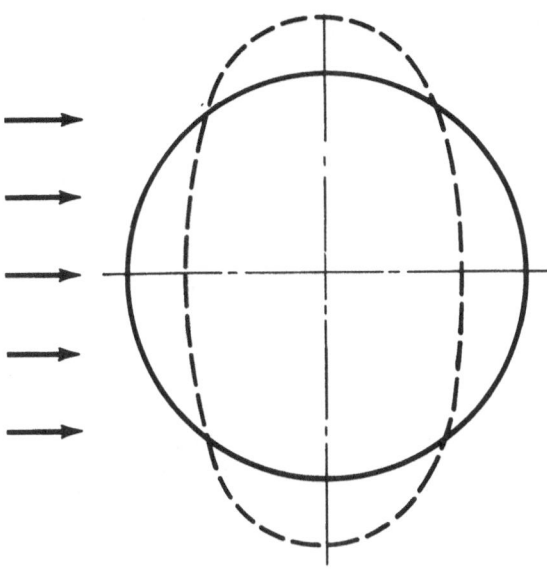

Fig. 3. The shell shape under its potential unseparated flow.

a shell decrease (when μ increases), its deflections
increase monotonically. Within the framework of
linear theory they grow without limit, if the aeroelastic parameter μ approaches the value

$$\mu_+ = 2.20 \qquad (1.27)$$

This is called the critical value of μ.

It should be noted, however, that the linear
theory ceases to be valid long before the parameter μ
reaches the critical value of (1.27). Therefore, the
value μ_+ is only some reference condition for the
order of the aeroelastic parameter values. The solution obtained is correct only for $\mu \ll \mu_+$. For $\mu > \mu_+$
the amplitudes W_2 and W_4 change their sign, which corresponds to stretching of the shell along the stream.
This result is physically invalid.

The linear approximation will be removed in the
next sections of this Chapter.

As one can see from (1.26), the series (1.21), (1.22), (1.23) quickly converge. If, for example, we consider the case of $N = 2$ or 3, then instead of (1.26) and (1.27) we obtain

$$W_2 = -2R\mu(9-4\mu)^{-1}, \qquad \mu_+ = 2.25 \qquad (1.28)$$

Thus, in the problem considered the harmonic $\cos 2\theta$ gives a rather complete picture of the flow and shell interaction.

3.2. Influence of distinction between deformed and undeformed surfaces in contact conditions

Let us estimate the contribution to the above solution of the distinction between deformed and undeformed surfaces when writing the contact conditions, that is of the last member on the right-hand side of (1.15) and the two last members on the right-hand side of (1.16). For this purpose we consider the relation of all these members to the main ones:

$$\frac{(\partial^2 \phi_1/\partial r^2)R^2 w}{(\partial \phi_1/\partial \theta)(\partial w/\partial \theta)} = \frac{\sum\limits_{s=1,3,\ldots} (W_{s+1} + W_{s-1})\cos s\theta}{\sum\limits_{s=1,3,\ldots} [(s+1)W_{s+1} - (s-1)W_{s-1}]\cos s\theta},$$

$$(2.1)$$

$$\frac{(\partial p_1/\partial r)w}{p_2} = \frac{\sum\limits_{s=2,4,\ldots} (2W_s - W_{s+2} - W_{s-2})\cos s\theta}{\sum\limits_{s=2,4,\ldots} [(s+1)W_{s+2} - 2sW_s + (s-1)W_{s-2}]\cos s\theta}$$

Here a resummation in series taking into account zero amplitudes W_n with the odd indexes and $W_0 = 0$ is performed. For example, the sum $(\partial^2 \phi_1/\partial r^2)w$ is written as follows

$$\sum_{n=1} W_n \cos n\theta \cos \theta = \frac{1}{2} \sum_{n=1} W_n [\cos(n-1)\theta + \cos(n+1)\theta]$$

$$= \frac{1}{2} [W_2 \cos \theta + (W_2 + W_4)\cos 3\theta, \ldots]$$

$$= \frac{1}{2} \sum_{s=1,3,\ldots} (W_{s-1} + W_{s+1})\cos s\theta$$

Because of the rapid convergence of the series herein, by taking into account only the principal amplitudes W_2 and W_4, we conclude that the terms mentioned in the bending of a shell along whose curvilinear coordinate a fluid flows are of the same order of magnitude as the main terms. In fact at $\theta = \pi/3$ the first of relations (2.1) equals $\sim 1/6$ (it is taken into account that $W_2 \gg W_4$). For some values of the angle θ it becomes zero, while at other values of θ it is infinity (in view of the fact that the tangential velocity component $\partial\phi_1/\partial\theta$ and also the deflection function w and its derivative become zero). The second relation at $\theta = 0$ equals ~ 1, while at $\theta = \pi/3$ it has the order of magnitude $(225-12\mu)/2\mu$.

The more readily seen are the estimates of the terms mentioned in their contribution to the contact conditions over the direct solution of problem. We begin by estimation of the influence of the last term on the right-hand side of equation (1.16). If it is neglected, then instead of (1.25) we obtain

$$W_1 - W_3 = 0,$$

$$(9/2\mu - 2)W_2 + 2W_4 = -R,$$

$$(n-2)W_{n-2} + [(n^2-1)^2/2\mu - 2(n-1)]W_n + nW_{n+2} = 0$$

$$(n \geqslant 3)$$

hence, at $N = 4$ ($W_1 = W_3 = 0$)

$$W_2 = - \frac{2R\mu}{9-4\mu(1 + \frac{4\mu}{225-12\mu})} ,$$

$$(2.2)$$

$$W_4 = \frac{8R\mu^2}{[9-4\mu(1 + \frac{4\mu}{225-12\mu})](225-12\mu)}$$

Comparing these values of W_2, W_4 with (1.26) shows that difference between them is small. In particular, at N=2 the value of W_2 exactly coincides with (1.28).

The critical value of μ in solution of (2.2) equals

$$\mu_+ = 2.16,$$

$$(2.3)$$

which is close to (1.27).

If we also discard the term $w\partial p_1/\partial r$ on the right-hand side of (1.16), then by reasoning analogous to the above one has

$$W_2 = - \frac{2R\mu}{9-8\mu(1 + \frac{9\mu}{550-32\mu})} ,$$

$$(2.4)$$

$$W_4 = \frac{12R\mu^2}{[9-8\mu(1 + \frac{9\mu}{550-32\mu})](225-16\mu)}$$

The corresponding critical value of μ is

$$\mu_+ = 1.12 ,$$

that is twice as small as (1.27).

Furthermore, we consider the influence of the last term in the kinematic condition (1.15). We shall not take it into account, but retaining the term $w\partial p_1/\partial r$ in (1.16), we obtain the following expressions for an additional pressure

$$P_2 = \rho_\infty V_\infty^2 R^{-1} \sum_{n=1}^{N} [(n+1)W_{n+1} - (n-1)W_{n-1}]$$

$$[\cos(n-1)\theta - \cos(n+1)\theta] \qquad (2.5)$$

while for the amplitudes of the deflection

$$W_2 = - \frac{2R\mu}{9-4\mu(1 + \frac{3\mu}{225-12\mu})} ,$$

$$(2.6)$$

$$W_4 = \frac{4R\mu^2}{[9-4\mu(1 + \frac{3\mu}{225-12\mu})](225-12\mu)}$$

The critical value of μ equals

$$\mu_+ = 2.18 \qquad (2.7)$$

and it is little different from the value of (1.27).

Thus, at the contact conditions the most essential of the terms generated by a distinction between the deformed and undeformed surfaces is the term $w\partial p/\partial r$ from dynamical condition (1.4) or $w\partial p_1/\partial r$ from (1.16).

If one neglects the last term in (1.15) and also the two last terms in (1.16) then p_2 is given by expression (2.5) while the amplitudes W_2, W_4 equal

$$W_2 = - \frac{2R\mu}{9-8\mu(1 + \frac{4\mu}{225-16\mu})} ,$$

$$(2.8)$$

$$W_4 = \frac{8R\mu^2}{[9-8\mu(1 + \frac{4\mu}{225-16\mu})](225-16\mu)}$$

Thus, the parameter of aeroelasticity has a critical value

$$\mu_+ = 1.10 \qquad (2.9)$$

instead of $\mu_+ = 2.20$ from (1.27).

It should be noted that in the case of flow over the surface along the rectilinear coordinate (flow around a plate parallel to its plane, or a cylindrical shell along its generatrix, etc.) the above mentioned terms in (1.8) and (1.9) vanish, and when writing of contact conditions, the deformed and undeformed surfaces are identical. Therefore we obtain the contact conditions used in classical linear theory of the divergence and the flutter of the thin lifting bodies. Consequently, in considering the problem of the deformation of a bluff body (cylinder), the classical interaction theory gives a critical value of μ_+ twice as small as that from the exact theory.

Finally we consider a problem with the greatest simplification. The pressure acting on the shell we will find from the assumption that it is undeformed. Consequently, all the pressure on the shell surface reduces to p_1 from (1.20). Then the shell deformation is determined from equation (1.16) with the known right-hand side. The given solution may be obtained from the next formulae for the deflection amplitudes by neglecting in the denominator terms containing the value μ:

$$W_2 = -2R\mu/9, \quad W_4 = W_6 = \ldots = 0 \qquad (2.10)$$

Since in the given approximation the problem ceases to be one of interaction, there is no critical aeroelasticity parameter. This solution may be made more precise by using the method of successive approximations, and there are efficient methods to accelerate the convergence [9].

Thus, we have obtained problem solutions with different levels of accuracy of the contact conditions on the shell and fluid surfaces. The values of dimensionless deflection $w = (W_2 + W_2)/R$ at frontal point A ($\theta = 0$) are shown in Fig. 4. There solution (1.26), obtained by satisfying the exact contact conditions, is denoted by number I. Solution (2.2), where in the dynamic contact condition the influence of the displacement tangential component is not taken into

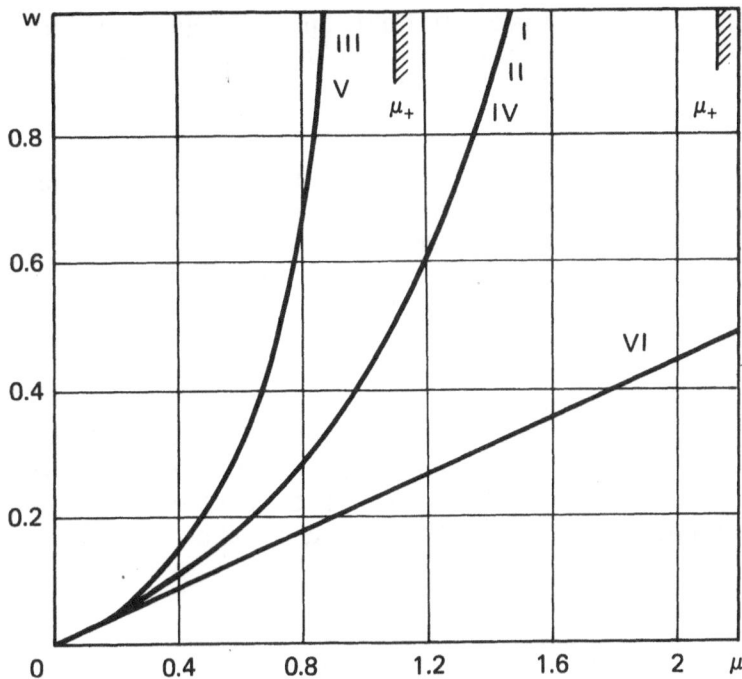

Fig. 4. Influence of the contact conditions on the solutions to various equations: I, (1.26); II, (2.2); III, (2.4); IV, (2.6), V, (2.8), VI, (2.10).

account, is denoted by number II; while III - solution (2.4) corresponds to assuming that in the dynamical condition the deformed and undeformed surfaces are identical (in the kinematic condition this distinction remains); number IV - solution (2.6) is obtained with assumptions that are opposite to solution (2.4). Solution (2.8), obtained by using the contact conditions of classical divergence theory, is denoted by number V; while solution (2.10), found from the hydrodynamical pressure on an absolutely rigid shell, is denoted by number VI.

3.3. Dynamical behavior of a shell in a flow

It is of great interest to consider the problem in a dynamical formulation. All the previous assumptions are retained without any changes. Therefore, equations and conditions (1.1), (1.2), (1.8), (1.9) with the connecting initial conditions are still valid.

The influence of fluid contained within the shell is neglected. This may be allowed when, for example, the fluid density is small as compared to the densities of the shell and external flowing liquid. This influence easily may be taken into account if necessary.

Dynamical phenomena may take place with shell static bending and a corresponding flow field of a liquid around it. Therefore the disturbed values of the velocity potential $\phi*$, the pressure $p*$ and the shell deflection $w*$ are assumed in the form

$$\phi* = \phi_0(r,\theta) + \phi(r,\theta,t) ,$$

$$p* = p_0(r,\theta) + p(r,\theta,t) , \qquad (3.1)$$

$$w* = w_0(\theta) + w(\theta,t)$$

In further approximate analysis of the system behavior we identify an initial static state of the shell with its undeformed state: $w_0 = 0$. Correspondingly, $\phi_0 = \phi_1 + \phi_2 \approx \phi_1$, $p_0 = p_1 + p_2 \approx p_1$, where according to designation of previous section, ϕ_1, p_1 are the flow parameters in the case of the absolute rigid shell, while ϕ_2, p_2 are their disturbances induced by the shell deformation. Influence of the shell deformed state and the corresponding changed flow field on dynamic behavior of the shell will be taken into account in section 3.7. Thus,

$$\phi* = \phi_1 + \phi, \quad p* = p_1 + p, \quad w* = w \qquad (3.2)$$

If in equations and conditions (1.1), (1.2), (1.8), (1.9) we ascribe the lower indices (stars) to parameters ϕ, p, w and take into account (3.2), then we obtain the necessary relations for disturbances ϕ, p, w. Dynamical deflections are considered to be small. Therefore, $\partial\phi/\partial\theta \ll \partial\phi_1/\partial\theta$, $p \ll p_1$ and the equations may be linearized. Thus, on the right-hand

sides of (1.8) and (1.9) we omit the last term.   As
is known from the previous sections, at linearization
the last term vanishes in (1.8), while the last term's
influence in (1.9) is negligible.

Consequently, we have the following equations for
the disturbances:

$$\nabla^2 \phi = 0,$$

$$p = -\rho_\infty \left(\frac{\partial \phi}{\partial t} + \frac{\partial \phi_1}{\partial r} \frac{\partial \phi}{\partial r} + \frac{1}{r^2} \frac{\partial \phi_1}{\partial \theta} \frac{\partial \phi}{\partial \theta} \right)$$

$$(3.3)$$

The condition at infinity is

$$\phi = 0 \quad (r \to \infty) \qquad (3.4)$$

The kinematic and dynamic conditions are

$$\frac{\partial w}{\partial t} - \frac{\partial \phi}{\partial r} - \frac{\partial^2 \phi_1}{\partial r^2} w + \frac{1}{r^2} \frac{\partial \phi_1}{\partial \theta} \frac{\partial w}{\partial \theta} = 0 \quad (r=R), \qquad (3.5)$$

$$\frac{\partial^6 w}{\partial \theta^6} + 2 \frac{\partial^4 w}{\partial \theta^4} + \frac{\partial^2 w}{\partial \theta^2} + \frac{\rho_s h R^4}{D} \frac{\partial^2}{\partial t^2} \left(\frac{\partial^2 w}{\partial \theta^2} - w \right)$$

$$= -\frac{R^4}{D} \frac{\partial^2}{\partial \theta^2} \left(p + \frac{\partial p_1}{\partial r} w \right) \quad (r=R) \qquad (3.6)$$

Without going into the details of initial conditions, we find the solution of the problem in the
form

$$w = e^{\lambda t} \sum_{n=0}^{N} W_n \cos n\theta, \qquad \phi = e^{\lambda t} \sum_{n=0}^{N} \Phi_n \cos n\theta \qquad (3.7)$$

As for the static bending, $W_0 = 0$ because of the shell
inextensibility, and for the angle θ we assume symmetry about the line passing across the points 0, π
(Fig. 1).

In the expression for the potential (3.7) the solution of the Laplace equation (3.3) has the form (1.18). Subjecting this solution to the condition (3.4) ($B_n = 0$) and substituting (3.7), (1.19) in (3.5), we write

$$A_n = - (\lambda/n)R^{n+1}W_n + V_\infty R^n(W_{n+1} - W_{n-1}) \quad (n>1)$$

Consequently,

$$\phi = e^{\lambda t} \sum_{n=1}^{N} [-(\lambda R/n)W_n + V_\infty(W_{n+1}-W_{n-1})](R/r)^n\cos n\theta \quad (3.8)$$

The formula to determine the pressure disturbance (3.3) on the shell surface (r=R) has the form

$$p = \rho_\infty V_\infty R^{-1} \sum_{n=1}^{N} [\lambda R W_n - nV_\infty(W_{n+1}-W_{n-1})][(\lambda R/nV_\infty)\cos n\theta$$

$$+ \cos(n+1)\theta - \cos(n-1)\theta]e^{\lambda t} \quad (3.9)$$

Here it is taken into account that $\partial\phi_1/\partial r = 0$ for r=R.

Substitution of (3.7), (3.9), (1.20) into (3.6) and coefficients of identical arguments in $\cos k\theta$ to zero yields

$$[(1 + \nu)X^2 - 1]W_1 - 2XW_2 + W_3 = 0 \quad (W_0 \equiv 0),$$

$$[(n^2 - 1)^2\mu^{-1} + (n^{-2} + 1)\nu X + n^{-1}X^2$$

$$- 2(n-1)]W_n + 4X(W_{n-1} - W_{n+1}) + 2(n-2)W_{n-2}$$

$$+ 2nW_{n+2} = 0 \quad (n>2) \quad (3.10)$$

where in addition to the aeroelastic parameter μ from (1.24) the following definitions are introduced:

$$\nu = \frac{2\rho_s h}{\rho_\infty R} , \quad X = \frac{R\lambda}{V_\infty} \quad (3.11)$$

Parameter ν is the ratio of the shell wall mass to the equivalent mass of the fluid.

According to (3.10) for N=2 from the condition of existence of nontrivial solutions for W_1, W_2 we obtain

$$(1 + \nu)[1 + (5/4)\nu]X^4 + [7 - (5/4)\nu$$

$$- (1 + \nu)(4 - 9\mu^{-1})]X^2 + (4 - 9\mu^{-1}) = 0 \quad (3.12)$$

Hence it is seen that for $X = 0$ ($\lambda = 0$) the value of μ for which unlimited stretching of the shell occurs is $\mu_+ = 2.25$, which agrees with (1.28).

Substituting $X = \varepsilon + i\omega$ into (3.12) and eliminating ω from the two equations obtained, we arrive at a biquadratic equation in ε. One of its roots will be positive if

$$[7 - (5/4)\nu - (1 + \nu)(4 - 9\mu^{-1})]^2$$

$$- 4(1 + \nu)[1 + (5/4)\nu](4 - 9\mu^{-1}) < 0$$

from which the value of μ corresponding to the onset of dynamic instability is

$$\mu_{++} = \frac{9(1 + \nu)}{\dfrac{1}{32 + 40\nu} + (11/4)\nu - 5} \quad (3.13)$$

A graph of the function (3.13) is presented in Fig. 5. It is seen that the smallest value of μ_{++} is in the range $3 < \nu < 10$ and is close to the critical parameter $\mu_+ = 2.16$ for the static case obtained when N = 2. The values of μ_+ are also plotted. The limiting values equal $\mu_{++} = 13.65$ (at $\nu = 0$) and $\mu_{++} = 3.27$ (at $\nu = \infty$).

In an analysis of the third-order determinant (3.10) (N = 3) calculation of the roots for different ν was carried out numerically. They are also given in Fig. 5 for comparison with the results of the two-term approximation (3.13) (interaction of oscillating

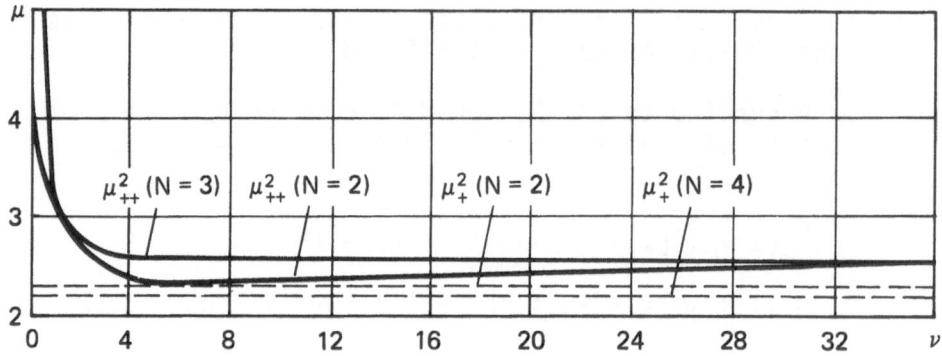

Fig. 5. Critical parameters of aeroelasticity.

shapes $\cos\theta$ and $\cos2\theta$). A further increase in the order of the determinant (3.10) (N = 4, for example, that is an analysis of interaction of the first four oscillating shapes) does not result in a significant change in the value of μ_{++}.

3.4. <u>Bending of a shallow cylindrical curved plate in a flow</u>

The plane deformation of an infinitely long shallow cylindrical curved plate 1, which is a part of a cylindrical surface with an unseparated transverse flow around it of a steady unbounded ideal incompressible fluid (Fig. 6), is considered. The remaining part 2 of the surface is considered to be absolutely rigid. Thus, the same problem as in section 3.1 is formulated. Hence, the assumptions concerning the fluid motion taken there are retained. As concerns the shallow shell deformation, the possibility of its "turning out" is allowed. Therefore, we consider the nonlinear equations in which a membrane stress and the shell curvature change are taken into account. In such a case the assumption of the middle surface inextensibility is inapplicable and it is not used here.

Taking into account the known simplifications for a shallow shell, namely, neglecting the shear force Q_θ

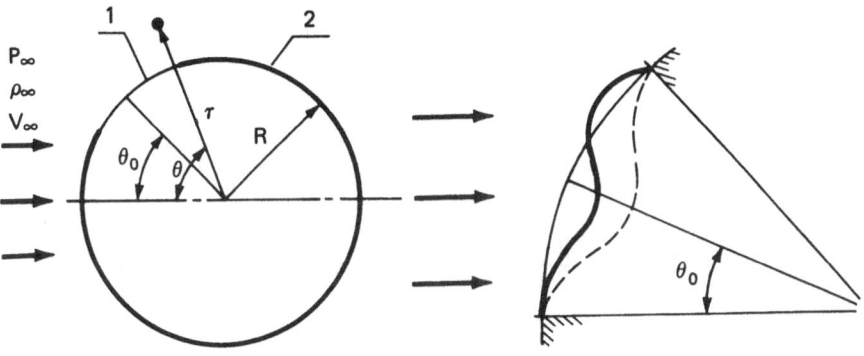

Fig. 6. Cylindrical curved plate in the flow.

in the first equation and the derivative of the tangential component v in the curvature change expression, we can write the bending equations as

$$\partial N_{\theta\theta}/\partial\theta = 0,$$

$$\frac{\partial Q_\theta}{R\partial\theta} - \frac{N_{\theta\theta}}{R^*} = - Z_r,$$

$$\frac{\partial M_{\theta\theta}}{\partial\theta} - RQ_\theta = 0,$$

$$\frac{1}{R^*} = \frac{1}{R} + X_{\theta\theta} = \frac{1}{R} - \frac{\partial^2 w}{R^2 \partial\theta^2}, \qquad (4.1)$$

$$N_{\theta\theta} = B\varepsilon_{\theta\theta} = B[\frac{\partial v}{R\partial\theta} + \frac{w}{R} + \frac{1}{2}(\frac{\partial w}{R\partial\theta})^2],$$

$$M_{\theta\theta} = -\frac{D}{R^2}\frac{\partial^2 w}{\partial\theta^2}, \quad B = \frac{Eh}{1-\nu^2}, \quad D = \frac{Eh^3}{12(1-\nu^2)}$$

Integrating the first equation, we find

$$\frac{CR}{B} = \frac{\partial v}{\partial \theta} + w + \frac{1}{2R} \left(\frac{\partial w}{\partial \theta}\right)^2 \qquad (4.2)$$

in which C is an integration constant. Excluding from the remaining relations $M_{\theta\theta}$, Q_θ and taking into account (4.2), we have

$$\frac{\partial^4 w}{\partial \theta^4} + \frac{CR^3}{D} \left(1 - \frac{\partial^2 w}{R \partial \theta^2}\right) = \frac{R^4}{D} Z_r \qquad (4.3)$$

Then we suppose that the curved plate is clamped (fastened) to the cylinder edges

$$v = w = \partial w/\partial \theta = 0 \qquad (\theta = \theta_1, \theta_2) \qquad (4.4)$$

If the flow field changes weakly in our problem (as may be concluded from section 3.1) for the case of shallow shell, the kinematic and dynamic conditions are

$$\frac{\partial \phi}{\partial r} = \frac{1}{r^2} \frac{\partial w}{\partial \theta} \frac{\partial \phi}{\partial \theta} - w \frac{\partial^2 \phi}{\partial r^2} \qquad (r=R), \qquad (4.5)$$

$$Z_r = P_i - p - w \frac{\partial p}{\partial r} \qquad (r=R) \qquad (4.6)$$

In the same sense as in (1.10), let us introduce the potentials ϕ_1, ϕ_2 and the pressures p_1, p_2. Then the problem for ϕ_1, p_1 has the form (1.11) and (1.12). As concerns the problem for ϕ_2, p_2, w, v, then (1.13) and (1.14) remain unchanged, but in the kinematic condition there is a difference, induced by the nonuniformity of the surface, i.e.,

$$\frac{\partial \phi_2}{\partial r} = \frac{1}{r^2} \frac{\partial w}{\partial \theta} \frac{\partial \phi_1}{\partial \theta} - w \frac{\partial^2 \phi_1}{\partial r^2} \qquad (r=R, \; \theta_1 < \theta < \theta_2), \qquad (4.7)$$

$$\partial \phi_2/\partial r = 0 \qquad (r=R, \; \theta_2 < \theta < 2\pi + \theta_1) \qquad (4.8)$$

Instead of (1.16) we have from (4.3) and (4.6) the equation

$$\frac{\partial^4 w}{\partial \theta^4} + \frac{CR^3}{D}(1 - \frac{\partial^2 w}{R \partial \theta^2}) = \frac{R^4}{D}(p_i - p_1 - p_2 - w \frac{\partial p_1}{\partial r})\ (\substack{r=R \\ \theta_1 < \theta < \theta_2})$$

$$(4.9)$$

which is considered in common with (4.2).

Solution of the problem for ϕ_1 and p_1 is known (1.19), (1.20). Solving the problem for ϕ_2, p_2, w, v, we present the function w, satisfying two last conditions of (4.4), in the form

$$-w = \frac{L^2}{R} \{ \sum_{i=1}^{I} f_i[1 + (-1)^{i-1} \cos i \pi \xi]$$

$$(4.10)$$

$$+ \sum_{j=1}^{J} q_j \sin j \pi \xi \cos j \frac{\pi}{2} \xi \}$$

in which the first sum gives the symmetrical form of the bending, while the second gives the nonsymmetrical form. Here $\xi = (\theta - \theta_0)R/L$, where $2L$ is the curved plate length.

The potential ϕ_2, satisfying Laplace equation (1.13) and condition (1.14), is written as

$$\phi_2 = \sum_{n=1}^{N} (\frac{R}{r})^n (A_n \cos n\theta + B_n \sin n\theta) \qquad (4.11)$$

Such a form for the potential is similar to expression (1.22). But unlike the problem previously considered the symmetry about the line passing across points $0; \pi$ is absent here, since the plate disposition with respect to the angle is arbitrary.

Substituting the expressions (1.19), (4.10), (4.11) into (4.7) and (4.8) and integrating them by the Galerkin method

$$\int_0^{2\pi} \frac{\partial \phi_2}{\partial r} \binom{\cos n\theta}{\sin n\theta} d\theta = \int_{\theta_1}^{\theta_2} (\frac{\partial w}{r^2 \partial \theta} \frac{\partial \phi_1}{\partial \theta} - w \frac{\partial^2 \phi_1}{\partial r^2}) \binom{\cos n\theta}{\sin n\theta} d\theta,$$

we find

$$A_n = \frac{2V_\infty L}{\pi n} \left(\sum_{i=1}^{I} C_{ni} f_i + \sum_{j=1}^{J} D_{nj} q_j \right),$$

$$(4.12)$$

$$B_n = \frac{2V_\infty L}{\pi n} \left(\sum_{i=1}^{I} F_{ni} f_i + \sum_{j=1}^{J} G_{nj} q_j \right)$$

Coefficients C_{ni}, D_{nj}, F_{ni}, G_{nj} entering here have the form

$$C_{ni} = (i\pi)^2 n \left[\frac{\cos(n+1)\theta_o \sin\beta_{n+1}}{(n+1)(\beta_{n+1}^2 - i^2\pi^2)} - \frac{\cos(n-1)\theta_o \sin\beta_{n-1}}{(n-1)(\beta_{n-1}^2 - i^2\pi^2)} \right],$$

$$D_{nj} = \frac{(-1)^j}{2} \left[\left(\frac{\ell\beta_{n+1} + \alpha_{j+1}^2}{\beta_{n-1}^2 - \alpha_{j+1}^2} - \frac{\ell\beta_{n-1} + \alpha_{j-1}^2}{\beta_{n-1}^2 - \alpha_{j-1}^2} \right) \sin(n-1)\theta_o \cos\beta_{n-1} \right.$$

$$\left. + \left(\frac{\ell\beta_{n+1} - \alpha_{j+1}^2}{\beta_{n+1}^2 - \alpha_{j+1}^2} - \frac{\ell\beta_{n+1} - \alpha_{j-1}^2}{\beta_{n+1}^2 - \alpha_{j-1}^2} \right) \sin(n+1)\theta_o \cos\beta_{n+1} \right],$$

$$F_{ni} = (i\pi)^2 n \left[\frac{\sin(n+1)\theta_o \sin\beta_{n+1}}{(n+1)(\beta_{n+1}^2 - i^2\pi^2)} - \frac{\sin(n-1)\theta_o \sin\beta_{n-1}}{(n-1)(\beta_{n-1}^2 - i^2\pi^2)} \right],$$

$$G_{nj} = \frac{(-1)^j}{2} \left[\left(\frac{\ell\beta_{n-1} + \alpha_{j-1}^2}{\beta_{n-1}^2 - \alpha_{j-1}^2} - \frac{\ell\beta_{n+1} + \alpha_{j+1}^2}{\beta_{n-1}^2 - \alpha_{j+1}^2} \right) \cos(n-1)\theta_o \cos\beta_{n-1} \right.$$

$$\left. + \left(\frac{\ell\beta_{n+1} - \alpha_{j-1}^2}{\beta_{n+1}^2 - \alpha_{j-1}^2} - \frac{\ell\beta_{n+1} - \alpha_{j+1}^2}{\beta_{n+1}^2 - \alpha_{j+1}^2} \right) \cos(n+1)\theta_o \cos\beta_{n+1} \right]$$

Here it is denoted

$$\alpha_{j\pm1} = (2j\pm1)\pi/2, \quad \beta_{n\pm1} = (n\pm1)\ell, \quad \ell = L/R$$

In the expression for C_{ni} the last term in the square brackets at n = 1 equals $-\ell/i^2\pi^2$, while in F_{ni} the last term at n = 1 vanishes.

According to (1.13), (1.19), (4.11), (4.12) the formula for the pressure disturbance at r=R has the form

$$P_2 = \frac{2\rho_\infty V_\infty^2 \ell}{\pi} \sum_{n=1}^{N} \{ (\sum_{i=1}^{I} C_{ni} f_i + \sum_{j=1}^{J} D_{nj} q_j)[\cos(n-1)\theta - \cos(n+1)\theta]$$

$$+ (\sum_{i=1}^{I} F_{ni} f_i + \sum_{j=1}^{J} G_{nj} q_j)[\sin(n-1)\theta - \sin(n+1)\theta]\} \quad (4.13)$$

In (4.9) and (1.20) the difference $p_i - p_\infty$ is the internal differential of the static pressure. In order to compare the calculations with experimental data we shall consider this differential to be induced by the hydrostatic pressure on the curved plate, immersed together with a rigid cylindrical part in water. If the depth of immersion of the plate center ($\xi = 0$) equals H, then

$$p_i - p_\infty = -\rho_\infty g\{H - R[\sin(\theta_0 + \ell\xi) - \sin\theta_0]\} \quad (4.14)$$

From (4.2) and (4.10), satisfying the condition v = 0 at $\xi = \pm 1$, we obtain

$$\sigma^2 = \frac{3K^2}{4} [\sum_{i=1}^{I} (1 - \frac{i^2\pi^2}{4} f_i)f_i - \frac{\pi^2}{32} \sum_{j=1}^{J} (4j^2+1)q_j^2] \quad (4.15)$$

where the definitions of the curvature dimensionless parameters and of the membrane stress are introduced

$$K = 4L^2/Rh, \quad \sigma^2 = -CL^2/D$$

Integrating equation (4.9) and taking into account (1.20), (4.10), (4.13), (4.14) by the Galerkin method, we find the following equation system:

$$m^4 \pi^4 f_m + (2 - m^2 \pi^2 f_m) \sigma^2$$

$$\text{(4.16)}$$

$$= - \gamma E_m + \mu [K_m + \sum_{n=1}^{N} (\sum_{i=1}^{I} P_{mni} f_i + \sum_{j=1}^{J} Q_{mnj} q_j)]$$

$$(m=1,2,\ldots,I),$$

$$\frac{\pi^2}{16} \{\frac{\pi^2}{4} [(2s+1)^4 + (2s-1)^4] - 2\sigma^2(4s^2+1)\} f_s$$

$$\text{(4.17)}$$

$$= - \gamma H_s + \mu [T_s + \sum_{n=1}^{N} (\sum_{i=1}^{I} R_{mni} f_i + \sum_{j=1}^{J} S_{mnj} q_j)]$$

$$(s=1,2,\ldots,J)$$

Here the parameters of the hydrostatic load and hydroelasticity are introduced:

$$\gamma = -\rho_\infty g H R L^2/D, \qquad \mu = \rho_\infty V_\infty^2 R L^2/2D \qquad \text{(4.18)}$$

The coefficients E_m, K_m, H_s, T_s are expressed by the relative curved plate width ℓ, its mounting angle θ_0, the immersion depth H, and the parameters $\alpha_{i\pm1}$. As concerns P_{mni}, Q_{mnj}, R_{mni}, S_{mnj}, in addition to ℓ, θ_0, $\alpha_{i\pm1}$ they also contain $\beta_{n\pm1}$ and C_{ni}, D_{nj}, F_{ni}, G_{nj}. These coefficients do not reduce further because of their cumbersome form [6].

The system of nonlinear algebraic equations (4.15)-(4.17) for f_i, q_i is solved for the given values of parameters of the hydrostatic loads and hydroelasticity γ, μ. This system is solved numerically by the Newton method [6]. From the found values of f_i, q_i the plate deflection w and the disturbance pressure p are determined according to (4.10), (4.13). The Newton iteration process is continued until the condition

$$|(x_k^{(m+1)} - x_k^{(m)})/x_k^{(m+1)}| < 10^{-5}$$

is reached, in which $x_k^{(m)}$ is the m-th approximation of one of the isolated roots corresponding to some values of parameters γ_k, μ_k.

3.5. <u>Research results</u>: <u>comparison with experiment</u>

The curved plate deformation depending on the hydroelasticity parameter μ may be characterized by the bending change at characteristic points along the plate length. Now we shall observe the dimensionless deflection $\overline{w}_0 = wR/L^2$ at its center at three values of the hydrostatic load parameter $\gamma = 0$, -6, -12. In Fig. 7 they are denoted by numbers 1, 2, 3, respectively.

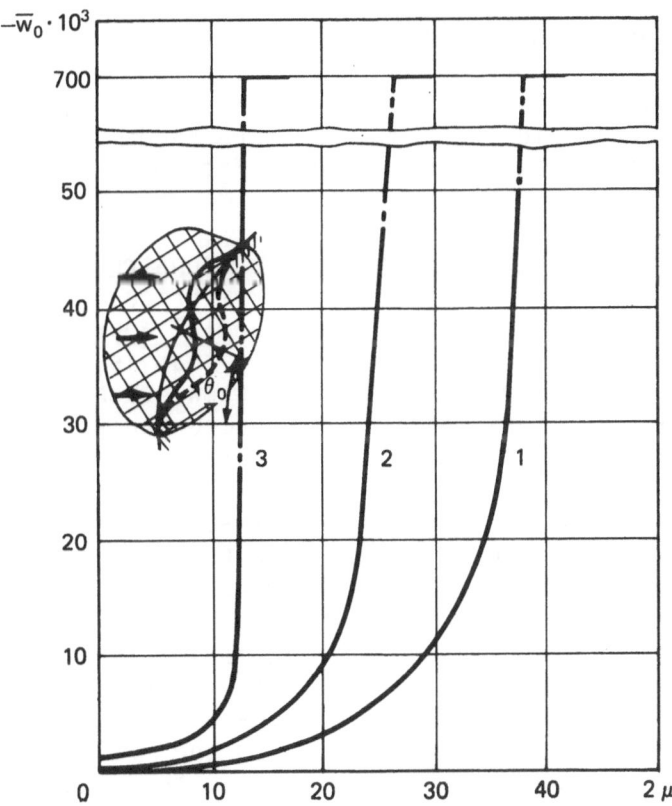

Fig. 7. Dependence of the curved plate bending on the hydroelasticity parameter and the hydrostatic load parameter.

These data are obtained for the version of contact conditions that was denoted in section 3.2 by number III. Hence in (4.7) all the terms are retained, while in (4.9) the last term on the right-hand side of the equation is omitted. Furthermore, we have taken: the curvature parameter k = 222, the mounting angle θ_0 = 13°, and L/R = 0.228.

As follows from the graph, if the parameter μ increases, then, as in the case of the closed cylindrical shell, the deflection increases. When μ approaches some value μ_+, depending on the parameter γ, then the bending increases suddenly (sharply). This part of the curve is plotted by a dot-dash-line. As in the analysis of the closed shell, we shall call this value critical. The deformation process is accompanied by the curved plate "turning out." When the external hydrostatic pressure γ increases, this state occurs at smaller values of μ_+.

Unlike the case of uniform external pressure under "dead" load [18], the plate bends in an unsymmetrical shape, as is shown in Fig. 6 by the continuous line.

The curves in Fig. 7 are constructed for the binomial approximation of the symmetrical part of the plate deflection and for the monomial approximation of the unsymmetrical part of (4.10), that is, I = 2 and J = 1. As numerous calculations show, such an approximation describes sufficiently accurately the plate bending. For example, for the plate with the parameter k = 175 at the mounting angle θ_0 = 30° and the parameter of the hydrostatic pressure differential γ = -10.34 the calculation for one term in (4.10) (I=1, J=1) gives the critical parameter μ_+ = 10.55; while for I=3, J=1 we obtain μ_+ = 10.40; for I=2, J=3 we obtain μ_+ = 10.38. Thus, the number of terms taken for the results plotted in Fig. 7 is quite enough. Note that we approach the true value of μ_+ monotonically from above, as occurs when using the Galerkin method in the problem of a curved plate with a uniform load [18]. Also the practical convergence was verified numerically.

In Fig. 8 comparison of the values for dimen_
sionless deflection at the plate center w_0 obtained by
using the pressure on a plate without taking into
account its deformation (version VI in section 3.2),
with the values from Fig. 7 (version III) for the
parameters $\gamma = 0$, -6 is plotted. The results corresponding to these values of γ are shown in Fig. 8 by
numbers 1 and 2. As in the case for small bending of
the shell considered in section 3.2, the version VI of
the contact condition leads to lower values of deflection. As a consequence of taking into account the
nonlinearities in the shell equations, the deflections
in this problem are finite. They are determined by
the turned out plate position. This state is stable.

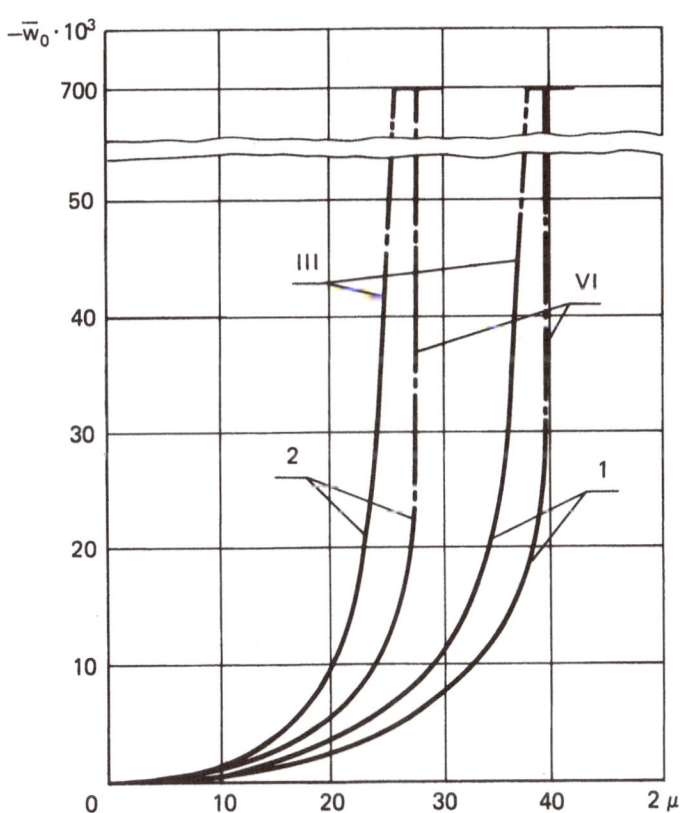

Fig. 8. Influence of the different contact conditions
on the value of a curved plate deflection.

If one assumes that the hydrostatic pressure
(4.14) does not depend on the angle θ (that is $p_i - p_\infty$
$= -\rho_\infty gH$), then for $\theta_0 = 0$ the plate loading about its
center will be symmetrical. In particular, in the
absence of the fluid flow ($\mu = 0$) we have the well-
known problem of a curved plate buckling under a uniform "dead" load. As is known [18], the buckling into
a symmetrical shape may take place only for very
shallow shells (k < 20.17).

Furthermore, we consider the results presented in
Fig. 9. Here the critical hydroelastic parameter μ_+
dependence on the mounting angle θ_0 of the plate on

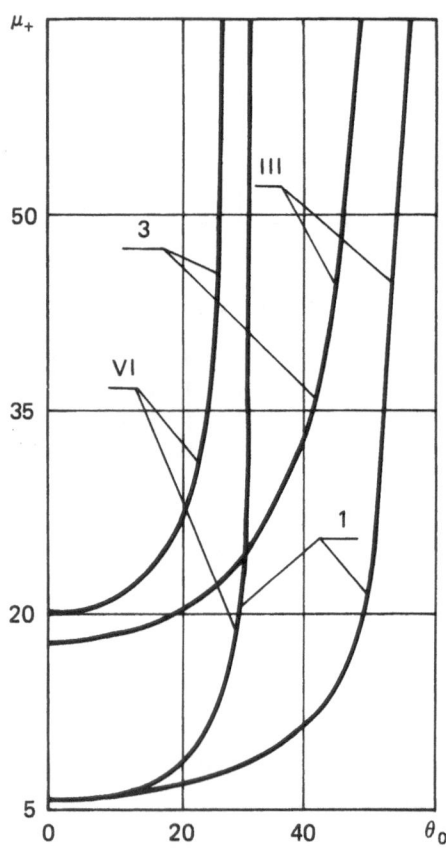

Fig. 9. Dependence of the critical hydroelasticity
parameter upon the curved plate angle of
mounting.

the cylinder for two parameter values $\gamma = 0; -12$ (corresponding curves are denoted by numbers 1 and 3) are shown. These data are just for the plates with the curvature parameter $k > 70$. As is seen, for small mounting angles ($\theta_0 < 15°$) and large $|\gamma|$ (analysis shows that for $-\gamma < 8$) both III and VI versions of the contact conditions give practically identical results. Consequently, in these cases the influence of interaction between the curved plate deformation and the fluid flow is absent. When the mounting angle increases results from version VI are considerably different from those from version III, the latter being more exact. Here the interaction influence is strong.

The increase of the critical values of hydro-elastic parameter μ_+ with growth of the mounting angle θ_0 may be explained by the fact that for motion from the frontal region of a cylinder to the section $\theta = \pi/2$, $3\pi/2$ the pressure falls, while the pressure differential may become negative because of the stream inleakage (the pressure distribution on a rigid cylinder is shown in Fig. 2). In these regions the plate tends to be deflected to one side of the external normal. Thus, in the case of zero hydro-static pressure differential there is no turning out of the curved plate, if the mounting angle reaches 55° (Fig. 9, curve 1). According to the approximate theory (version VI) this angle equals $\theta_0 = 30°$.

Now consider the comparison with the experimental results. At first we consider the stability of a curved plate under the uniform "dead" load ($\mu = 0$). In Fig. 10 the critical values of the dimensionless pressure or the hydrostatic load parameter $-\gamma_+ = (p_i - p_\infty)RL^2/D$ are shown with their dependence on the curvature parameter $k = 4L^2/Rh$. Here the experimental results from [18] are indicated by points, while from [3] they are indicated by crosses. Good agreement between the computational curve and experimental data marked by the crosses, verifies the appropriateness of the computational model for the test specimen and the conditions of the experiment (insignificant initial deviations of the specimen geometrical shape, strict observance of the rigid fastening conditions, etc.).

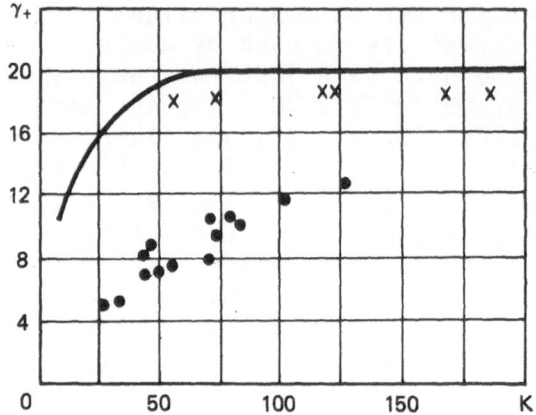

Fig. 10. Comparison of the computational and experimental values of the hydrostatic load
critical parameter.

In Fig. 11 the dependence of the critical flow
velocity V_∞^+ (in m/s) corresponding to the parameter

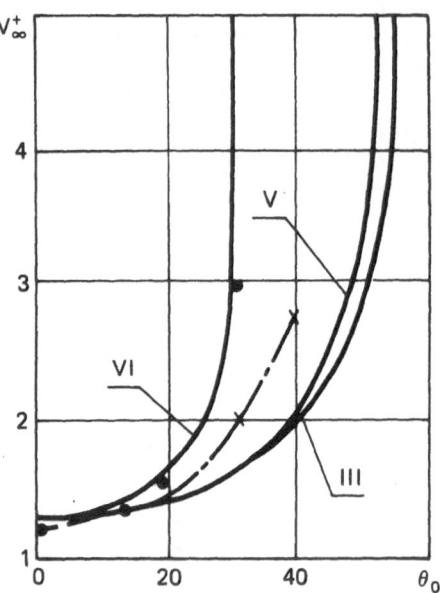

Fig. 11. Influence of the curved plate angle of
mounting on the critical value of a flow
velocity.

μ_+, on the angle of plate mounting θ_0 [3,6] is shown. Here the experimental results obtained in testing of copper specimens M1 with dimensions: $2L = 49$ mm, $R = 107.8$ mm, $h = 0,1$ mm are plotted by points. The curved plate length is $L_1 = 250$ mm and its depth of immersion in water $H = 200$ mm. Experimental data for longer plates ($L_1 = 400$ mm) are plotted in the figure by crosses. The values for experimental data for the specimens, produced from steel 1Kh18N10T having dimensions $2L = 49$ mm, $R = 107.8$ mm, $h = 0.127$ mm, $L_1 = 400$ mm for $H = 600$ mm are denoted in Fig. 12 by crosses.

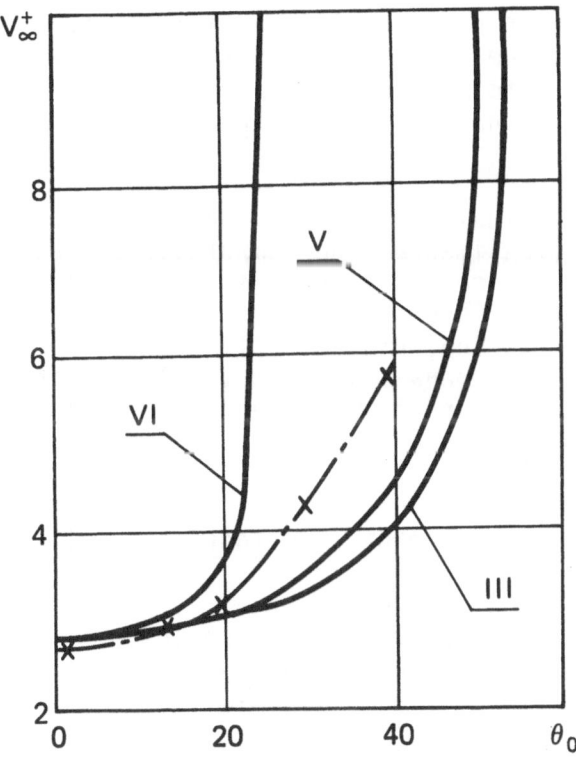

Fig. 12. Influence of the curved plate angle of mounting on the critical value of a flow velocity.

Analysis of the results obtained shows that at
small angles of plate mounting and in the presence of
a hydrostatic pressure both versions of the contact
conditions (version III and VI) give identical results
in practice. Their good agreement with experimental
results is also observed. When the mounting angle
increases, version VI is strongly different from
version III. This difference begins to be seen for
angles $\theta_0 > \pi/9$. Also the results of computation by
using the contact conditions from the classical
divergence theory (version V in section 3.2) are
plotted here, which are little different from the
results of version III.

The value of the plate elongation parameter
$L_1/2L$, for which the plate may be considered to be
infinitely length, is far greater than it is in the
case of a uniform loading (in calculations [18] it is
obtained when $L_1/2L \geqslant 3$). In fact, the experimental
critical value of V_∞^+ for L = 400 mm ($L_1/2L$ = 8.15) is
much closer to more exact computational values (curves
III and V), than V_∞^+ for the plate having a length L =
250 mm.

In conclusion we note that we did not present
here the experimental and computational data for that
case when a curved plate is located in the bottom
region of cylinder ($\theta_0 > \pi/2$). In this region fluid
stream separation, plate oscillations, and other complicated phenomena take place. Taking into account
separation of the flow around the cylinder will be
discussed in section 3.8.

3.6. Large bending of a cylindrical shell in a flow

The plane deformation of an infinitely long thin-
walled shell with a transverse flow around it of an
unbounded ideal incompressible fluid with parameters
at infinity p_∞, ρ_∞, V_∞, is considered. The model of
unseparated fluid flow along the surface is taken.
One assumes that the pressure p_i within it is constant. Thus, the part of the problem concerning fluid
flow is the same as in section 3.1. The difference is

connected with shell deformation. We consider the shell displacement along a normal to be comparable with its radius. Therefore one has to invoke the large bending equations. Moreover, as in section 3.1, we use an assumption on inextensibility for the shell middle surface.

The problem concerning the velocity potential ϕ and the hydrodynamic pressure p reduces to the solution of system (1.1), (1.2). The kinematic condition is

$$\frac{\partial w}{\partial t} - \frac{1}{r^2}(v - \frac{\partial w}{\partial \theta})[\frac{\partial \phi}{\partial \theta} + v\,(\frac{\partial^2 \phi}{r \partial \theta^2} + \frac{\partial \phi}{\partial r}) + w\,\underline{(\frac{\partial^2 \phi}{\partial \theta \partial r} - \frac{\partial \phi}{r \partial \theta})}]$$

$$- (1 + \frac{\partial v}{r \partial \theta} + \frac{w}{r})[\frac{\partial \phi}{\partial r} + \frac{v}{r}\,\underline{(\frac{\partial^2 \phi}{\partial \theta \partial r} - \frac{\partial \phi}{r \partial \theta})} + w\,\underline{\frac{\partial^2 \phi}{\partial r^2}}]$$

$$- \underline{\frac{w^2}{2}\frac{\partial^3 \phi}{\partial r^3}} = 0 \qquad (r = R) \qquad (6.1)$$

The dynamic condition gives

$$Z_r^* = p_i - p - \frac{v}{r}\frac{\partial p}{\partial \theta} - \underline{w\frac{\partial p}{\partial r}} - \underline{\frac{w^2}{2}\frac{\partial^2 p}{\partial r^2}} \qquad (r=R) \qquad (6.2)$$

In (6.1) and (6.2) the underlined terms occur as a result of the contact condition transfer from deformed surface into the surface before deformation. In the third terms of a Taylor series expansion only the most important contributions $(w^2/2)\partial^3\phi/\partial r^3$ and $(w^2/2)\partial^2 p/\partial r^2$ remain. The less essential terms of the same order of magnitude $(v^2/2r^2)\partial^3\phi/\partial\theta^2\partial r, \ldots$ are not included.

The large bending equation of an inextensive shell (ring) may be written for the function $U = R/R^*$ [37], where R, R* are the radii of curvature before and after deformation.

The connection between the bending moment and curvature change gives

$$M_{\theta\theta} = Dx_{\theta\theta} = \frac{D}{R} \left(\frac{R}{R*} - 1\right) = \frac{D}{R} (U - 1) \qquad (6.3)$$

Therefore, the bending equations are written as follows:

$$\partial N_{\theta\theta}/\partial\theta + UQ_\theta = 0,$$

$$\partial Q_\theta/\partial\theta - UN_{\theta\theta} + RZ_r^* = 0, \qquad (6.4)$$

$$\partial M_{\theta\theta}/\partial\theta - RQ_\theta = 0$$

Excluding from (6.3), (6.4) $M_{\theta\theta}$, Q_θ, $N_{\theta\theta}$, one obtains

$$\frac{\partial}{\partial\theta} \left[\frac{1}{U} \left(\frac{\partial^2 U}{\partial\theta^2} + \frac{1}{2} U^3 + \frac{R^3}{D} Z_r^*\right)\right] = 0,$$

from which, taking into account (6.2),

$$\frac{\partial^2 U}{\partial\theta^2} + \frac{1}{2} U^3 + CU$$

$$= -\frac{R^3}{D} \left(p_i - p - \frac{v}{r}\frac{\partial p}{\partial\theta} - w\frac{\partial p}{\partial r} - \frac{w^2}{2}\frac{\partial^2 p}{\partial r^2}\right) \qquad (6.5)$$

where C is the constant of integration. An equation similar to (6.5) was used for an analysis of the shell large bending in [37].

The constant C is determined by integration of equation (6.5). We have to take into account that $Rd\theta = R*d\theta*$, in which $d\theta*$ is the angle change induced by the curvature change, and

$$\int\limits_{0}^{2\pi} Ud\theta = \int\limits_{0}^{2\pi} (R/R^*)d\theta = \int\limits_{0}^{2\pi} (d\theta^*/d\theta)d\theta = 2\pi$$

$$\int\limits_{0}^{2\pi} (d^2U/d\theta^2)d\theta = (dU/d\theta)_0^{2\pi} = 0$$

Then

$$C = -\frac{1}{4\pi} \int\limits_{0}^{2\pi} U^3 d\theta$$

$$-\frac{R^3}{2\pi D} \int\limits_{0}^{2\pi} (p_i - p - \frac{v}{r}\frac{\partial p}{\partial \theta} - w\frac{\partial p}{\partial r} - \frac{w^2}{2}\frac{\partial^2 p}{\partial r^2})d\theta \quad (6.6)$$

Since in the right-hand side of equation (6.5) and (6.6) there are the displacement components v, w, it is necessary to determine a connection between them and function U. For this purpose we consider an expression of the total (complete) dimensionless curvature $U = 1 + Rx_{\theta\theta}$. In the case of a plane deformation we have (Chapter I, 1.2)

$$x_{\theta\theta} = -\frac{1}{R} (E_2 \frac{\partial e_{\theta\theta}}{\partial \theta} + E_3 \frac{\partial \omega_\theta}{\partial \theta}),$$

$$e_{00} = \frac{\partial v}{R\partial \theta} + \frac{w}{R}, \quad \omega_0 = \frac{\partial w}{R\partial \theta} - \frac{v}{R},$$

$$E_2 = -\omega_\theta, \quad E_3 = 1 + e_{\theta\theta}$$

Consequently,

$$U = 1 + \frac{1}{R} [(1 + \frac{\partial v}{R\partial \theta} + \frac{w}{R})(\frac{\partial v}{\partial \theta} - \frac{\partial^2 w}{\partial \theta^2})$$

$$\qquad\qquad (6.7)$$

$$- \frac{1}{R} (v - \frac{\partial w}{\partial \theta})(\frac{\partial^2 v}{\partial \theta^2} + \frac{\partial w}{\partial \theta})]$$

The connection between the components v and w is determined using the condition of inextensibility of the middle surface

$$\varepsilon_{\theta\theta} = \frac{1}{R} \left(\frac{\partial v}{\partial \theta} + w\right) + \frac{1}{2R^2} \left(\frac{\partial v}{\partial \theta} + w\right)^2$$

$$+ \frac{1}{2R^2} \left(v - \frac{\partial w}{\partial \theta}\right)^2 = 0 \tag{6.8}$$

Before beginning to solve the formulated inter-action problem, we consider an auxiliary problem to determine the needed number of terms in the expression for the function U. For this purpose we consider the large static bending of a shell and the pressure, given as follows.

$$(R^3/D)(p_i - p) = - \bar{p}_0(1 + q\cos2\theta) \tag{6.9}$$

Such a distribution of the pressure occurs for the unseparated flow of an ideal incompressible fluid around the rigid cylinder.

Let us consider equations (6.5), (6.6) without the underlined terms on their right-hand sides. In order to solve them with the help of the Galerkin method we take the trinomial approximation

$$U = U_0 + U_2\cos2\theta + U_4\cos4\theta \tag{6.10}$$

Because of the assumption of inextensibility of the shell middle surface, $U_0 = 1$. The system of two nonlinear algebraic equations obtained for U_2, U_4 is solved numerically. The curvature change (6.10) at points $\theta = 0$ and $\pi/2$ as depending on $\bar{p}_0$, q is shown in Figs. 13 and 14 by the hatched lines.

Equation (6.5) for given pressure (6.9) was solved with great accuracy by the finite difference method in [37]. Corresponding results are plotted in Fig. 13 and 14 by the continuous lines. As may be seen approximation (6.10) is sufficient even when the

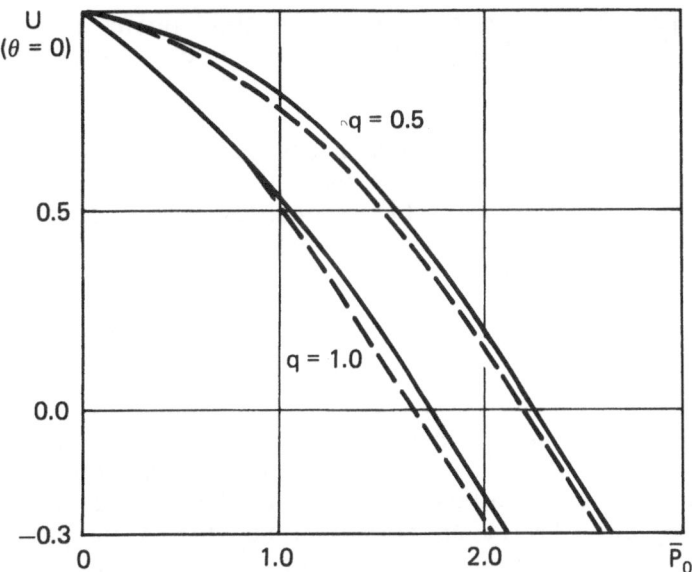

Fig. 13. Dependence of a relative curvature of a frontal point of the shell on the load parameters.

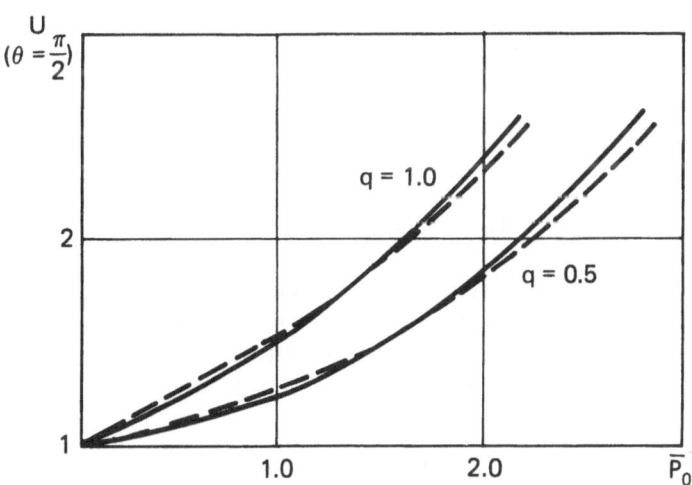

Fig. 14. Dependence of a relative curvature of a periphery point of the shell on the load parameters.

negative curvature occurs near the point $\theta = 0$ and upon the more than double increase of curvature near the point $\theta = \pi/2$. Corresponding maximum deflection of the shell reaches 1/3 of its radius. The term U_2 is dominant, while U_4 has a correcting character. If, for example, $U_2 = 1$, then $U_4 \approx 0.05$. Proceeding from solution of this auxiliary problem, in the future we shall take the curvature change in form (6.10) for $U_0 = 1$.

In previous sections the field of the velocity and the pressure was described by sums (1.10), and, moreover, it was assumed that $p_2 \ll p_1$, $\partial\phi_2/\partial\theta \ll \partial\phi_1/\partial\theta$, etc., and the contact conditions about p_2, ϕ_2, w, v were linearized. Such an assumption is justified, if the disturbances in the flow induced by shell bending are small as compared to the flow parameters around the undeformed cylinder. This occurs if the shell deflections are small compared to the radius. In this problem the deflections are comparable with the cylinder radius; hence, one must modify the approach mentioned and apply conditions in the form (6.1), (6.2).

Taking into account symmetry of the flow and of the shell deformations about the lines $\theta = 0$, π and $\theta = \pi/2$, $3\pi/2$, we assume

$$v = V_2\sin2\theta + V_4\sin4\theta,$$

$$w = W_0 + W_2\cos2\theta + W_4\cos4\theta, \qquad (6.11)$$

$$\phi = \Phi_0 + \Phi_1\cos\theta + \Phi_3\cos3\theta$$

Substituting the expression for ϕ from (6.11) into the Laplace equation (1.1) one obtains the ordinary differential equations for $\Phi_0(r)$, $\Phi_1(r)$, $\Phi_3(r)$. Their solutions are contained in (1.18).

$$\Phi_0 = A_0 \ell nr + B_0,$$

$$\Phi_1 = A_1 r^{-1} + B_1 r, \qquad \Phi_3 = A_3 r^{-3} + B_3 r^3 \qquad (6.12)$$

From condition (1.2) we have $B_0 = B_3 = 0$, $B_1 = -V_\infty$. The kinematic condition (6.1) serves to determine A_0, A_1, A_3. First of all it is necessary to determine the dependence among U_2, U_4, V_2, V_4, W_0, W_2, W_4.

Substituting (6.11) into (6.8) and comparing coefficients of the terms 1, $\cos\theta$, $\cos4\theta$ we express W_0, V_4, W_4 in terms of V_2, W_2. Proceeding from the above mentioned estimates of U_2 and U_4 we conclude that the values V_2, W_2 are dominant, while W_0, V_4, W_4 have a correcting character. Therefore, in expressions for W_0, V_4, W_4 we take into account only the main terms, omitting W_0, $4V_4 + W_4$ as compared to R and the value $(2V_2 + W_2)^2$ as compared to $(V_2 + 2W_2)^2$. Thus

$$4w_0 = -(v_2 + 2w_2),$$

$$v_4 + 4w_4 = -2(2v_2 + w_2)(v_2 + 2w_2)^{-1}, \qquad (6.13)$$

$$4(4v_4 + w_4) = (v_2 + 2w_2)^2$$

Here the definitions $w_0 = W_0R^{-1}$, $v_2 = V_2R^{-1}$, $v_4 = V_4R^{-1}$, $w_2 = W_2R^{-1}$, $w_4 = W_4R^{-1}$ are introduced.

The connection between coefficients U_2, U_4 from (6.10) and w_0, v_2, v_4, w_2, w_4 is determined from the formula for the curvature (6.7). Substituting expressions (6.10), (6.11) into (6.7) and separating the terms corresponding to 1, $\cos2\theta$, $\cos4\theta$, using (6.13) one may obtain the approximate expressions

$$w_0 = -\frac{1}{16} U_2^2 \left(1 - \frac{1}{16} U_4\right),$$

$$v_2 = -\frac{1}{6} U_2 \left(1 - \frac{1}{8} U_2^2 + \frac{7}{32} U_4\right),$$

$$w_2 = \frac{1}{3} U_2 \left(1 - \frac{1}{32} U_2^2 + \frac{1}{32} U_4\right), \qquad (6.14)$$

$$v_4 = -\frac{1}{60} U_4, \quad w_4 = \frac{1}{15} U_4$$

Here it is taken into account that $U_0 = 1$. From the expressions for w_2 in (6.14) it follows that for $U_2 = -1$ and $U_4 \ll 1$ [when at points $\theta = 0; \pi$ the total curvature (6.10) vanishes, while at points $\theta = \pi/2$, $3\pi/2$ it is doubled] the displacement along a normal w reaches 1/3 of the cylinder radius, that is, large bending of the shell takes place. In the future we shall assume the deflections to be no more than R/3, while $|U_2| < 1$ and $|U_4| < 0.1$.

Now from condition (6.1) for $\partial w/\partial t = 0$ and taking into account (6.11), (6.12), and (6.14) the coefficients A_0, A_1, A_3 may be expressed in terms of U_2, U_4:

$$A_1 = - V_\infty R(1 - \frac{1}{3} U_2 - \frac{9}{10} U_2^2 + \frac{2}{47} U_2^3$$

$$- \frac{1}{28} U_2 U_4 + \frac{1}{48} U_2^2 U_4),$$

$$A_3 = - \frac{1}{3} V_\infty R(U_2 - \frac{21}{72} U_2^2 - \frac{2}{3} U_2^3 - \frac{1}{5} U_4$$

$$+ \frac{1}{60} U_2 U_4 + \frac{3}{28} U_2^2 U_4)$$

These coefficients we substitute into the expressions for the potential (6.11), (6.12) and the pressure (1.1). In the latter it is necessary to assume $\partial\phi/\partial t = 0$. After that the right-hand side of the bending equation (6.5) and coefficient (6.6) may be calculated. In consequence, equation (6.5), taking into account (6.10), leads to two nonlinear equations

$$(\frac{3}{8} + \frac{19}{20} \mu)U_2^3 + \frac{11}{30} \mu U_2^2 + (3 + \gamma - \frac{1}{3} \mu)U_2 - \frac{9}{10} \mu U_2^2 U_4$$

$$- (\frac{3}{2} - \frac{1}{9} \mu)U_2 U_4 + \frac{3}{10} \mu U_4 + 2\mu = 0,$$

$$\tag{6.15}$$

$$\frac{13}{10} \mu U_2^3 + \frac{1}{4} (3-\mu)U_2^2 - \mu U_2 - \frac{17}{20} \mu U_2^2 U_4 + \frac{4}{5} \mu U_2 U_4$$

$$- (15 + \gamma + \frac{13}{15} \mu)U_4 = 0$$

in which the following definitions are introduced:

$$\mu = \rho_\infty V_\infty^2 R^3/2D, \qquad \gamma = (p_i - p_\infty)R^3/D \qquad (6.16)$$

Obtaining equations (6.15), one takes into account the terms with U_4 only in the first degree, while with U_2 no more than third degree.

Solution of the system (6.15) is obtained numerically. In Figs. 15 and 16 curve 1 shows the dependence of the total dimensionless curvature $U = 1 + U_2 + U_4$ and $U = 1 - U_2 + U_4$ at points $\theta = 0$, $\pi/2$ on the hydroelastic parameter μ for zero hydrostatic pressure differential ($\gamma = 0$).

Using the known values U_2, U_4 the displacement components are determined according to formulae (6.11), (6.13), (6.14).

The value of parameter γ is subjected to the following restrictions. From the stability condition of the uniform external pressure difference it should be considered that $\gamma > -3$, while from the strength condition on the shell discontinuity, $(p_i - p_\infty)R < \sigma_n h$, in which σ_n is the strength limit of material. Therefore,

$$-3 < \gamma < 12(\sigma_n/E)(R/h)^2$$

Restrictions also have to be put on the values of μ. One of them is the above taken restriction on the deflections and the curvature change ($w < R/3$, $|U| < 1$), while the second is the retention of a fluid continuity (the absence of negative pressures) near the regions $\theta = \pi/2$; $3\pi/2$, where large velocities and stream suction occur.

Furthermore, we shall again consider some estimates for the influence of the underlined nonlinear terms in (6.1), (6.5), (6.6), induced by the contact conditions transfer to the surface before deformation. Equations (6.15) and the above mentioned results concern that case, when on the right-hand sides of equations (6.5), (6.6) as well as in the kinematic condition (6.1) all the terms are retained. This most nearly exact solution is denoted by number I.

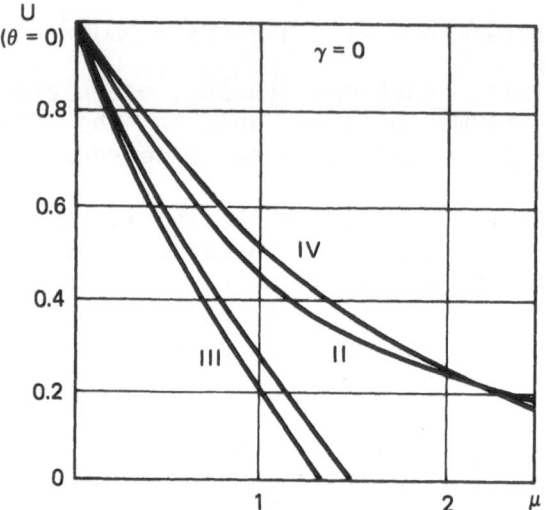

Fig. 15. Dependence of a relative curvature of a frontal point of the shell on the hydro-elasticity parameter.

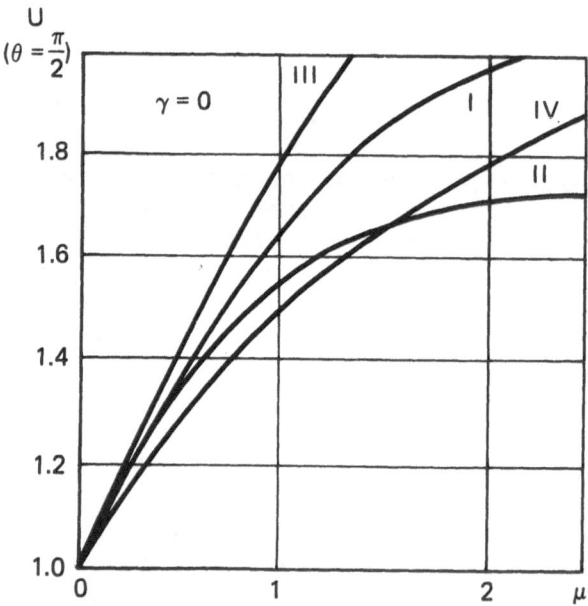

Fig. 16. Dependence of a relative curvature of a periphery point of the shell on the hydro-elasticity parameter.

If on the right-hand sides of (6.5), (6.6) one omits the term $(w^2/2)\partial^2 p/\partial r^2$ and the kinematic condition (6.1) is used in a complete form, then instead of system (6.15) one obtains

$$(\frac{3}{8} + \frac{93}{35}\mu)U_2^3 + \frac{1}{30}\mu U_2^2 + (3 + \gamma - \frac{1}{3}\mu)U_2 - \frac{20}{25}\mu U_2^2 U_4$$

$$- (\frac{3}{2} - \frac{1}{3}\mu)U_2 U_4 + \frac{3}{10}\mu U_4 + 2\mu = 0,$$

$$\frac{16}{7}\mu U_2^3 + \frac{1}{4}(3 - \mu)U_2^2 - \mu U_2 - \frac{19}{50}\mu U_2^2 U_4 \qquad (6.17)$$

$$+ \frac{13}{15}\mu U_2 U_4 - (15 + \gamma + \frac{13}{15}\mu)U_4 = 0$$

This less exact version of solution we denote by number II. In Figs. 15 and 16 a comparison of this solution with the version I is shown. As μ increases, the difference between them also increases, and for μ = 2 it reaches a considerable value.

The solution (6.17) dependence (version II) on hydrostatic load parameter, γ, (6.16) is shown in Fig. 17. The shell curvature, for positive values of γ (internal difference) with increasing μ, increases smoothly. When the external pressure differential (γ < 0) occurs and especially when it approaches its critical value γ = -3 as the static buckling of a circular shape under the uniform pressure occurs, then this dependence becomes sharper. Even for small flow velocities, elastic displacements of the shell points occur.

Omitting in (6.1), (6.5), (6.6) all the underlined terms and taking the expressions for U, v, w, ϕ as before in a form (6.10), (6.11) instead of systems (6.15), (6.16), we obtain

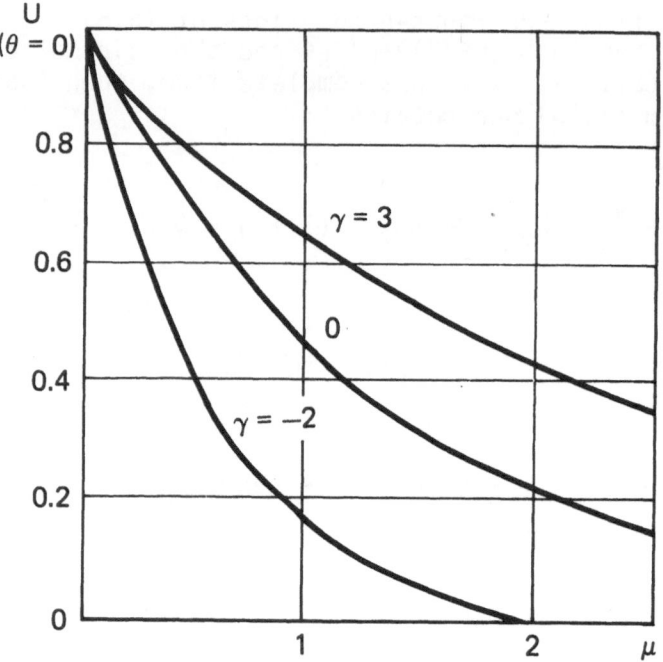

Fig. 17. Influence of the hydrostatic load parameter on the change of relative curvature.

$$\frac{1}{4}(3 + \mu)U_2^3 - \frac{1}{3}\mu U_2^2 + 2(3 + \gamma - \mu)U_2 - \frac{2}{75}\mu U_2^2 U_4$$

$$- (3 - \frac{27}{50}\mu)U_2 U_4 + \mu U_4 + 4\mu = 0,$$

$$\frac{1}{24}\mu U_2^3 + (\frac{3}{4} + \frac{1}{3}\mu)U_2^2 - \mu U_2 - \frac{10}{25}\mu U_2^2 U_4$$

$$+ \frac{22}{25}\mu U_2 U_4 - (15 + \gamma + \frac{1}{2}\mu)U_4 = 0$$

$$(6.18)$$

The solution of system (6.18) we denote by number III. It is obtained for the case of identical deformed and undeformed surfaces when writing the contact conditions.

If one produces the further simplification of the contact conditions, namely, (6.1) reduces to condition $\partial\phi/\partial r = 0$ (r=R), while in (6.5), (6.6) one omits the underlined terms, then according to (1.1) one obtains the pressure on an absolutely rigid cylinder (1.20). Hence, the shell bending under the given pressure may be determined from the system of equations

$$(3/8)U_2^3 + (3 + \gamma + \mu)U_2 - (3/2)U_2U_4 + 2\mu = 0,$$

$$\tag{6.19}$$

$$(15 + \gamma + \mu)U_4 - (3/4)U_2^2 = 0$$

Here, as it was above in obtaining the equations (6.15), (6.17), (6.18), the terms with U_4 are taken into account only to the first degree, while with U_2 no more than third degree is considered. The solution of the system of equations (6.19) is denoted by number IV.

Dimensionless pressure differential on the absolutely rigid cylinder equals $\gamma + \mu - 2\mu\cos 2\theta$. Therefore the approximation coincides with an auxiliary problem for a given pressure (6.9). Only here the coefficients γ, μ are expressed by the flow parameters.

The results of numerical solution of I, II, III, IV systems of equations (6.15), (6.17), (6.18), (6.19) for zero average pressure differential (γ = 0) are shown in Fig. 15 (for point θ = 0) and Fig. 16 (for point $\theta = \pi/2$). One can see that all four versions of the solution for small values of the parameter μ and the curvature change U give practically identical results. When μ and U increase, the versions II, III, IV are much different from version I, the latter being more exact.

In conclusion we shall return to estimates of the terms U_2 and U_4 and the assumptions concerning their orders. For this purpose we consider an approximate analytical solution of system (6.19). Deleting from it U_4, we obtain

$$\frac{12 + \gamma + \mu}{15 + \gamma + \mu} U_2^3 + \frac{8}{3} (3 + \gamma + \mu)U_2 = -\frac{16}{3}\mu$$

Substituting the first approximation for the root

$$U_2 = - 2\mu(3 + \gamma + \mu)^{-1} \qquad (6.20)$$

into the cube term we find the second approximation in the form

$$U_2 = - \frac{2\mu}{3+\gamma+\mu} + \frac{3(12+\gamma+\mu)\mu^3}{(15+\gamma+\mu)(3+\gamma+\mu)^4} ,$$

$$\qquad (6.21)$$

$$U_4 = \frac{3\mu^2}{(15+\gamma+\mu)(3+\gamma+\mu)^2}$$

These approximations converge very quickly. Indeed, the second approximation (6.21) gives a good agreement with the curves IV in Figs. 15 and 16. From (6.21) we shall construct a relation

$$|U_4/U_2| \approx (3/2)\mu(15+\gamma+\mu)^{-1}(3+\gamma+\mu)^{-1}$$

For example, for $\mu = 2$, $\gamma = 0$ this relation gives $|U_4/U_2| \approx 1/30$. Consequently, in the above solutions retaining terms with U_4 only to the first degree was justified.

Solution (6.20) has to be compared with (2.10). Unlike (2.10), here the influence of the hydrostatic pressure differential γ is taken into account. But, also for $\gamma = 0$, solution of (6.20), unlike (2.10), does not increase infinitely with μ having a finite limit. This fact is explained by the external flow influence, making the shell more rigid, which is taken into account in equations (6.5), (6.6) by the term CU, namely, that inleakage increases when the flow velocity increases.

3.7. Dynamical behavior of a system: Taking into account initial bending of the shell

In section 3.3 interaction of the shell with a fluid flow was analyzed in a dynamical formulation. However, the initial bending state of the shell was identified with its undeformed state. In this section such an assumption is removed. Thus, in representation (3.1) to which one now has to join $v* = v_0(\theta) + v(\theta,t)$, $U* = U_0(\theta) + U(\theta,t)$, the functions $w_0(\theta)$, $v_0(\theta)$, $U_0(\theta)$ are taken from the previous section. In other respects the physical formulation of the problem does not differ from the formulation in section 3.3. Disturbances of the velocity potential and pressure are substituted into equation (3.3) and into the condition at infinity (3.4), but in (3.3) ϕ_0 is used instead of ϕ_1.

The kinematic condition on the surface r=R has a very complicated form:

$$\frac{\partial w}{\partial t} - \frac{\partial \phi}{\partial r} + \frac{\partial w_0}{R^2 \partial \theta} \frac{\partial \phi}{\partial \theta} + \frac{\partial \phi_0}{R^2 \partial \theta} \frac{\partial w}{\partial \theta} - \frac{1}{R} \left(\frac{\partial v_0}{\partial \theta} + w_0\right) \frac{\partial \phi}{\partial r}$$

$$- \frac{\partial \phi_0}{R \partial r} \left(\frac{\partial v}{\partial \theta} + w\right) - w_0 \frac{\partial^2 \phi}{\partial r^2} - \frac{\partial^2 \phi_0}{\partial r^2} w \qquad (7.1)$$

$$- \frac{v_0}{R} \frac{\partial^2 \phi}{\partial \theta \partial r} - \frac{\partial^2 \phi_0}{\partial \theta \partial r} \frac{v}{R} + F = 0$$

An expression for F contains the terms consisting of the products of functions v_0, w_0, ϕ and their derivatives, for example, $v_0 w_0 / \partial \phi / \partial \theta$; moreover, it contains the products of functions v_0, ϕ_0, w and w_0, ϕ_0, v and their derivatives. The obvious form of F is easily obtained from (6.1) and the representation $\phi*$, $v*$, $w*$ as the summations (3.1). It should be noted that in (7.1) effects of change of the normal to the middle surface under a shell deformation and of transfer of the kinematic condition to the surface before deformation are partially taken into account when for simplification of computations we assume $F = 0$.

We shall suppose that the inertia of the system is determined only by the inertia of the liquid. This assumption is true if the coefficient $\nu = 2\rho_s h / \rho_\infty R$, introduced in (3.11) is small. For thin-walled shells produced from nonmetallic materials and for the case of water, ν is always a small value compared to unity. Then in our problem one may invoke the equations of statical bending (6.5), (6.6). So, the motion equation for an initial statically deformed non-inertia shell being under flow action has the form

$$\partial^2 U/\partial\theta^2 + (3/2)U_0^2 U + C_0 U + U_0 C \tag{7.2}$$

$$= \frac{R^3}{D} \left(p + \frac{\partial p_0}{\partial\theta} \frac{v}{R} + \frac{v_0}{R} \frac{\partial p}{\partial\theta} + \frac{\partial p_0}{\partial r} w + w_0 \frac{\partial p}{\partial r} \right)$$

Here the coefficient C_0 is given by expression (6.6), in which instead of p, v, w, U it is necessary to substitute p_0, v_0, w_0, U_0. The coefficient disturbance equals

$$C = \frac{R^3}{2\pi D} \int_0^{2\pi} \left(p + \frac{\partial p_0}{\partial\theta} \frac{v}{R} + \frac{v_0}{R} \frac{\partial p}{\partial\theta} + \frac{\partial p_0}{\partial r} w + w_0 \frac{\partial p}{\partial r} \right) d\theta \tag{7.3}$$

$$- \frac{3}{4\pi} \int_0^{2\pi} U_0^2 U \, d\theta$$

On the right-hand side of (7.2) and in (7.3) the terms $(w*^2/2)\partial^2 p*/\partial r^2$ are omitted.

The dimensionless curvature change according to (6.7) equals

$$U = \frac{1}{R^2} \left[\left(R + \frac{\partial v_0}{\partial\theta} + w_0 \right) \left(\frac{\partial v}{\partial\theta} - \frac{\partial^2 w}{\partial\theta^2} \right) \right.$$

$$+ \left(\frac{\partial v_0}{\partial\theta} - \frac{\partial^2 w_0}{\partial\theta^2} \right) \left(\frac{\partial v}{\partial\theta} + w \right) - \left(\frac{\partial^2 v_0}{\partial\theta^2} + \frac{\partial w_0}{\partial\theta} \right) \left(v - \frac{\partial w}{\partial\theta} \right) \tag{7.4}$$

$$\left. - \left(v_0 - \frac{\partial w_0}{\partial\theta} \right) \left(\frac{\partial^2 v}{\partial\theta^2} + \frac{\partial w}{\partial\theta} \right) \right]$$

while the expression for the deformation is

$$\varepsilon_{\theta\theta} = \frac{1}{R^2} [(R + \frac{\partial v_0}{\partial \theta} + w_0)(\frac{\partial v}{\partial \theta} + w)$$

$$+ (v_0 - \frac{\partial w_0}{\partial \theta})(v - \frac{\partial w}{\partial \theta})]$$

(7.5)

From the previous section one can see that for the corresponding small curvature changes solution IV differs little from the more exact one (Figs. 15 and 16). Therefore, for an approximate analysis of the dynamical phenomena the values with zero index may be taken for the simplest case of static bending, when the pressure on the shell is determined without taking into account its deformation (solution IV). Moreover, in (6.10), (6.11), (6.14) we omit the terms with U_4, V_4, W_4 because of their smallness. Thus, the solution for U_0 may be taken in a form $U_0 = 1 + U_2\cos2\theta$, in which U_2 is determined from (6.21); solutions for v_0, w_0 are determined from (6.14).

The solution for the disturbances is found in the class of functions

$$v = e^{\lambda t} \sum_{n=1}^{N} V_n\sin n\theta, \quad w = e^{\lambda t} \sum_{n=1}^{N} W_n\cos n\theta,$$

(7.6)

$$U = e^{\lambda t} \sum_{n=1}^{N} \bar{U}_n\cos n\theta, \quad \phi = e^{\lambda t} \sum_{n=1}^{N} \Phi_n\cos n\theta$$

From the Laplace equation (3.3) it follows that functions Φ_n have the same form (1.18) as in the static problem. Here according to condition at infinity (3.4), $B_n = 0$.

The further course for the solution is the same as in section 3.3. In our calculations we restrict the number of terms in (7.6), which equals $N = 3$. Introducing the parameter X according to (3.11) (parameter $\nu = 0$ because of the assumption that the shell is inertialess), we obtain the following condition for the existence of the nonzero solution:

$$K_0 X^6 + K_2 X^4 + K_4 X^2 + K_6 = 0 \qquad (7.7)$$

in which coefficients K_i equal

$$K_0 = \frac{1}{2} - \frac{1}{12} U_2 + \frac{3}{4} U_2^2 \, ,$$

$$K_2 = 14 - \frac{11}{3} U_2 - \frac{25}{6} U_2^2 - \frac{19}{4} U_2^3 \, ,$$

$$K_4 = 12 + 35 U_2 + \frac{12}{5} U_2^2 + 22 U_2^3 - \frac{3}{\mu} [20 + 6 U_2 + 5 U_2^2$$

$$+ (\gamma + \mu)(7 + \frac{5}{9} U_2 - \frac{20}{9} U_2^2)] + \frac{3}{\mu^2} [144 - 24 U_2 + 59 U_2^2$$

$$+ (\gamma + \mu)(66 - 11 U_2 + \frac{18}{10} U_2^2) + (\gamma + \mu)^2 (6 - U_2 - \frac{3}{40} U_2^2)],$$

$$K_6 = -9 U_2 + 30 U_2^2 + 14 U_2^3 + \frac{9}{\mu} [12 + 5 U_2 + 9 U_2^2 + \frac{33}{7} U_2^3$$

$$+ (\gamma + \mu)(4 + \frac{1}{2} U_2 - \frac{19}{6} U_2^2)] - \frac{18}{\mu^2} [24 + 8 U_2 + 38 U_2^2$$

$$+ (\gamma + \mu)(11 - \frac{5}{3} U_2 + 4 U_2^2) + (\gamma + \mu)^2 (1 - \frac{1}{3} U_2 - \frac{15}{12} U_2^2)]$$

Here the terms $A U_2^3$ with the coefficient A, which is small compared to the main terms in K_i, are omitted. Moreover, rounding-off of some coefficients with insignificant change of their values is carried out. Contained in K_i are values of U_2 that are functions of μ and γ, as is seen from (6.21).

The real part of one root of equation (7.7) will be positive if in the solution of the cubic equation

$$z^3 + az^2 + bz + c = 0$$

$$z = X^2, \quad a = K_2/K_0, \quad b = K_4/K_0, \quad c = K_6/K_0$$

the condition

$$\left(\frac{b}{3} - \frac{a^2}{9}\right)^3 + \left(\frac{a^3}{27} - \frac{ab}{6} + \frac{c}{2}\right)^2 > 0 \qquad (7.8)$$

is satisfied.

From this condition are found the critical values of hydroelastic parameter μ_{++} depending on the pressure differential-γ (minus sign indicates an external differential). They are shown in Fig. 18 by the curve 1. The value of a pressure differential $\gamma = -3$, for which μ_{++} vanishes, corresponds to the critical pressure of the shell static buckling, with two waves on the circle of the shell. Let us note that for $\gamma > 0$ the curve continues to the left part of a figure, while the values of μ_{++} increase, which is explained by the stiffening influence of the internal pressure differential.

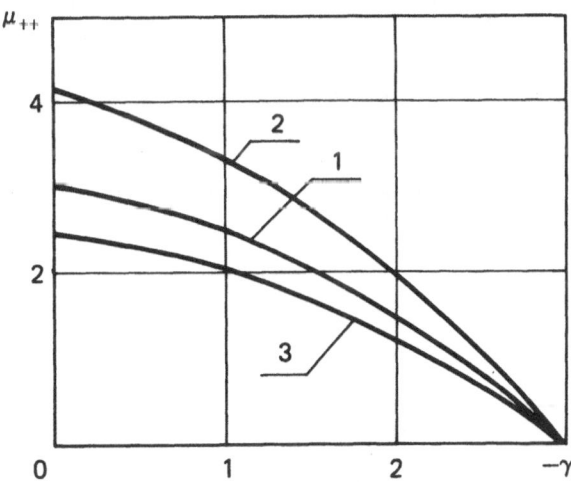

Fig. 18. Dependence of a dynamic critical value of hydroelasticity parameter on the hydrostatic parameter

It is of interest to know what influence is exerted by taking into account the static deformation of a shell and the membrane stresses on the dynamical stability of a shell. If the initial static state of the shell is identified with its undeformed state (U_2 = 0 in coefficients K_i, hence, U_0 = 1, v_0 = 0, w_0 = 0), then we come to a problem analogous to that considered in section 3.3. The corresponding curve μ_{++} is designated in Fig. 18 by the number 2.

Thus, a difference between the curves 1 and 2 characterizes the influence of an initial static deformation of the shell on the critical value of the hydroelastic parameter. Taking into account such a deformation leads to a decrease of the value of the critical hydroelastic parameter.

Curves 1 and 2 are obtained for the case of retaining the terms $C_0U + U_0C$ in equation (7.2), characterizing an interaction between the membrane stresses and the shell curvature. If one does not take them into account, as has been done in section 3.3, then one obtains the curve 3 in Fig. 18 (but there is taken into account, $v_0 \neq 0$, $w_0 \neq 0$, $U_2 = 0$). Let us note that in order to neglect the influence of initial deformation in construction of K_i the parameter μ inside of the brackets has to vanish. The increase of the critical value of μ_{++} by taking into account the membrane stresses (curves 1 and 3) is induced by an increase of the average pressure differential at the expense of an inleakage of the flow around the cylinder (decrease of an average pressure in the stream along the external surface of a shell).

3.8. Influence of flow separation from the shell surface

If in considering the transverse flow around a shell one attempts to take into account the flow separation from its surface, then the problem is more complicated.

In previous sections the principal assumption was that the fluid flow did not separate from the shell surface. Such an assumption is quite justified for many practically important and commonly occurring cases. However, the circular cylinder is bluff body and, as experiments show, flow separation takes place from the surface. The reason is that the boundary layer formed because of an internal friction in fluid does not permit the pressure reestablishment in the part $\theta \approx 80° - 120°$ of the cylinder and here there is flow separation.

Let us introduce the Reynolds number and the pressure coefficient

$$Re = 2RV_\infty/\nu_0, \quad C_p = 2(p - p_\infty)/\rho_\infty V_\infty^2$$

in which ν_0 is the coefficient of viscosity.

It is determined experimentally that there are two principal regimes of circular cylinder flow, occurring at various Reynolds numbers. For $Re < 2.10^5$ a subcritical regime of flow with separation angle $\theta_s \approx 80°$ and pressure coefficient $C_{pb} = -0.96$ in the bottom region takes place, while for $Re > 2.10^5$ a supercritical regime of flow with separation angle $\theta_s \approx 120°$ and bottom pressure $C_{pb} = -0.38$ takes place. The bottom region $\theta_s < \theta < \pi$, containing the field between the separation lines, is shown in Fig. 19. Also shown is the separation angle $\theta_s \approx 100°$ (its more exact value equals to 104°) under the transcritical flowing regime ($C_{pb} = -0.86$).

In Fig. 20 the theoretical curve C_p is shown for the surface of an undeformed cylinder upon the unseparated flow model according to (1.20) and also the pressure distribution in the indicated regimes of separated flow. The experimental results are taken from [35,36].

There exist various theories for describing separated flow around a circular cylinder. Computations with the Parkinson-Jandali model [35] give satisfactory agreement for the pressure distribution with the experimental values, if the separation angle

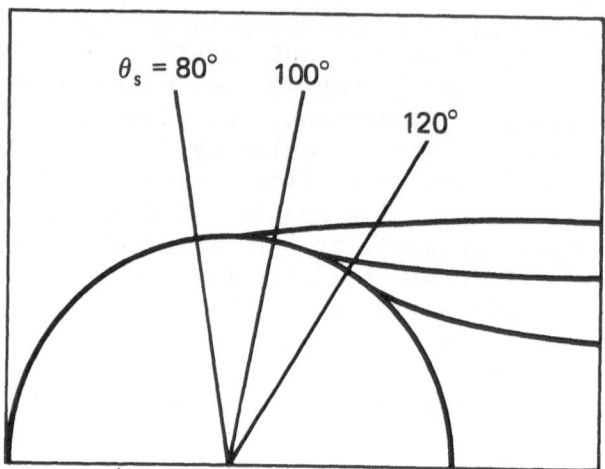

Fig. 19.   Angles of the separation of flow.

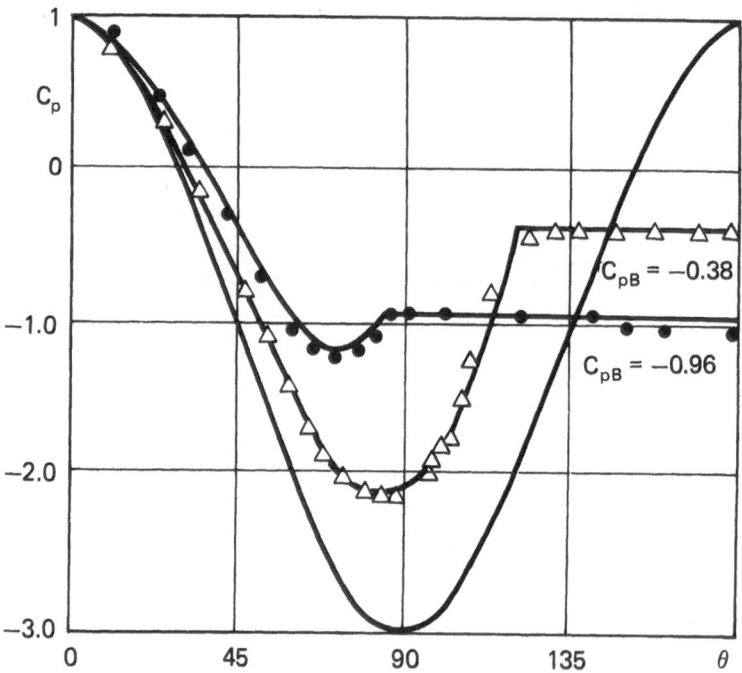

Fig. 20.   The pressure distribution over the surface
of a rigid cylinder.

and the pressure value in the bottom (base) region are taken from experiment.

The expression for the pressure on the cylinder in the range $0 < \theta < \theta_S$ has the form [35]

$$p = p_\infty + \frac{\rho_\infty V_\infty^2}{2} C_p ,$$

$$(8.1)$$

$$C_p = 1 - \frac{A^6 \sin^2 \theta/2}{(\cos\alpha\cos\delta - 1 + A\cos\theta/2)^2}$$

Here

$$A = \cos \frac{\theta}{2} + \frac{1}{\sqrt{2}} \sqrt{\cos\theta - \cos\theta_S}$$

$$\cos\delta = \cos\alpha + K^{-1}\sin^3\alpha, \qquad 2\alpha = \pi - \theta_S$$

while the coefficient $K = 1.40$ at the separation angle $\theta_S = 80°$, $K = 1.36$ at $\theta_S = 100°$ and $K = 1.17$ at $\theta_S = 120°$.

In order to obtain the pressure in the bottom part $\theta_S < \theta < \pi$ in (8.1) instead of C_p it is necessary to substitute the mentioned values of coefficients $C_p = C_{pb}$ for the various flow regimes.

Upon the separated flow of an ideal fluid around the body the resultant force on it is not equal to zero as it is in the case of an unseparated flow. Therefore, one has to consider the supports holding the shell in the flow. In [29-31] a support at a frontal point of the shell has been introduced. The Parkinson-Jandali model in combination with a Galerkin method and also with the successive perturbations approximations method has been applied. There was considered the case of small and large static displacements of the shell contour in a plane formulation. In experimental work [4] the holding supports were at points $\theta = \pi/2$ and $3\pi/2$.

In [5] there was considered experimentally and theoretically the bending of a cylindrical shell having finite length. The shell was fixed along its ends into the two rigid thin pylons (Fig. 21) moving together with the towing carriage in a tank with water. Measurement of the relative elongations of the shell was produced by strain gauge sensing elements glued on its inner surface. Curves 1 in Fig. 22 show the indications of these strain gauges for the various velocities of a shell motion.

In the approximate theoretical description of this experiment one does not take into account the fluid flow along a shell induced by its deformation and also by disturbances due to the pylons. In the case of small shell bending and small pylon thickness this does not produce a visible error. Thus, the hydrodynamical part of the computations may be reduced to a plane problem, while the computation of a finite length shell has to be considered as a spatial structure. The corresponding computations are produced in [5] in two approximations.

In the first approximation, the pressure on the shell is determined with the assumption of its absolute rigidity. Consequently, it is given by formulae (8.1). Furthermore, the linear bending of a shell under the action of this fluid pressure is considered. In the second approximation the flow field is determined taking into account the deformation of the shell contour and the displacements and stresses in a shell are defined more precisely. It is shown [9] that such a successive approximations method converge and there are methods of convergence acceleration. In Fig. 22 the values of circular deformation on an internal shell surface are plotted. Curve 2 indicates the result of the first approximation, while curve 3, the second approximation. These data are plotted for an average section of the shell (that is, for the section C-C in Fig. 21).

The stated computation for the two approximations gives the qualitatively true picture of the shell strain and the fluid flow. Differences between the experiment and the computation are conditioned by both

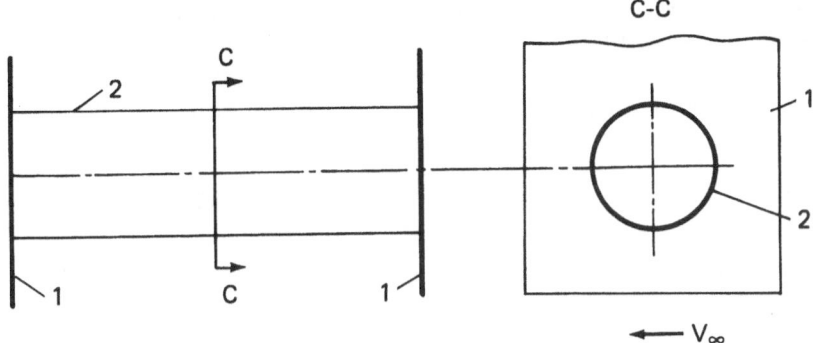

Fig. 21.   Scheme of an experiment.

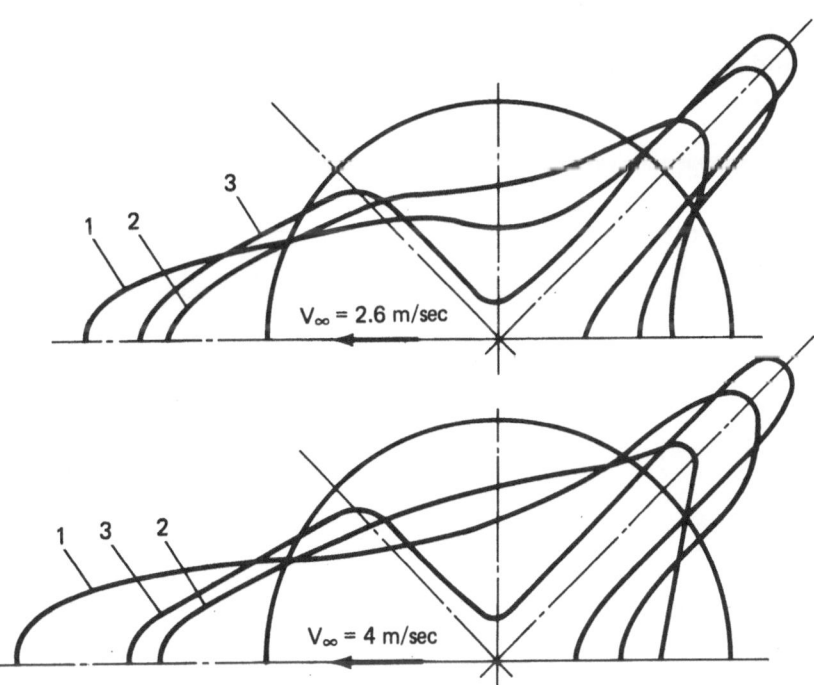

Fig. 22.   Relative elongations of the shell bending in
           a flow.

the error in the measurement and approximations in the theory as well as incomplete conformity between the conditions in which the experiment was produced and the theoretical model of the flow and structural deformation.

The static buckling of a cylindrical shell having finite length induced by bending in a fluid flow is considered experimentally and theoretically in [2].

In [34] the authors consider the dynamic stability of a cylindrical shell in flow in the same formulation as in section 3.3, but they did take into account flow separation. Separation angle, the pressure distribution upon the surface of a rigid shell, are taken into account according to the Parkinson-Jandali model (8.1). As in [5] a distribution (8.1) is approximated by trigonometric functions, then in [34] it is approximated by polynomials of fourth degree. The theoretical part is carried out in a plane formulation; however, the experiment for flutter is produced with the cylindrical shell having a finite length. Qualitative agreement between computational and experimental data is obtained.

References

1. Barmina, L. A.,   Force Acting on the Undeformed
    Contour which Moves in an Arbitrary Fluid Flow,
    Izv. Akad. Nauk SSSR, Mekh. Zhidk. Gaza, 1973,
    No. 1, pp. 4-8.

2. Gafurov, M. B., Stability of the Cylindrical
    Shell having a Finite Length upon the Transverse
    Flow Around it of a Liquid, Proceedings of a
    Shell Theory Seminar,   Kazan Physicotechnical
    Inst., USSR Academy of Sciences, 1975, No. 6, pp.
    341-352.

3. Gafurov, M. B., Experimental Analysis of Cylin-
    drical Plate Stability in a Fluid Flow, Investi-
    gations on a Shell Theory, Proceedings of a
    Seminar, Kazan: Physicotechnical Inst.,   USSR
    Academy of Sciences, 1976, No. 7, pp. 89-97.

4. Gafurov, M. B., Bending of the Cylindrical Shell
    Fastened Along the Opposite Generatrices upon its
    Transverse Flow of a Liquid, Statics and Dynamics
    of Shells,   Proceedings of a Seminar, Kazan:
    Physicotechnical Inst., USSR Academy of Sciences,
    1977, No. 8, pp. 154-161.

5. Gafurov, M. B., Ilgamov, M. A., Bending of the
    Cylindrical Shell Having a Finite Length upon the
    Transverse Flow Around it of a Liquid, Prikl.
    Mekh., V. XIV, No. 3, 1978, pp. 60-67.

6. Gafurov, M. B., Ilgamov, M. A., Bending and Sta-
    bility of the Shallow Cylindrical Curved Plate
    upon the Transverse Flow Around it of a Liquid,
    Nonlinear Problems of Aerohydroelasticity,   Pro-
    ceedings of a Seminar, Kazan: Physicotechnical
    Inst., USSR Academy of Sciences, 1979, No. 11,
    pp. 59-69.

7. Gafurov, M. B., Ilgamov, M. A., Dynamical Sta-
    bility of Cylindrical Shell upon the Transverse
    Flow Around it of Liquid, Interaction between the
    Shells and a Liquid, Proceedings of a Seminar,

Kazan: Physicotechnical Inst., USSR Academy of Sciences, 1981, No. 14, pp. 81-95.

8. Joukowski, N. E., Determination of a Liquid Motion at Some Condition Given on the Current Line, The Collected Works of N. E. Joukowski, V. 3, M.L., ONTI, 1936, pp. 45-55.

9. Ilgamov, M. A., Dynamics of Elastic Shells Containing an Acoustic Medium with Sources, In: Theory of Plates and Shells, M., Nauka, 1971, pp. 90-122.

10. Ilgamov, M. A., Bending and Stability of a Cylindrical Shell upon the Transverse Flow Around it of a Liquid, Prikl. Mekh., V. XI, No. 3, 1975, pp. 12-19.

11. Ilgamov, M. A., Bending and Stability of a Cylindrical Shell upon the Transverse Flow around it of a Liquid, Proceedings of XIII All-Union Conference on the Theory of Plates and Shells, Tallin, 1983, Part II, pp. 124-129.

12. Ilgamov, M. A., Sabitov, M. Z., Interaction of Elastic Permeable Cylindrical Shell with a Flow of Incompressible Liquid, Nonlinear Problems of Aerohydroelasticity, Proceedings of a Seminar, Kazan: Physicotechnical Inst., USSR Academy of Sciences, 1979, No. 11, pp. 70-81.

13. Kiseljov, O. M., On a Problem Concerning the Gas Bubble in a Plane Flow of Ideal Liquid, Izv. Akad. Nauk SSSR, Mekh. Zhidk. Gaza, 1969, No. 4, pp. 13-23.

14. Kiseljov, O. M., On a Problem Concerning the Flow Around the Shell, Containing the Gas by the Plane Flow of an Ideal Liquid, Izv. Akad. Nauk SSSR, Mekh. Zhidk. Gaza, 1971, No. 3, pp. 108-118.

15. Kiseljov, O. M., Unseparated Flow of a Cylindrical Shell, Proceedings of a Seminar on the Boundary Value Problems, Kazan: University, 1982, No. 18, pp. 104-115.

16. Kiseljov, O. M., Rapoport, E. F., On a Problem of the Stream Flow of an Elastic Shell, Izv. vuzov. Mathematics, 1976, No. 10, pp. 97-100.

17. Kiseljov, O. M., Rapoport, E. F., On the Stream Flow of an Elastic Shell, Izv. Akad. Nauk SSSR, Mekh. Zhidk. Gaza, 1977, No. 2, pp. 24-32.

18. Kornishin, M. S., Isanbajeva, F. S., Flexible Plates and Panels, M., Nauka, 1968, 260 p.

19. Legler, E. L., Computational and Experimental Investigation of Self Oscillations of a Membrane Cylindrical Shell, Proceedings of TCAGI, No. 1253, 36 p.

20. Legler, E. L., Experimental Determination of a Static Form of the Membrane Shell in Air Flow, Proceedings of TCAGI, No. 1382, 35 p.

21. Mavljutov, R. R., Khakimov, A. G., Large Displacements of the Cylindrical Shell in a Plane Flow of Ideal Liquid, Proceedings of Ufa aviatz. Inst. Prochn. Konstr., 1973, No. 76, pp. 7-12.

22. Petrova, S. I., Form of Equilibrium of the Strip which is Limited by an Elastic Film in Homogeneous Fluid Flow, Izv. Acad. Nauk SSSR, Mekh. Zhidk. Gaza, 1971, No. 1, pp. 120-127.

23. Sljozkin, N. A., Flow of the Shell Containing a Gas by the Plane Ideal Liquid, Utchjonye Zapiski MGU, v. 3, No. 152, 1951, pp. 103-110.

24. Faddejev, Ju.I., Nonstationary Motion of a Plane Deformed Contour being in Inhomogeneous Flow of Inviscid Liquid, Proceedings of Leningr. corabl. inst., 1968, No. 63, pp. 75-81.

25. Faddejev, Ju.I., Listovsky, M. K., Computation of the Pressure Distribution and of Determination of the Inertia Forces at Deformed Contour Motion, Proceedings of Leningr. corabl. Inst., 1969, No. 65, pp. 61-69.

26. Fotinitch, D. I., Solution of a Plane Aeroelas-
    ticity Problem for the Air-Supported Membrane
    Shell, Izv. vuzov, Mashinostr, 1976, No. 8, pp.
    12-16.

27. Khakimov, A. G., Flow Around the Flexible Cylin-
    drical Shell of a Plane Ideal Liquid, Izv. Acd.
    Nauk SSSR, Makh. Zhidk. Gaza, 1975, No. 6, pp.
    147-150.

28. Khakimov, A. G., Gafurov, M. B., Large Displace-
    ments of an Elliptical Cylindrical Shell in a
    Plane Flow of Ideal Liquid, Proceedings of a
    Seminar on the Shell Theory, Kazan: Physicotech-
    nical Inst. USSR Academy of Science, 1974, No. 4,
    pp. 226-236.

29. Shagidullin, R. R., Linear Bending of an Elastic
    Cylindrical Shell at Various Regimes of the
    Transverse Flow by Small-Viscous Liquid, Proceed-
    ings of a Seminar on the Shell Theory, Kazan:
    Physicotechnical Inst. USSR Academy of Science,
    1974, No. 4, pp. 214-225.

30. Shagidullin, R. R., Large Bending of an Elastic
    Cylindrical Shell upon the Transverse Flow Around
    it of a Liquid, Investigations on the Shell
    Theory, Proceedings of a Seminar, Kazan: Physi-
    cotechnical Inst. USSR Academy of Science, 1976,
    No. 7, pp. 75-83.

31. Shagidullin, R. R., Investigation of Problem of
    the Shell Flow over the Model of Parkinson-
    Jandali Trace, Statics and Dynamics of Shells,
    Proceedings of a Seminar, Kazan: Physicotechni-
    cal Inst. USSR Academy of Science, 1977, No. 8,
    pp. 130-137.

32. Jakimov, Ju.L., Cylinder Motion in an Arbitrary
    Plane Flow of Ideal Incompressible Liquid, Izv.
    Akad. Nauk SSSR, Mekh. Zhidk. Gaza, 1970, No. 2,
    pp. 201-204.

33. McLeod, E. B., The Explicit Solution of a Free
    Boundary Problem Involving Surface Tension, J.

Rational Mechanics and Analysis, Vol. 4, 1955, No. 4, pp. 45-49.

34. Paidoussis, M. P., Wong, D.T.-M., Flutter of Thin Cylindrical Shells in Cross Flow, J. Fluid Mech., 1982, V. 11, pp. 411-426.

35. Parkinson, G. V., Jandali, T., A Wake Source Model for Bluff Body Potential Flow, J. Fluid Mech., 40, pt. 3, 1970, pp. 577-594.

36. Roshko, A., Experiments on the Flow Past a Circular Cylinder at Very High Reynolds Number, J. Fluid Mech., 1961, V. 10, pt. 3, pp. 345-356.

37. Seide, P., Jamjoom, T. M. M., Large Deformations of Circular Rings Under Nonuniform Normal Pressure, J. Appl. Mech., No. 1, 1974, pp. 192-196.

Chapter IV

INTERACTION OF A PERMEABLE SHELL
WITH A FLUID FLOW

The general relations for the problem of interaction of a permeable shell with a fluid flow are considered. They may be used to determine both the forms (shapes) and stress state of perforated shells and plates, parachute, networks (nets), etc., as well as the pressure and velocity field in the flow around them.

Formulating the conditions on the contact surface we accept the following principal assumptions: the shell wall is thin, in consequence of which the contact surface of the shell with the fluid is identified with its middle surface; the shell displacements may be comparable with its characteristic dimensions, but the strains are small as compared to unity; the fluid permeation across (through) a shell occurs only along the normal to the deformed surface; geometrical and mechanical parameters and properties of permeability are continuous on the surface; the fluid is ideal and compressible.

Conditions imposed on the fluid motion, on the uncontact surfaces, and also on the boundary edges of a shell do not differ from the generally accepted ones, and they are not reproduced here. Some of them were considered in the previous chapter.

At the end of this chapter we obtain the solution for the problem of the transverse flow around an elastic cylindrical shell. There the permeability influence on the flow picture (pattern) and on the shell bending is demonstrated.

4.1. Conditions on the contact surface

At time t = 0 the shell is in an undeformed state under the action of surface and volume forces. If the shell does not resist both bending and compression (membrane shell), then its initial state is that in which the shell is in some unbent "frozen" state. At t = 0 one has to prescribe also the conditions on the motion velocity of both the shell and the liquid.

As in Chapter II, the fluid motion is described here by a system of orthogonal curvilinear Euler coordinates x_1, x_2, x_3. Axes x_1,x_2 being on the middle surface of the shell σ at time t = 0 are directed along the lines of principal curvatures, while axis x_3 is along the outward normal (Fig. 1). Thus, the shell occupies in space a position $x_3 = x_3^0$ = const. The shell deformation is described by a system of curvilinear Lagrange coordinates α_1, α_2, which coincide with axes x_1, x_2 on the surface σ at t = 0. Therefore, in both coordinate systems the Lamé coefficients and the unit vectors of the local thrihedral $\vec{K}_1$, $\vec{K}_2$, $\vec{K}_3$ are identical (for more detail about that see Chapter II, 2.2).

Unit vectors $\vec{K}_i(\alpha_1, \alpha_2)$ and unit vectors on the deformed surface $\vec{K}_j^*(\alpha_1, \alpha_2, t)$ are related by formulae (2.2) of Chapter I.

Let us denote the values of density, pressure and velocity vector on two sides of surface σ (Fig. 1) by ρ, p, $\vec{v}$ and ρ', p', $\vec{v}'$, respectively. For the fixed Euler coordinate system the velocities along the normal on the two sides of σ equal $\vec{v}\,\vec{K}_3^*$ and $\vec{v}'\vec{K}_3^*$. The normal velocity with which the fluid particles intersect the surface σ moving with velocity $\partial\vec{u}/\partial t$ is $(\vec{v} - \partial\vec{u}/\partial t)\vec{K}_3^*$. The corresponding mass flow rate $\rho(\vec{v} - \partial\vec{u}/\partial t)\vec{K}_3^*$ equals $\rho\vec{V}\,\vec{K}_3^*$, where $\vec{V}$ is the velocity vector of the fluid flow across the shell at point M*. By the assumption made at the beginning of this chapter, this vector is always perpendicular to the moving surface: $\vec{V} = V_3^*\vec{K}_3^*$. Thus, $\vec{v}$, $\vec{v}'$ are the

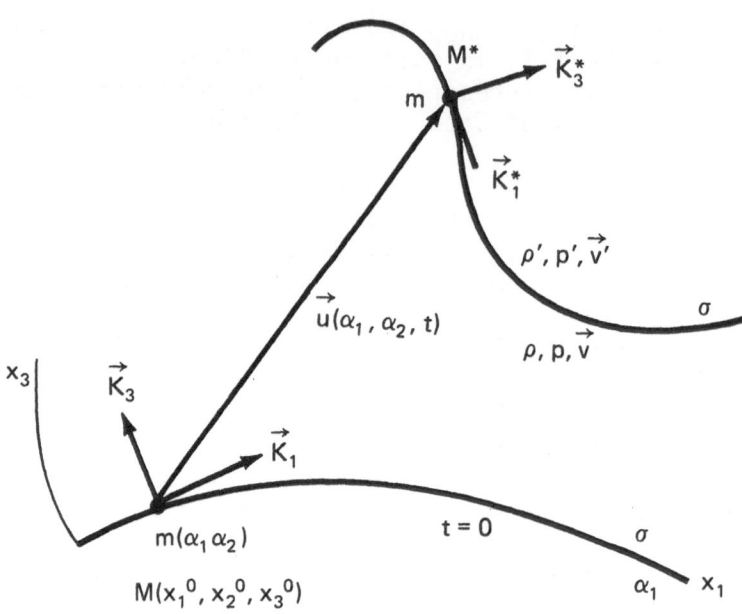

Fig. 1. Two positions of shell in a space. The Euler
         and Lagrange coordinate systems coincide at
         an initial moment of time.

vectors of absolute velocities, while $\vec{V}$ is the vector
of relative velocity of the fluid (relative to the
moving shell).

     The flow velocity $\vec{V}'$ also may be introduced.   It
is obvious that the equality $\rho\vec{V}\,\vec{K}_3* = \rho'\vec{V}'\vec{K}_3*$ holds.
We shall use the parameters $\vec{V}$ and V_3*.

     Thus, we have an equation

$$\rho(\vec{v} - \partial\vec{u}/\partial t)\vec{K}_3* = \rho'(\vec{v}' - \partial\vec{u}/\partial t)\vec{K}_3* = \rho\vec{V}\,\vec{K}_3* \quad (M*)$$

$$(1.1)$$

which represents the mass conservation law for a
stream of ideal fluid flowing across the unit area per
time unit [equation (4.4) of Chapter I].

Transferring across the surface σ, the tangential components of velocity, in general, undergo the discontinuity

$$(\vec{v} - \partial\vec{u}/\partial t)\vec{K}_i{}^* \neq (\vec{v}' - \partial\vec{u}/\partial t)\vec{K}_i{}^* \qquad (M^*) \qquad (1.2)$$

or $\vec{v}\,\vec{K}_i{}^* \neq \vec{v}\,\vec{K}_i{}^*$ ($i = 1,2$).

In (1.1) and (1.2) et. seq., the sign (M^*) indicates that the corresponding equations are written for a point M^* of space that coincides with the fixed point m of the shell on surface σ at a given moment of time.

Equation (1.1) may be written in scalar form in different ways. Representing $\vec{v}$, $\vec{v}'$, $\vec{V}$, $\vec{u}$ in terms of components with respect to a main basis $\vec{K}_1$, $\vec{K}_2$, $\vec{K}_3$

$$\vec{v} = v_i\vec{K}_i, \quad \vec{v}' = v_i'\vec{K}_i, \quad \vec{V} = V_i\vec{K}_i,$$

$$\vec{u} = u_i\vec{K}_i, \quad \partial\vec{u}/\partial t = (\partial u_i/\partial t)\vec{K}_i \qquad (1.3)$$

and taking into account (1.1) from Chapter II, we obtain the following equations in scalar form:

$$[\rho(v_i - \partial u_i/\partial t) - \rho'(v_i' - \partial u_i/\partial t)]E_i = 0,$$

$$(M^*) \qquad (1.4)$$

$$(v_i - V_i - \partial u_i/\partial t)E_i = 0$$

In formulae (1.3), (1.4) summation is performed by the repeated indices.

The system of equations (1.4) represents the kinematic conditions for the interaction problem of a permeable shell with an ideal compressible liquid.

Furthermore, we obtain the kinematic conditions for the components of fluid velocity in the deformed basis $\vec{K}_1*$, $\vec{K}_2*$, $\vec{K}_3*$. Let us use the representations $\vec{v}$ = $v_i*\vec{K}_i*$, $\vec{v}' = v_i'*\vec{K}_i*$, $\vec{V} = V_3*\vec{K}_3*$. In the latter expression one takes into account that the flow velocity across the surface σ is always directed along the normal. In this connection, however, we shall operate with the components of shell displacement u_i in the main basis $\vec{K}_1$, $\vec{K}_2$, $\vec{K}_3$, because in the shell theory the displacement components in deformed coordinates are not applied. Their application in this case would lead to a complication of the time derivative $\partial\vec{u}/\partial t = \partial(u_i* \vec{K}_i*)/\partial t$, in which $u_i*(\alpha_1, \alpha_2, t)$ and $\vec{K}_i*(\alpha_1, \alpha_2, t)$.

The first terms in (1.1) are presented in a form $v_i*\vec{K}_i*\vec{K}_3*$, while the second ones, as well as in obtaining (1.4), are presented in a form

$$(\partial u/\partial t)\vec{K}_i(\vec{E}_j\vec{K}_j)$$

In consequence, the kinematic conditions containing the components of fluid velocity in the shell deformed coordinates, take the form

$$\rho V_3* - \rho'v_3'* - (\rho - \rho')E_i\partial u_i/\partial t = 0,$$

$$(M*) \qquad (1.5)$$

$$v_3* - V_3* - E_i\partial u_i/\partial t = 0$$

They are especially simple in the case when the bending of a shell may be considered as static $(\partial u_i/\partial t = 0)$.

The dynamical conditions on the contact surface we obtain from equation (4.5) of Chapter I. The equation of momentum conservation of an ideal compressible fluid flowing across the unit area of a shell at a point M* per time unit is as follows

$$p\vec{K}_3* + \rho\vec{v}[(\vec{v} - \partial\vec{u}/\partial t)\vec{K}_3*] - \vec{Z}$$

$$(1.6)$$

$$= p'\vec{K}_3* + \rho'\vec{v}'[(\vec{v}' - \partial\vec{u}/\partial t)\vec{K}_3*] \qquad (M*)$$

Here $(-\vec{Z})$ is the vector of forces acting on the fluid from the shell. The latter is under the action of a force $\vec{Z}$.

Hence, we obtain the expressions for the force components Z_m and Z_m* in projections on the axis at an initial $(\vec{K_1}, \vec{K_2}, \vec{K_3})$ and actual $(\vec{K_1}*, \vec{K_2}*, \vec{K_3}*)$ instant of time. The latter enter on the right-hand sides of the shell motion equations (1.2) and (1.3) in Chapter II. Taking into account the first conditions in (1.4) and (1.5) we have

$$Z_m = (p-p')E_m + \rho(v_m-v_m')(v_i-\partial u_i/\partial t)E_i, \qquad (1.7)$$

$$(M*)$$

$$Z_m* = (p-p')\delta_{3m} + \rho(v_m*-v_m'*)(v_3*-\partial u_i/\partial t \cdot E_i) \quad (1.8)$$

Here as well as above, one carries out the summation over the double indices $(i=1,2,3)$; δ_{3m} is Kronecker's symbol.

According to (1.7) and (1.8), the right-hand sides of the above mentioned bending equations include the components of velocity associated with the shell displacement u_i. They may be excluded from the composition of Z_m and Z_m* with the help of the last conditions in (1.4) and (1.5). Then (1.7) and (1.8) will be expressed by the components of flow velocity through the shell:

$$Z_m = (p-p')E_m + \rho v_i E_i(v_m-v_m'), \qquad (1.9)$$

$$Z_m* = (p-p')\delta_{3m} + \rho v_3*(v_m*-v_m'*) \quad (M*) \quad (1.10)$$

Thus, in equations of shell bending, $L_m = R_m + Z_m$, which follows by projections of all the forces on $\vec{K_1}, \vec{K_2}, \vec{K_3}$, and the terms Z_m may be employed in form (1.7) as well as in form (1.9). Analogously, the bending equations $L_m* = R_m* + Z_m*$, obtained by projec-

136

tion of all the forces on $\vec{K_1}^*$, $\vec{K_2}^*$, $\vec{K_3}^*$, are considered in common with expression (1.8) or (1.10). The advantage of the various versions of the kinematical and dynamical conditions depends on the concrete problem to be solved.

As follows from (1.7)-(1.10), the shell permeability reduces to an appearance of tangential forces acting on the shell even in the case of an ideal liquid. They are equal to zero only in flowing conditions as the values of tangential components of a fluid velocity on the surface σ are identical (v_1^*-$v_1'^* = 0$, $v_2^*-v_2'^* = 0$). This is the principal pecularity of an interaction problem with a permeable shell as compared to the case of an impermeable shell.

In conclusion we consider one generalization of the problem formulation. It consists in the following.

For the solution of some concrete problems it is more convenient to operate with an Euler coordinate system x_1, x_2, x_3, which does not coincide with the system α_1, α_2, α_3 (Fig. 2) at the initial instant of time $t = 0$. Let us denote the unit vectors along the curvilinear orthogonal lines x_1, x_2, x_3 at point M of a space by $\vec{a_1}$, $\vec{a_2}$, $\vec{a_3}$. Their connection with unit vectors $\vec{K_1}$, $\vec{K_2}$, $\vec{K_3}$ along the lines α_1, α_2, α_3 at the same point M is expressed by the formula

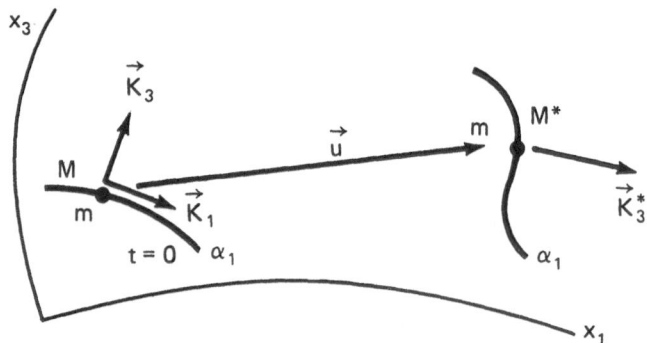

Fig. 2. Two positions of shell in a space. The Euler and Lagrange coordinate systems do not coincide at an initial moment of time.

$$\vec{a}_j = c_{ji}\vec{K}_i \qquad (1.11)$$

Then, if one represents $\vec{v}$, $\vec{v}'$ in terms of components of a basis $\vec{a}_1$, $\vec{a}_2$, $\vec{a}_3(\vec{v} = v_j\vec{a}_j)$ and also takes into account that $\vec{u} = u_i\vec{K}_i$, instead of (1.4), one may obtain (summation on i and j)

$$[\rho(c_{ji}v_j - \partial u_i/\partial t) - \rho'(c_{ji}v_j' - \partial u_i/\partial t)]E_i = 0,$$

$$(c_{ji}v_j - \partial u_i/\partial t)E_i = 0 \qquad (M^*) \qquad (1.12)$$

By analogy one can write the other versions of kinematical conditions and also of expressions for the force on the shell. For example, instead of (1.9) we have

$$Z_m = (p-p')E_m + \rho V_3{}^*c_{jm}(v_j - v_j') \qquad (M^*) \qquad (1.13)$$

Here it is also taken into account that $V_3{}^* = V_iE_i$.

In the case when at $t = 0$ the shell takes a place $x_3 = x_3{}^0$, and coordinates x_1, x_2 coincide with α_1, α_2, then we have $c_{ii} = 1$, $c_{ij} = 0$ $(i \neq j)$ and $\vec{a}_i = \vec{K}_i$. In this connection we obtain the conditions (1.4) and (1.9).

4.2. Representation of the flow velocity through the shell and the shell rigidity

The question of an experimental determination of the flow velocity of both air and water across the fabric, perforated metal plates, wire nets, etc., has attracted significant attention in the literature [1-11]. This velocity depends on a number of factors, namely, the shell perforation, the pressure differential p-p', the liquid density and viscosity, the shell deformation, etc.

The most often used representations are as follows

$$p-p' = aV_3* + bV_3*^2, \quad V_3* = \alpha(p-p')^\beta \quad (M*) \quad (2.1)$$

where a, b, α, β are experimental coefficients.

In a table we record the values of coefficients a, b, α, β for three kinds of a capron material thread. The values for β agree with the general estimates of [6], where one indicates that for natural fabrics β = 0.59 ÷ 0.7, while for the artificial ones - β = 0.55. Based on these data one can confirm that for the fabrics $1/2 < \beta < 2/3$. The other coefficients range more widely, especially a and b.

Table

	a $\dfrac{kg \cdot sec}{m^3}$	b $\dfrac{kg \cdot sec^2}{m^4}$	α $\dfrac{m}{sec}(\dfrac{m}{kg})^\beta$	β dimensionless
1	3.7	2.0	0.49	0.55
2	21.7	3.7	0.39	0.50
3	55.0	76.0	0.09	0.52

Relations (2.1) have the defect that when the sign of the pressure differential changes, the velocity's V_3* sign does not change and vice versa. Therefore, one may assume the following form:

$$p-p' = aV_3* + bV_3*^2 signV_3*,$$

$$(M*) \quad (2.2)$$

$$V_3* = \alpha|p-p'|^\beta sign(p-p')$$

The more general permeability relations, in which the influence of the deformation of the shell middle surface is also taken into account, are presented in [12-15].

It should be noted that relationships of the type (2.1) between the fluid flow velocity and the pressure differential are often used in other fields of mechanics as well. For example, in the theory of liquid and gas filtration in porous media, the Darcy law is applied, which is obtained from (2.1) if one assumes that b = 0, β = 1. In this case the value a is called the filtration coefficient and it characterizes the medium porosity and the fluid properties.

V_3* is also determined from the relation of the hole (aperture, orifice) area S_0 of the perforated plate to its common (general) area S.

Let us consider one of the first examples of the definition of such dependence for the perforated plate and the metal net, placed across the flow [2,3]. A linear relationship is taken for V_3* with respect to the velocity V_∞ at infinity:

$$V_3* = 4V_\infty/(4 + K) \qquad (2.3)$$

In Fig. 3 taken from [3], the experimental points are plotted, and the analytical dependences K = $f(S/S_0)$ are shown by the dashed lines. The latter are based on certain hypotheses. However, none of them describes satisfactorily an experiment. In the article [17] there was assumed another dependence

$$K + 1 = \frac{1}{3} \left(4 \frac{S^2}{S_0^2} - 1 \right) \qquad (2.4)$$

which is shown in Fig. 3 by the continuous line.

From (2.3) and (2.4) we have

$$V_3* = \frac{3V_\infty}{2 + (S/S_0)^2} \qquad (2.5)$$

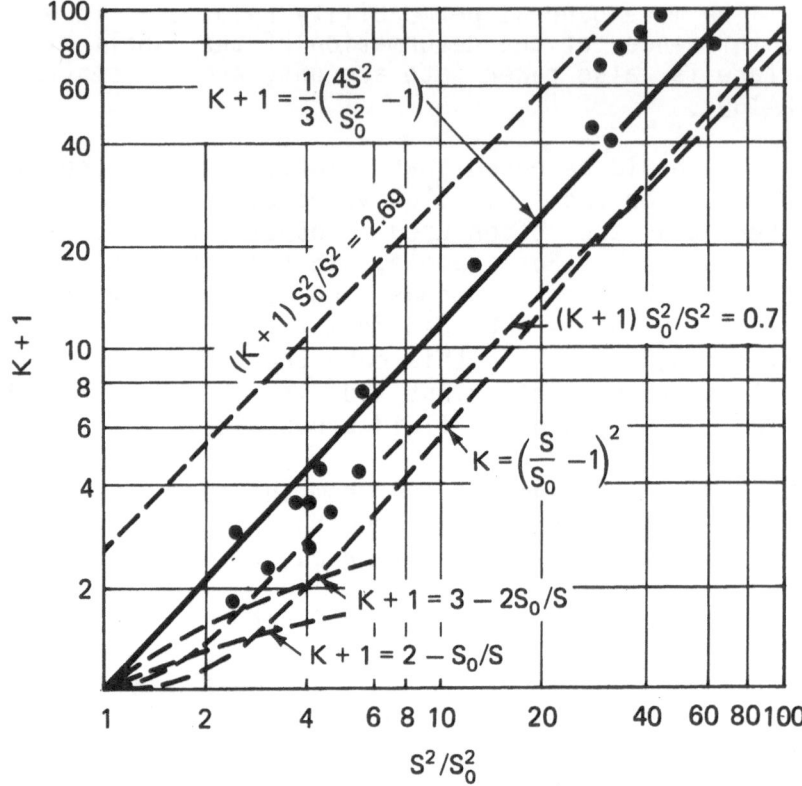

Fig. 3. Dependence of permeability characteristics of
the shell on its perforation degree.

According to (2.5) for an impermeable plate $(S_0/S \rightarrow 0)$
the flow velocity equals zero, while for the other
limiting case $(S_0/S \rightarrow 1)$ it is $V_3^* = V_\infty$. If $S_0/S = 1/2$, then $V_3^* = (4/7)V_\infty$. It should be noted that V_3^*
is the velocity per unit of the plate common area, but
it is not the real flow velocity across the hole.

Based on the experimental data of [3], the following formula for the pressure differential was
assumed in the article [17]:

$$p - p' = \frac{2}{3} \rho_\infty V_\infty^2 \left(1 - \frac{S_0^2}{S^2}\right) \qquad (2.6)$$

In this connection one has assumed that the pressure on an impermeable plate equals $(2/3)\rho_\infty V_\infty^2$.

The perforated plate is bent under the action of the pressure Z_3*, approximately equal to $(1 - S_0/S)$ $(p - p')$, that is, in accordance with (2.6)

$$Z_3{}^* = \frac{2}{3} \rho_\infty V_\infty^2 (1 - \frac{S_0^2}{S^2})(1 - \frac{S_0}{S}) \qquad (2.7)$$

By analogy with (2.7) we introduce in the general formulae for the forces acting on the shell (1.7)-(1.10) the coefficient ε, which takes into account the relation of the unflowing part of the plate area to its whole common area, $(1 - S_0/S)$:

$$Z_m = \varepsilon[p-p')E_m + \rho V_i E_i (v_m - v_m')], \qquad (2.8)$$

$$(M*)$$

$$Z_m{}^* = \varepsilon[(p-p')\delta_{3m} + \rho V_3*(v_m{}^* - v_m'{}^3)] \qquad (2.9)$$

Strictly speaking, along each direction there should be different values of ε (that is, ε_m; moreover, all the values of ε_m are less than unity). Moreover, in equations (2.8) and (2.9) they also have to differ (in the latter, ε_m*). However, we limit ourselves to the case $\varepsilon_m = \varepsilon_m* = \varepsilon$. It should be noted that the coefficient ε in the general case is connected with parameters a, b, α, β in relations (2.1) and (2.2).

Let us consider the question of the shell rigidity. As compared with the case of a heat-isolated shell with a dense wall, here the change is connected with both the material heating up and a perforation.

Let us introduce the following definitions: $\tilde{W}$ is the energy transfer from the shell to a fluid; $\tilde{U}$, $\tilde{q}$, $\tilde{\sigma}$ are the specific internal energy, energy transfer, and entropy in a fluid, $\tilde{U}'$, q', σ' are the same values from the other side of the shell. Therefore, the equation of energy conservation of a fluid stream, flowing across the unit area at a point M* of space, has the form

$$\rho v \, \vec{K}_3{}^* + q + \rho[(\vec{v} - \frac{\partial \vec{u}}{\partial t})\vec{K}_3{}^*](\frac{v^2}{2} + \tilde{U}) + \vec{Z}(\vec{v} - \frac{\partial \vec{u}}{\partial t}) - \tilde{W}$$

(2.10)

$$= \rho'\vec{v}'\vec{K}_3{}^* + q' + \rho'[(\vec{v}' - \frac{\partial \vec{u}}{\partial t})\vec{K}_e{}^*](\frac{v'^2}{2} + \tilde{U}') \quad (M^*)$$

As in the previous section, we operate with both the components of fluid motion velocity on the Euler coordinates x_1, x_2, x_3 and the components of a shell displacement vector on the Lagrange coordinates α_1, α_2, α_3 in an initial position. Thus, it is more convenient to employ the value of the flow velocity along the normal to the deformed surface $V_3{}^*$, because it is connected directly with the pressure differential. Then, taking into account (1.1), (1.4), one can write equation (2.10) as follows:

$$\tilde{W} = Z_m(v_m - \frac{\partial u_m}{\partial t}) + (\rho v_i - \rho'v_i')E_i + q - q'$$

(2.11)

$$+ \rho V_3{}^* [\frac{1}{2}(v^2 - v'^2) + \tilde{U} - \tilde{U}']$$

The entropy jump in a fluid stream equals

$$\Omega = \rho V_3{}^*(\tilde{\sigma} - \tilde{\sigma}') \quad (M^*)$$

(2.12)

It is provided by an entropy change due to heat transfer from the shell and nonreversible entropy increase. In (2.11) and (2.12) a correcting coefficient may also be introduced, which is connected with surface porosity.

From (2.11) and the heat-conduction equation, the temperature of the shell wall is determined. Furthermore, from a knowledge of the temperature, one finds the corresponding value of the elastic modulus of the shell.

A sizable literature [16] is devoted to an analysis of the extension and bending of perforated plates and shells.

Reduction of the shell bending stiffness is characterized by a coefficient $\mu = D^*/D$, where D^* is the reduced cylindrical stiffness of the perforated shell, while D is a stiffness of the dense unperforated shell with the same thickness and material. Special investigations show that the coefficient μ depends weakly on the perforation form. In Fig. 4 two forms (1 and 2) of perforation are shown, while in Fig. 5 the graphical dependence of μ on the ratio of the holes' diameter to the distance between them d/ℓ [17] is shown. The computational values are shown by continuous lines, and the experimental values by points.

The growth of a shell perforation leads to reduction of both the force acting from the flow and the stiffness. In order to evaluate which two of these factors dominates, we consider the cylindrical bending of a long plate having free supported edges placed across the flow. For the pressure on the plate we make use of formula (2.7). The bending equation is

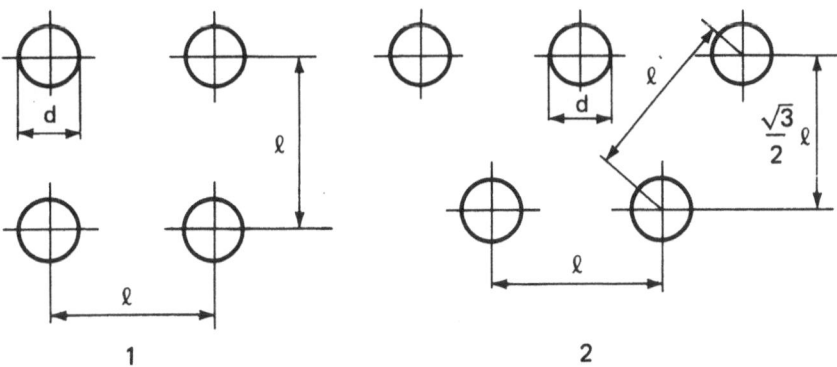

Fig. 4. Shapes of perforation.

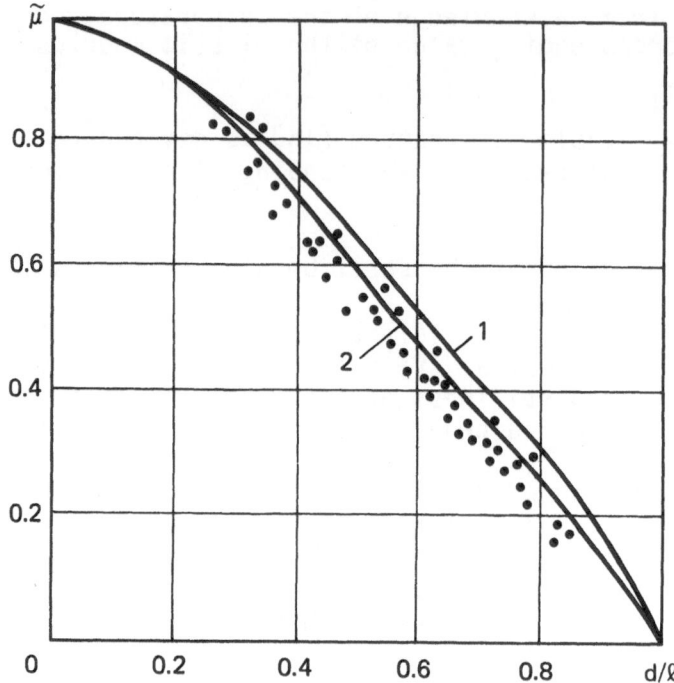

Fig. 5. Change of a bending rigidity depending on the perforation degree. (Lines 1 and 2 correspond to the perforation shapes from Fig. 4.)

$$D\tilde{\mu}d^4w/dx^4 = Z_3* \qquad (2.13)$$

Assuming $w = W\sin(\pi x/L)$ and integrating (2.13) by the Galerkin method, we obtain the value of W. The ratio of the deflection of a perforated to an unperforated plate is [in (2.7) one has taken into account that $S_0/S = \pi d^2/4\ell^2$]

$$\frac{W_0}{W} = \frac{1}{\tilde{\mu}(d/\ell)} (1 - \frac{\pi^2}{16} \frac{d^4}{\ell^4})(1 - \frac{\pi}{r} \frac{d^2}{\ell^2}) \qquad (2.14)$$

Here $\tilde{\mu}(d/\ell)$ is taken from Fig. 5.

145

In Fig. 6 this deflection ratio is plotted. It is seen that the plate deflection increases with increasing perforation. For $d/\ell \to 1$ according to Fig. 5 and to the formulae (2.7), (2.14) the pressure on the plate has a finite value, while bending stiffness tends to zero. Therefore the ratio w_0/w tends to infinity.

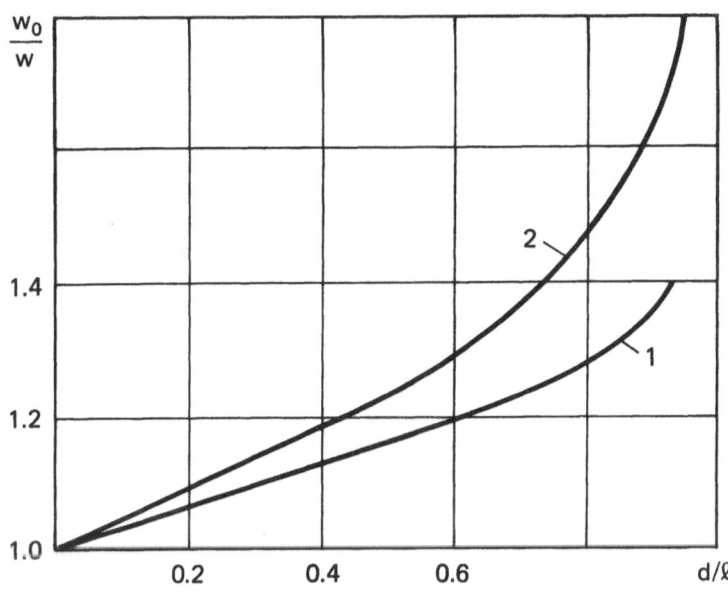

Fig. 6. Change of a plate deflection depending on the perforation degree. (Lines 1 and 2 correspond to the perforation shapes from Fig. 4.)

4.3. Particular cases

Conditions (1.4), (1.5), (1.7)-(1.10), (2.1), (2.2) are for a point M* of space which coincides with the fixed point m of the shell. The reduction to a point M in space, which has known Euler coordinates on the undeformed surface, may be made as in Chapter II by means of an analytical continuation of functions that describe the liquid motion. However, such a reduction in general form causes excessive complication in the interaction problem of a permeable shell. It may be more expedient to do this in concrete problems. The manner of reduction and the analysis in detail has been stated in Chapter II.

However, even in particular cases the problem solution with the above mentioned reduction is not always possible by analytical methods. The principal methods are numerical ones, as conditions on the contact surface may be taken into account at every step of time. Below we shall limit ourselves to consideration of some particular cases without the mentioned reduction.

1. A significant particular case is an incompressible liquid. Accepting that $\rho = \rho'$, from (1.4) we obtain the kinematic conditions

$$(v_i - v_i')E_i = 0$$

$$(M\star) \qquad (3.1)$$

$$(v_i - V_i - \partial u_i/\partial t)E_i = 0$$

while from (1.5) we obtain

$$v_3{}^\star - v_3{}'^\star = 0,$$

$$(M\star) \qquad (3.2)$$

$$v_3{}^\star - V_3{}^\star - E_i \partial u_i/\partial t = 0$$

These conditions may be easily written in terms of the flow parameters in the domain on each side of the shell. For example, from the latter conditions it follows that

$$v_3{}^\star = V_3{}^\star + E_1 \frac{\partial u_1}{\partial t} + E_2 \frac{\partial u_2}{\partial t} + E_3 \frac{\partial u_3}{\partial t} ,$$

$$(M\star)$$

$$v_3{}'^\star = V_3{}^\star + E_1 \frac{\partial u_1}{\partial t} + E_2 \frac{\partial u_2}{\partial t} + E_3 \frac{\partial u_3}{\partial t}$$

The relations of permeability (2.1), (2.2) and expressions for the forces acting on the shell (2.8), (2.9) remain the same.

2. For slow shell deformation one can assume that $\partial u_i/\partial t \ll v_i$, v_i', and the contact conditions have the form

$$(\rho v_i - \rho' v_i')E_i = 0, \qquad (v_i - V_i)E_i = 0,$$

$$\rho v_3^* - \rho' v_3'^* = 0, \qquad v_3^* - V_3^* = 0,$$

$$\text{(M*)} \qquad (3.3)$$

$$Z_m = \varepsilon[(p - p')E_m + \rho v_i E_i(v_m - v_m')],$$

$$Z_m^* = \varepsilon[(p - p')\delta_{3m} + \rho v_3^*(v_m^* - v_m'^*)]$$

The expressions for Z_m, Z_m^* (2.8), (2.9) and also the relations of permeability remain unchanged.

The question of the neglect of the inertia terms in the equations for the shell deformation requires additional analysis.

Unlike the displacement components u_i entering in E_i, in $\partial u_i/\partial t$ one can also imagine the motion velocity of the shell as a rigid body about the (fixed in space) Euler coordinate system. Therefore, for the uniform motion of a shell as a solid body, the inertia terms in its bending equations are equal to zero, but the terms $\partial u_i/\partial t$ remain in the kinematic conditions.

3. With assumption of absolute rigidity we have $E_1 = E_2 = 0$, $E_3 = 1$, $v_i^* = v_i$, $V_3^* = V_3$, $Z_m^* = Z_m$. In this connection equations (1.4), (2.8) coincide with (1.5) and (2.9) and take the form

$$\rho v_3 - \rho' v_3' - (\rho - \rho')\partial u_3/\partial t = 0,$$

$$v_3 - V_3 - \partial u_3/\partial t = 0, \qquad \text{(M*)} \qquad (3.4)$$

$$Z_m = \varepsilon[(p - p')\delta_{3m} + \rho V_3(v_3 - v_3')]$$

Here $\partial u_3/\partial t$ is the component of motion velocity normal to the middle surface of the absolutely rigid shell.

4. The shell is impermeable: $V_3^* = 0$ or $a = \infty$, $\alpha = 0$ in relations (2.1), (2.2). The kinematic conditions take the form

$$(v_i - \partial u_i / \partial t)E_i = 0, \quad (v_i' - \partial u_i / \partial t)E_i = 0,$$

$$(M^*) \quad (3.5)$$

$$v_3^* - E_i \partial u_i / \partial t = 0, \quad v_3'^* - E_i \partial u_i / \partial t = 0$$

that, is they are formulated separately for each domain on the two sides of the shell. In this connection the densities undergo the arbitrary discontinuity $(\rho \neq \rho')$.

The pressure on the shell from the liquid equals $(\varepsilon = 1)$

$$Z_m = (p - p')E_m, \quad Z_m^* = (p - p')\delta_{3m} \quad (M^*) \quad (3.6)$$

Here the tangential force components to the deformed surface are absent.

The detailed analysis of this case is discussed in Chapter II.

5. Now we consider the case of $\rho = \rho'$; $(v_i, v_i') \ll \partial u_i / \partial t$, which occurs, for example, for the motion of a shell having large permeability (for instance, nets) in water at rest. From (3.1), (3.2) we obtain

$$(v_i - v_i')E_i = 0, \quad (V_i + \partial u_i / \partial t)E_i = 0,$$

$$(M^*) \quad (3.7)$$

$$v_3^* - v_3'^* = 0, \quad V_3^* + E_i \partial u_i / \partial t = 0$$

Expression for Z_m has the form

$$Z_m = \varepsilon[(p-p')E_m - \rho E_i \partial u_i / \partial t \cdot (v_m - v_m')] \quad (M^*) \quad (3.8)$$

Significant simplifications may be made under the restrictions on separate components of the shell displacement vectors and on fluid velocities and also on their change characteristics with respect to coor-

dinates, etc. Some other simplifications are described in Chapters II and III.

Let us consider an example of the interaction of
a shallow curved plate with a subsonic flow of compressible fluid under small angle of attack (that is,
the angle between the plate chord and the velocity
direction far from it). In this case a smooth change
of the flow parameters with respect to coordinates
takes place. The numbers in (5.1) of Chapter II equal
$m = 1$, $n = 1/2$, $k = -1/2$. Therefore, as one simplifies the kinematic and dynamic conditions (1.4),
(1.5), (2.8), (2.9) and also the permeability relations (2.2) to an undeformed surface, the second terms
in the Taylor series expansion $\vec{v}(M*) = \vec{v}(M) +$
$[\vec{u}\cdot\vec{\nabla})\vec{v}](M)$, $\rho(M*) = \rho(M) + [(\vec{u}\cdot\vec{\nabla})\rho](M)$,...are small
compared to the first ones. Because of that fact when
writing the contact conditions, the deformed surface
is identified with the undeformed one (or with the
chord).

As is known from Chapter II, in the case of a
shallow shell $u_1 \sim \omega_1 u_3$, $u_2 \sim \omega_2 u_3$, $\omega_1 = \partial w/\partial x$, $\omega_2 =
\partial w/\partial y$, $E_1 = -\omega_1$, $E_2 = -\omega_2$, $E_1 = 1$. Taking into ac-
count that $V_i = V_3*E_i$ and evaluating the terms (1.4)
as in Chapter II, we obtain the kinematic conditions

$$V_3 = \frac{\partial w}{\partial t} + V_1 \frac{\partial w}{\partial x} + V_2 \frac{\partial w}{\partial y} + V_3*, \qquad \text{(M)}$$

$$\rho V_3 - \rho' V_3' - (\rho - \rho') \frac{\partial w}{\partial t}$$

$$\qquad\qquad\qquad (3.9)$$

$$- (\rho V_1 - \rho' V_1') \frac{\partial w}{\partial x} - (\rho V_2 - \rho' V_2') \frac{\partial w}{\partial y} = 0$$

The last condition also may be represented in the form
of the law of mass conservation for the fluid stream
transporting across the shell, viz.

$$\rho(v_3 - \frac{\partial w}{\partial t} - v_1 \frac{\partial w}{\partial x} - v_2 \frac{\partial w}{\partial y})$$

$$= \rho'(v_3' - \frac{\partial w}{\partial t} - v_1' \frac{\partial w}{\partial x} - v_2' \frac{\partial w}{\partial y}) \qquad (3.10)$$

From (2.8), (2.9) we obtain

$$Z_1 = Z_2 = 0, \quad Z_3 = Z_3^* = \epsilon(p - p') \quad (M) \quad (3.11)$$

while the permeability relations (2.2) retain their form.

In the case of an incompressible liquid ($\rho = \rho'$) one must add to the first condition of (3.9), an equation, i.e.

$$v_3 - v_3' = (v_1 - v_1') \frac{\partial w}{\partial x} + (v_2 - v_2') \frac{\partial w}{\partial y} \qquad (M)$$

or

$$v_3' = \frac{\partial w}{\partial t} + v_1' \frac{\partial w}{\partial x} + v_2' \frac{\partial w}{\partial y} + V_3^* \quad (M)$$

Relations (3.11) do not change.

4.4. Bending of a cylindrical shell in an incompressible liquid

The plane problem for static bending of a thin elastic cylindrical shell with a transverse flow around it of an ideal incompressible liquid is considered. Such a problem in a linear formulation was solved in section 3.1. The difference here is the consideration of the (uniform) permeability of the shell. Therefore, one accepts the linear connection of flow velocity and pressure differential, that is, $\beta = 1$ in the second relation of (2.1).

We suppose that introduction of permeability does not invalidate the flow potential; parameters of the flow and shell are such that the shell is far from its elastic buckling. Furthermore, the limit cases of the problem are the uniform fluid flow at absolute shell permeability [in the law $V_3^* = \alpha(p - p')$, the coefficient $\alpha = \infty$] and the unseparated flow of an impermeable shell ($\alpha = 0$). In the latter case a separated flow model would be more realistic, but here we do not consider it. By writing the conditions of solutions conjugation on the shell middle surface, the deformed surface is identified with the undeformed one. The influence of their difference was considered in Chapter III. Here we concentrate our attention on the influence of permeability.

We denote the pressure and the velocity potential outside the shell by index 1, while inside it is denoted by index 2. At a great distance from the shell the parameters are p_∞, ρ_∞, V_∞. According to the Bernoulli equation for the steady motion of an incompressible liquid

$$\frac{1}{2} (\nabla\phi)^2 + \frac{p}{\rho} = C \qquad (4.1)$$

from which we obtain

$$P_1 = P_\infty + \frac{\rho_\infty}{2} [V_\infty^2 - (\nabla\phi_1)^2], \qquad (4.2)$$

$$P_2 = P_{02} - \frac{\rho_\infty}{2} (\nabla\phi_2)^2 \qquad (4.3)$$

in which the parameter p_{02} is determined from the problem solution.

Solutions of equations

$$\nabla^2\phi_1 = 0, \qquad \nabla^2\phi_2 = 0 \qquad (4.4)$$

have to satisfy the conditions

$$\phi_1 = -V_\infty r\cos\theta \qquad (r \to \infty),$$
$$\phi_2 \neq \infty \qquad (r = 0) \qquad (4.5)$$

and in accordance with the previous sections - ($\partial u_i/\partial t$ = 0, $v_\theta = \partial\phi/r\partial\theta$, $v_r = \partial\phi/\partial r$)

$$\frac{\partial\phi_1}{\partial r} - \frac{\partial\phi_1}{r\partial\theta}\,\omega_\theta = \frac{\partial\phi_2}{\partial r} - \frac{\partial\phi_2}{r\partial\theta}\,\omega_\theta \qquad (r=R) \qquad (4.6)$$

As is known from Chapter I, in the actual case the rotation of the normal to the shell surface equals $\omega_\theta = R^{-1}(\partial w/\partial\theta - v)$, where v, w are the displacement components.

The permeability law (p = p_2, p' = p_1) is

$$V_3^* = \frac{\partial\phi_1}{\partial r} - \frac{\partial\phi_1}{r\partial\theta}\,\omega_\theta = \alpha\,(p_2 - p_1), \qquad (r=R) \quad (4.7)$$

while the equations of shell bending are

$$\frac{\partial^2 v}{\partial\theta^2} + \frac{\partial w}{\partial\theta} = -\frac{b^2 R^4}{D}\,Z_\theta \qquad (r=R), \qquad (4.8)$$

$$\frac{\partial v}{\partial\theta} + w + b^2\,(\frac{\partial^2}{\partial\theta^2} + 1)^2 w = \frac{b^2 R^4}{D}\,Z_r \quad (r=R) \qquad (4.9)$$

Here α is the coefficient of proportionality, b^2 = $h^2/12R^2$, D = $(\tilde{\mu}/\varepsilon)Eh^3/12(1-\nu^2)$. The coefficients $\tilde{\mu}$ and ε are known from section 4.2.

From the previous sections of this chapter

$$Z_\theta = \alpha\rho_\infty(p_2-p_1)(\frac{\partial\phi_2}{R\partial\theta} + \frac{\partial\phi_2}{\partial r}\,\omega_\theta - \frac{\partial\phi_1}{R\partial\theta} - \frac{\partial\phi_1}{\partial r}\,\omega_\theta),$$
$$\qquad (4.10)$$
$$Z_r = p_2 - p_1$$

in which one has taken into account (4.7) and the inclusion of ε in D.

We obtain the approximate solution of the problem in two stages. First we determine the forces acting on the shell with the assumption of its absolute rigidity. At the second stage we determine the shell bending under the action of these forces.

As is known from Chapter III, the solution of Laplace's equation (4.4) for the external domain [which satisfies the first condition of (4.5)] has the form

$$\phi_1 = - V_\infty r \cos \theta + \sum_{n=1}^{N} B_n r^{-n} \cos n\theta \qquad (4.11)$$

For the internal domain we write

$$\phi_2 = K_0 + \sum_{n=1}^{N} K_n r^n \cos n\theta \qquad (4.12)$$

In the case of an absolutely rigid ($v = w = \omega_\theta = 0$) permeable shell, from (4.11), (4.12) we obtain

$$K_n = - (\delta_{1n} V_\infty + B_n R^{-2n}) \qquad (4.13)$$

where δ_{1n} is the Kronecker's symbol.

Substituting (4.11), (4.12), (4.13), (4.2), (4.3) into (4.7) and comparing the coefficients of cosines of identical arguments, we obtain the following system of algebraic equations:

$$P_{02} = P_\infty + \frac{1}{2} \rho_\infty V_\infty^2 (1 + 2D_1) \qquad (n=0),$$
$$(4.14)$$

$$\gamma D_1 - D_2 = -\gamma \ (n=1), \quad D_{n-1} + \gamma D_n - D_{n+1} = 0 \ (2 \leqslant n \leqslant N-1)$$

$$D_{N-1} + \gamma D_N = 0 \qquad (n=N)$$

Here one has denoted

$$\gamma = \frac{1}{\alpha \rho_\infty V_\infty} , \qquad D_n = \frac{n B_n}{R^{n+1} V_\infty} \qquad (4.15)$$

154

Solution of system (4.14) gives

$$D_1 = - \gamma \ \gamma_N,$$

$$D_n = (-1)^n \ \gamma \prod_{K=N-(n-1)}^{N} \gamma_k, \ldots, \qquad (4.16)$$

$$D_N = (-1)^N \gamma \prod_{k=1}^{N} \gamma_k,$$

in which γ_k is the sequence

$$\gamma_1 = \gamma^{-1}, \quad \gamma_k = (\gamma + \gamma_{k-1})^{-1} \quad (k \geqslant 2) \qquad (4.17)$$

It may be shown [14] that the series (4.11), (4.12) with the coefficients (4.16), (4.17) converge absolutely and uniformly. The series convergence is better when the value γ (4.15) is larger, that is, when the shell is less permeable.

According to (4.10), (4.2), (4.3), (4.11), (4.12), (4.14), (4.15) the forces acting on the shell (without taking into account the coefficient ε) equal

$$Z_\theta = -\rho_\infty V_\infty^2 \{ \sum_{n=1}^{N} D_n[\sin(n-1)\theta + \sin(n+1)\theta]$$

$$+ \sum_{m=1}^{N} \sum_{s=1}^{N} D_m D_s \sin(m+s)\theta \}, \qquad (4.18)$$

$$Z_r = p_2 - p_1 = -\gamma \rho_\infty V_\infty^2 (\cos\theta + \sum_{n=1}^{N} D_n \cos n\theta)$$

For the absolutely impermeable shell ($\gamma = \infty$) from (4.16), (4.17) it follows that $D_1 = -1$, $D_2 = \ldots = 0$, while for the absolutely permeable one ($\gamma = 0$) $D_1 = D_2 = \ldots = D_N = 0$. In the first limiting case we obtain

$$\frac{\partial \phi_1}{\partial r} = - V_\infty \left(1 - \frac{R^2}{r^2}\right)\cos\theta, \quad \frac{\partial \phi_1}{r \partial \theta} = V_\infty \left(1 + \frac{R^2}{r^2}\right)\sin\theta,$$

$$(4.19)$$

$$\frac{\partial \phi_2}{\partial r} = 0, \quad \frac{\partial \phi_2}{r \partial \theta} = 0, \quad p_1 = p_\infty - \rho_\infty V_\infty^2 \frac{R^2}{r^2}\left(\frac{R^2}{2r^2} - \cos 2\theta\right),$$

$$p_2 = p_{02} = p_\infty - \frac{1}{2}\rho_\infty V_\infty^2,$$

$$Z_\theta = 0, \quad Z_r = - \rho_\infty V_\infty^2 \cos 2\theta$$

while in the second one we obtain the plane-parallel fluid flow across the shell ($\partial \phi_1/\partial r = \partial \phi_2/\partial r, \ldots, p_1 = p_2 = p_\infty$, $Z_\theta = Z_r = 0$).

It should be noted that from (4.19) one obtains for the pressure p_2 inside the shell a quite definite connection with the external flow parameters. This is the consequence of the limiting method of the problem solution; because of the fact that for $\alpha \to 0$ ($\gamma \to \infty$) relation (4.7) degenerates to the impermeability condition $\partial \phi_1/\partial r = 0$, there is no influence of p_2 on the external flow. In the case of the solution of the problem with condition $\partial \phi_1/\partial r = 0$, the pressure inside the shell may be arbitrary.

In Fig. 7 the angular distribution of dimensionless pressure differential $\bar{Z}_r = (p_2 - p_1)\rho_\infty V_\infty^2$ is plotted, while in Fig. 8 - the distribution of tangential forces $\bar{Z}_\theta = Z_\theta/\rho_\infty V_\infty^2$ is plotted. The values of the permeability parameter γ were taken as 0.1, 1, 10, ∞.

From these values it follows, that for $\gamma > 10$, both the normal and tangential forces are approximately equal to their corresponding values in the case of an impermeable shell (the latter tends to zero as $\gamma \to \infty$). With increasing permeability (decreasing γ) the normal forces monotonically fall, while the tangential forces first increase and then also fall to

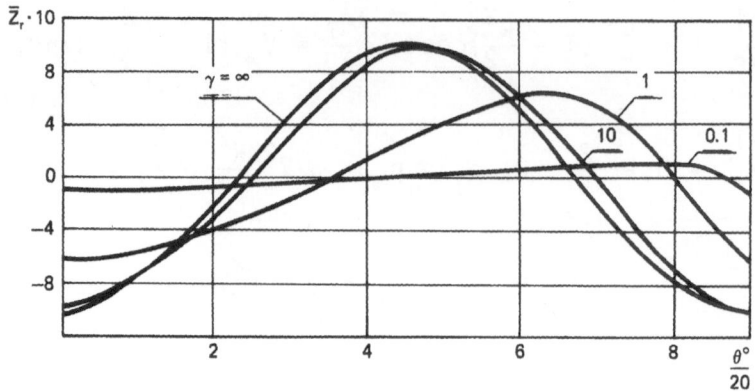

Fig. 7. Distribution of the pressure overfall over
the cylinder for the various values of
permeability parameter.

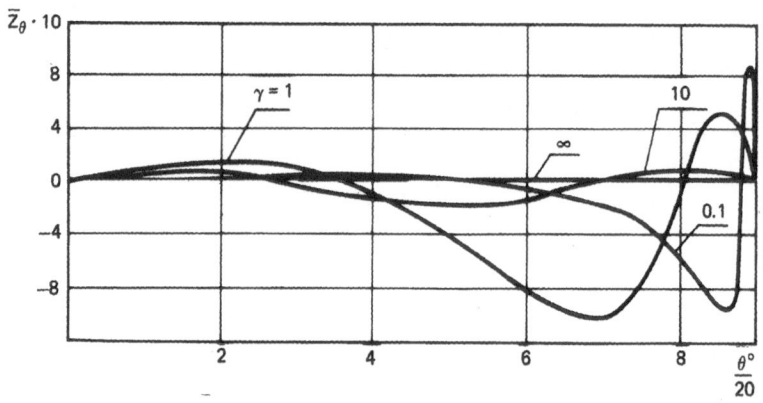

Fig. 8. Distribution of the tangential forces over
the cylinder for the various values of
permeability parameter.

zero. It should be noted that these results are presented for $\varepsilon = 1$.

In order to construct the stream lines we use the Cauchy-Riemann relations

$$\frac{\partial \phi_i}{\partial r} = \frac{\partial \psi_i}{r \partial \theta} \ , \qquad \frac{\partial \phi_i}{r \partial \theta} = - \frac{\partial \psi_i}{\partial r} \qquad (i=1,2)$$

Taking into account the above obtained values of potentials ϕ_1, ϕ_2 and accepting that the zero line $\psi_1 = \psi_2 = 0$ corresponds to angle $\theta = 0$, we have

$$\psi_1 = - RV_\infty [\frac{r}{R} \sin \theta + \sum_{n=1}^{N} \frac{D_n}{n} (\frac{R}{r})^n \sin n\theta] \qquad (4.20)$$

One obtains an identical expression for the flow function ψ_2.

Accepting that $\psi_1 = \psi_2 = Q = $ constant in which the absolute value of Q determines the fluid flux between the analyzed and zero flow line, one can construct the stream lines. For some values of $\bar{Q} = Q/RV_\infty$ they are shown in Fig. 9, 10. In this connection the values of permeability parameter γ were taken to be 0.1;1. The dimensionless radius $\bar{r} = r/R$ is introduced here.

From the above figures one can see a considerable influence of the surface permeability on the flow. For $\gamma > 10$ the flow lines that do not intersect the shell surface are close to the corresponding lines in the case of an impermeable cylinder. For the small values of γ a small disturbance to the given uniform flow occurs.

In the bottom part of a shell there is a region in which the particles move over the crossed trajectories. This is the consequence of our accepting the unseparated flow model (in the limiting case of an impermeable surface), when in the bottom part of a

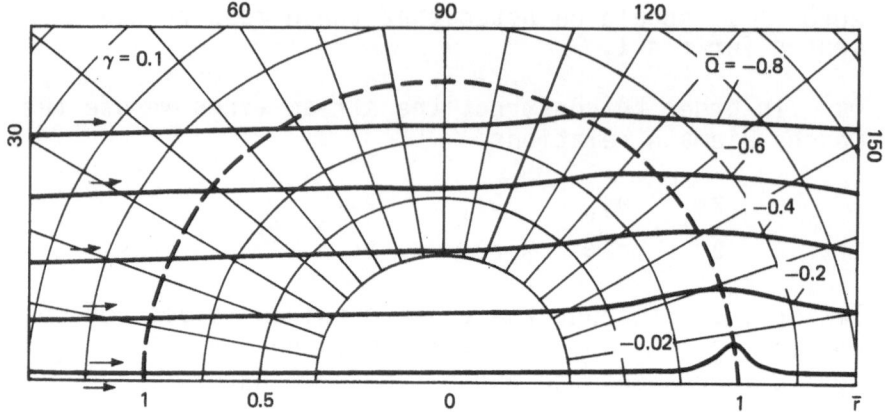

Fig. 9. Lines of flow at the large permeability of a
         shell.

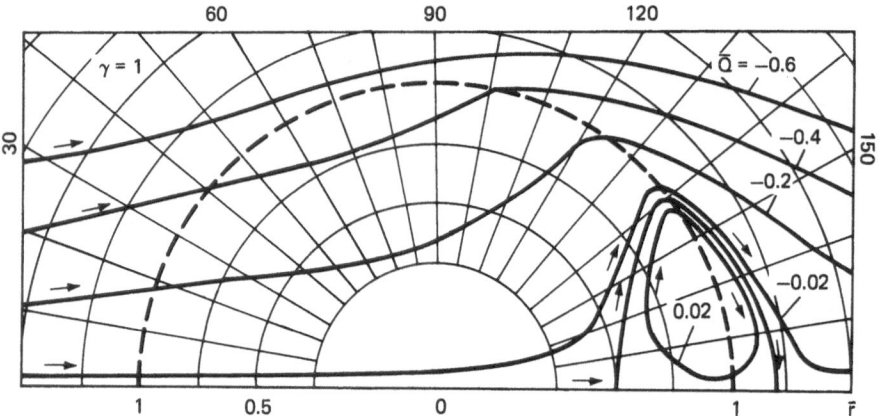

Fig. 10. Lines of flow at the medium permeability of
          a shell.

shell the pressure rises. On the contact surface the tangential velocity components from the two sides undergo a discontinuity. This induces the possibility of the existence of closed stream lines in spite of the potentiality of the flow field.

Projections of forces acting on the shell, on the x axis equal

$$N_x = \int_0^{2\pi} (-Z_r\cos\theta + Z_\theta\sin\theta)Rd\theta \quad (r=R)$$

Substituting here expressions (4.18), (4.19) we obtain $N_x = 0$ (D'alembert paradox). It should be noted that ignoring the tangential forces acting on a shell produces a nonzero force, N_x.

Now we consider the second stage of the problem solution. The components of the shell displacement are presented in the form

$$v = \sum_{n=1}^{N} V_n\sin n\theta,$$

$$(4.21)$$

$$w = \sum_{n=0}^{N} W_n\cos n\theta$$

Substituting (4.21) and (4.18) into (4.8), (4.9) and further comparing the coefficients of sine and cosine terms of identical arguments, we are led to an equations system for V_n, W_n. Its solution is

$$W_0 = 0,$$

$$V_n = -\frac{1}{n} W_n - \frac{2b^2 R\mu}{n^2} [D_{n-1} + \sum_{s=1}^{n-1} D_{n-s}D_s + (1-\delta_{nN})D_{n+1}]$$

$$W_n = \frac{2R\mu}{(n^2-1)^2} \{\frac{1}{n} [D_{n-1} + \sum_{s=1}^{n-1} D_{n-s}D_s \qquad (4.22)$$

$$+ (1-\delta_{nN})D_{n+1}] - \gamma D_n\} \qquad (2<n<N)$$

Here, as well as in Chapter III, the hydroelastic parameter $\mu = \rho_\infty V_\infty^2 R^3/2D$ is introduced. Now the value

D includes the parameters μ and ε, which characterize the surface permeability.

For the case $\gamma = \infty$ as was shown above, $D_1 = -1$, $D_2 = D_3 = \ldots = 0$ and instead of (4.22) we obtain

$$W_0 = 0, \qquad V_2 = R\mu/9, \qquad W_2 = -2R\mu/9,$$

$$V_3 = \ldots = 0, \qquad W_3 = \ldots = 0$$

This solution coincides with case (2.10) of Chapter III.

Displacement components v and w are determined from (4.21) and (4.22).

References

1. Rakhmatulin, Kh.A.   Theory of parachute opening.
    J. "Technica vozdushnogo flota," 1940, No. 8, pp.
    79-89.

2. Taylor, G. I. Air resistance of a flat plate of
    very porous material.   Reports and Memoranda of
    the Aeronautical Research Council, No. 2236,
    London, 1944, pp. 383-386.

3. Taylor, G. I., Davies, R. M. The aerodynamics of
    porous sheets.   Reports and Memoranda of the
    Aeronautical Research Council, No. 2237, London,
    1944, pp. 391-405.

4. Khanjonkov, V. I.   The nets resistance.   In:
    Promyshlennaja aerodynamica.   Central Aerohydro-
    dynamics Institute, 1944, pp. 35-42.

5. Rakhmatulin, Kh.A.   Flow of an impermeable body.
    Vestnic MGU, Ser. phys.-math. i jestestv. nauk,
    1950, No. 3, pp. 41-56.

6. Brown W. D. Parachutes, London, 1951, p. 85.

7. Heinrich, H. G.   The effective porosity of para-
    chute cloth.   Z. Flugwiss., 1963, 11, pp. 389-
    397.

8. Smetana, F. O. On the determination of parachute
    cloth permeability.   Z. Flugwiss., 1966, 14, pp.
    429-435.

9. Armour, J. C., Cannon, J. N.   Fluid flow through
    woren screens.   AIChE Journ., 1968, vol. 14, No.
    3, pp. 415-420.

10. Uljanov, G. S.   Supersonic gas flow around
    permeable plates.   Nauchnyje trudy instituta
    mechaniki MGU, 1975, No. 35, pp. 99-126.

11. Ilgamov, M. A., Shirjaev, A. P.   On permeability characteristics of the parachute fabrics. Strength and stability of shells. Proceedings of a Seminar. Kazan: Physicotechnical Inst., USSR Academy of Sciences, 1977, No. 9, pp. 116-124.

12. Ilgamov, M. A.   Conditions on a contact surface of the permeable shell with the ideal liquid. Strength and stability of shells. Proceedings of a Seminar. Kazan: Physicotechnical Inst., USSR Academy of Sciences, 1977, No. 9, pp. 125-135.

13. Ilgamov, M. A. Interaction of a fluid flow and a permeable membrane shell. Abstract of Seventh Canadian Congress of Appl. Mech. (CANCAM-79), Sherbrooke, 1979, pp. 143-144.

14. Ilgamov, M. A., Sabitov, M. Z. Interaction of an elastic permeable cylindrical shell with incom- pressible fluid flow.   Nonlinear problems of aerohydroelasticity.   Proceedings of a Seminar. Kazan: Physicotechnical Inst., USSR Academy of Sciences, 1979, No. 11, pp. 70-81.

15. Ilgamov, M. A.   General formulation of the prob- lem of the interaction of an ideal compressible liquid with a permeable shell upon its finite displacements and strains.   Proceedings of XIIth All-Union conf. on theory of the shells and plates, Jerevan, 1980, v. II, pp. 177-183.

16. Grigolyuk, E. I., Filschtinskii, L.A. Perforated plates and shells.   M.:   Physmatgiz, 1970, p. 556.

17. Gulin, B. V., Ilgamov, M. A., Ridel, V. V. Dynamics of interaction of a membrane shell with a flowing gas - Interaction of shells with fluid. Proceedings of a Seminar. Kazan: Physicotech- nical Inst., USSR Academy of Sciences, 1981, No. 14, pp. 96-117.

CHAPTER V

SELF-EXCITED NONLINEAR OSCILLATIONS OF ELASTIC BODIES IN A FLOW: AN INTRODUCTION

Summary

Three nonlinearities are considered. They are structural nonlinearities for plates and shells, fluid mechanical nonlinearities for bluff body oscillators, and aerodynamic nonlinearities for airfoils in transonic flow. For aeroelasticity of plates and shells, the equations of motion are well established. Results obtained by numerical time integration may be compared to those obtained by topological theories of dynamics and also from experiment. All of these suggest that the chaotic limit cycle oscillations may occur for this deterministic system. In a later chapter, chaotic oscillations in mechanical systems are discussed in a broader context. For bluff body oscillators, the equations of motion themselves are still in an emerging stage. Here a qualitative theory may be useful in establishing a generic model that describes the essential features of the physical phenomena. For aerodynamic nonlinearities in transonic flow effective methods of analysis are under development based on first principles of fluid and solid mechanics, and results from exploratory studies are now becoming available. An introductory discussion is contained in this chapter and a fuller treatment is provided in the following chapter.

Introduction

Nonlinear aeroelasticity [1-4] is a rich source of static and dynamic instabilities and associated limit cycle motions. Fundamentally, aeroelasticity combines the classical fields of fluid and solid mechanics. Usually, as the name implies, it does so within the context of the subfields of aerodynamics combined with elasticity. The aerodynamic forces almost invariably are nonconservative. When acting on a resonant elastic structure whose motion modifies these forces in a feedback sense, they lead to a complex and fascinating variety of dynamical behavior.

A further word should be said about the determination of the aerodynamic forces and the consequent implications for the mathematical modeling of aeroelastic phenomena. Even in the simplest, general theoretical model of small perturbation (inviscid, irrotational) potential flow, the aerodynamic forces due to the structural motion are related to the motion through a convolution integral. Hence, the ultimate structural equations of motion are nonlinear integro-differential equations. Moreover, the kernel of the convolution integral is frequently only known numerically. Hence, these equations are often attacked numerically to obtain time histories of motion. The method of harmonic balance is also occasionally used. Of course, the great bulk of the theoretical literature in aeroelasticity is devoted to completely linear structural and aerodynamic models [1,5-8], which may be solved by Fourier methods in the frequency domain. Here, however, the emphasis is on nonlinear behavior.

In the present chapter three aspects of the subject are addressed. First, in Section 5.1 (a simplified aerodynamic and structural representation of) the nonlinear aeroelasticity of plates and shells is considered. The principal focus here is on the evolution in parameter space of chaotic motion from simple deterministic motion and its possible implications for the methods and results of differential dynamics as typified by Holmes' work [9,10]. The results discussed have been obtained by numerical simulation of time histories. For this class of problems, qualitative methods of the sort studied by Holmes may suggest effective ways of interpreting the results of numerical studies. There is a comprehensive theoretical and experimental literature [2], which the reader interested in the physical background of this problem also may wish to consult.

In Section 5.2 a physical, intuitive approach is used to construct a theoretical nonlinear aeroelastic model based on experimental fluid mechanical information. This is done in the context of bluff body oscillators where no theoretical relationship is presently known from first principles that describes the aerodynamic forces on such bodies. It is suggested that the qualitative methods of differential dynamics may be able to place the construction of such theoretical models on a more rigorous and systematic basis.

In Section 5.3, the effect of nonlinear aerodynamic forces at transonic Mach numbers on airfoil flutter is addressed. Both the questions of under what conditions the aerodynamic forces are nonlinear and how to conduct a flutter analysis when they are nonlinear are considered.

5.1. Flutter of a buckled plate as an example of chaotic motion of a deterministic autonomous system

It has been known for some time that a plate under a compressive in-plane load with a fluid flow over its (upper) surface may undergo complex motions [2]. See the sketch of plate geometry in Fig. 1.

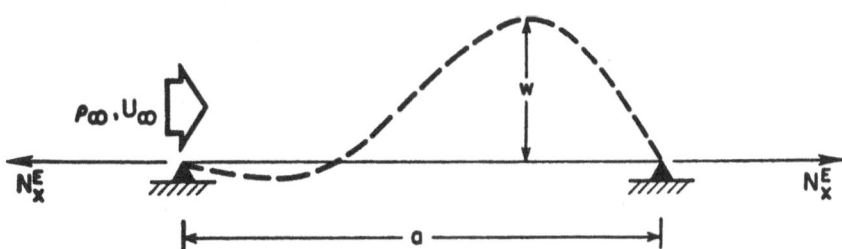

Fig. 1. Sketch of Plate Geometry. Note: Plate deflection, w, is shown to a greatly exaggerated scale for clarity. w is the order of a plate thickness.

For no fluid flow, but with a sufficiently large compressive load, $-N_x^E$, the plate will buckle into a statically deformed shape. By contrast, for no compressive load, but with a fluid flow of sufficiently large velocity, U_∞, the plate will flutter with a periodic, nearly harmonic motion. However, with both compressive load and a fluid flow, chaotic motion may occur as determined from numerical simulations, though no formal mathematical proof has yet been given [9,10].

The problem is considered here in its simplest formulation, i.e., a one-dimensional structural model (albeit nonlinear) and (linear) piston theory aerodynamics (appropriate to high supersonic Mach numbers). The governing partial differential equation is reduced to a system of ordinary differential equations by using a modal expansion and Galerkin's method. For this relatively simple model, Holmes [9,10] has obtained a number of interesting results using the methods of differential dynamics (for a two mode expansion). Here results are obtained (for two to six mode expansions) using numerical integration in time. Motivated by Holmes' results, the resulting data are examined in the phase plane and also by Poincaré plots. By changing systematically the compressive load and fluid flow parameters, the evolution of chaotic motion from simple, deterministic motion is studied.

It is worthy of emphasis that at lower Mach numbers the aerodynamic model needed becomes more elaborate and leads to integrodifferential equations. These have successfully been solved by numerical integration techniques [2]. However, it is unclear to what extent the methods of differential dynamics may be used for such equations. Hence, for the present class of problems, these latter methods appear primarily to be of value in explaining the qualitative character of the motion and in suggesting effective formats for presentation of the results of numerical solutions and their interpretation.

Equations of Motion. Here only an abbreviated account is given as the underlying theory is discussed thoroughly elsewhere [2,13]. The governing partial differential equation is

$$D \frac{\partial^4 w}{\partial x^4} - (N_x + N_x^E) \frac{\partial^2 w}{\partial x^2} + m \frac{\partial^2 w}{\partial t^2}$$

$$+ \frac{\rho_\infty U_\infty^2}{M} \left[\frac{\partial w}{\partial x} + \frac{1}{U_\infty} \frac{\partial w}{\partial t} \right] = \Delta p$$

(1.1)

where

w — plate transverse deflection
x — streamwise spatial coordinate
t — time
D — $Eh^3/12(1 - \nu^2)$, plate bending stiffness
N_x — $(Eh/2a) \int_0^a (\frac{\partial w}{\partial x})^2$ dx; physically this is the tension created by stretching of the plate due to bending

N_x^E — externally applied in-plane load; positive in tension
E — modulus of elasticity
ν — Poisson's ratio
a — plate length
h — plate thickness
m — mass/per unit length
ρ_∞ — flow density
U_∞ — flow velocity
M — flow Mach number
Δp — static pressure difference across the plate

A set of ordinary differential equations is obtained by using Galerkin's method with the modal expansion,

$$w = \Sigma a_n(t) \sin \frac{n\pi x}{a}$$

(1.2)

The result is (in nondimensional notation)

$$A_n(n\pi)^4/2 + 6(1-\nu^2) \left[\sum_r A_r^2(r\pi)^2/2\right] A_n(n\pi)^2/2$$

$$+ R_x A_n(n\pi)^2/2 + a_n''/2 \qquad (1.3)$$

$$+ \lambda\left\{\sum_m [nm/(n^2-m^2)][1 - (-1)^{n+m}]A_m + (\mu\lambda/M)^{1/2} A_n'\right\}$$

$$= P[1 - (-1)^n]/(n\pi) \qquad n = 1,2,\ldots, \infty$$

where

$A_n \equiv a_n/h$ and later $W \equiv w/h$
$\lambda \equiv \rho_\infty U_\infty^3 \, a^3/MD$
$\mu \equiv \rho_\infty \, a/m$
$R_x \equiv N_x E \, a^2/D$
$P \equiv \Delta p \, a^4/Dh$
$\tau \equiv t(D/ma^4)^{1/2}$
$' \equiv \partial(\ )/\partial\tau$

These equations may be numerically integrated to obtain time histories of motion. These are used to construct phase plane plots and Poincare plots. Systematic numerical studies are discussed in the following section. The choice of parameters is guided by the earlier results of Dowell [2,13] and Holmes [9,10]. Among the parameters studied are

λ - a nondimensional flow velocity parameter
R_x - a nondimensional in-plane load parameter

Initial conditions

P - a nondimensional static pressure
  differential

All results were obtained using a four mode expansion except for a few two and six mode calculations done for comparison.

To make the nature of the mathematical model as transparent as possible, consider a two mode representation from (1.3), where various numerical constants are omitted for clarity. One has

$$A_1'' + A_1 (1 + R_x) - \lambda A_2 + \lambda^{1/2} \zeta_1 A_1' + (A_1^2 + A_2^2)A_1 = P$$

$$\text{(1.4)}$$

$$A_2'' + A_2 (4 + R_x) + \lambda A_1 + \lambda^{1/2} \zeta_2 A_2' + (A_1^2 + A_2^2)A_2 = 0$$

The skew-symmetric terms involving λ are responsible for the dynamic instability, flutter, while the R_x terms (when $R_x < 0$) can cause a static instability, buckling. The form of flutter modeled by (1.4) is called coupled mode flutter and is so named because a minimum of two modes is required to produce it. If one examines a root locus of (1.4) for infinitesimal perturbations about the trivial equilibrium, $A_1 = A_2 = 0$, then the two eigenvalues, which were the system natural frequencies at $\lambda \equiv = 0$, approach each other as λ increases and after near coincidence one of them passes into the unstable half plane of the root locus. Hence this form of flutter is also sometimes called merging frequency flutter [1,2]. Only this type of plate flutter has been treated by the qualitative theory of differential equations in the literature [9,10].

By contrast another form of flutter is called single mode flutter [1,2], and it arises because the aerodynamic forces create negative damping. It is discussed briefly here, before returning to (1.3) and (1.4). Analogous to (1.4) a single equation for A_1 displays the essential features of this type of flutter.

$$A_1'' + \int_{-\infty}^{\tau} K_1 [(\tau-\sigma)\lambda^{1/2}] A_1'(\sigma) \, d\sigma$$

$$\text{(1.5)}$$

$$+ A_1 (1 + R_x) + A_1^3 = P$$

The convolution integral is a mathematical representation of the physical fact that at time τ, the aerodynamic forces depend in general on the entire past history of the plate motion. This effect is important for Mach numbers, M, near unity, but is unimportant at large M where (1.3) and (1.4) apply.

A simple explanation of the nonlinear limit cycle behavior of (1.5) for $R_x = 0$, $P = 0$ may be obtained by using the method of harmonic balance.

Assume

$$A_1 = \bar{A}_1 \cos \Omega \tau \qquad (1.6)$$

Substitute (1.6) into (1.5) and integrate over one period of motion (as the steady state is approached, $\tau \to \infty$). Then one obtains

$$K_1^*[\Omega/\lambda^{1/2}] = 0 \qquad (1.7)$$

and

$$\Omega^2 = 1 + \bar{A}_1^2 \qquad (1.8)$$

where nonessential constants have been dropped. K_1^* is the (real part of the) Fourier transform of K_1 and physically is a damping coefficient. It is shown in Fig. 2. The infinitesimal stability is obtained when $\bar{A}_1 \equiv 0$. From (1.8) the frequency of the flutter oscillation is $\Omega_f = 1$ (i.e., the single mode natural frequency) while from (1.7) the flutter velocity parameter, λ_f, is found from the condition that $K_1^* = 0$ at $\lambda = \lambda_f$; see the above sketch.

Now consider what happens when $\lambda > \lambda_f$, $\bar{A}_1 > 0$. One still requires that $K_1^* = 0$, thus

$$\Omega/\lambda^{1/2} = \Omega_f/\lambda_f^{1/2} \qquad (1.9)$$

and

$$\Omega^2 = 1 + \bar{A}_1^2 \qquad (1.10)$$

Solving (1.9) and (1.10), one has

$$\Omega^2 = \lambda/\lambda_f \qquad (1.11)$$

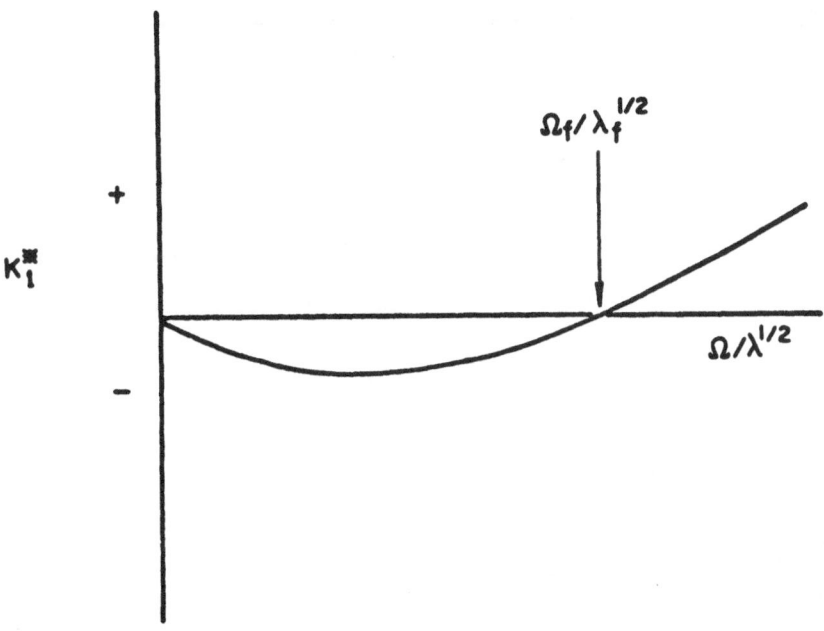

Fig. 2. Sketch of aerodynamic damping coefficient.

and

$$A_1 = [\lambda/\lambda_f - 1]^{1/2} \qquad (1.12)$$

(1.11) gives the limit cycle frequency and (1.12) the amplitude for $\lambda > \lambda_f$.

When $R_x < 0$ one anticipates that chaotic motion may occur [2], but we do not pursue that issue here.

Numerical Studies. For all results reported below, $\mu/M = 0.01$, $x/a = 0.75$, $\nu = 0.3$ and $P = 0$ unless otherwise noted. Equations (1.3) were used.

From previous numerical simulation studies [2], it is known that the parameters λ and R_x govern the type of motion which may occur. Fig. 3 displays a map in λ, R_x space that identifies the various types of motion which may occur. Only $R_x < 0$ is considered,

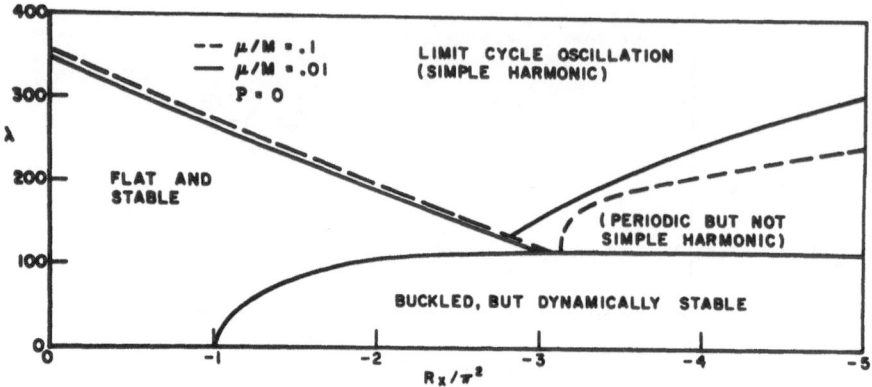

Fig. 3. Sketch of stability regions.

since this proves to be the interesting case. For small λ and R_x, the steady state motion is a flat, undeformed plate. For small λ, but moderate R_x, the plate buckles. For small R_x, but moderate λ, the plate flutters with simple harmonic motion. For moderate λ, R_x a more complicated periodic limit cycle motion occurs and for sufficiently large R_x and moderate to large λ, chaotic motions ensues.

The results will be presented in the phase plane. Anticipating these, the distinctive types of motion which may occur are sketched in Fig. 4 as follows. The buckled plate corresponds to two (nonlinear static equilibria) points in the phase plane. Simple harmonic motion flutter is an ellipse. The more complicated periodic limit cycle motion flutter comprises a smaller orbit about each buckled state and a larger orbit that evolves from the flutter motion. The chaos is beyond the author's ability to sketch simply.

To investigate the possibility of chaotic motion, two trend studies were made. First R_x was held fixed at $-4\pi^2$ and λ was varied: 300, 250, 200, 175, 150, 130, 115, 100. See Figs. 5-8. At $\lambda = 300$ and 250 the phase plane plot, W' vs. W, of the limit cycle shows an ellipse typical of pure flutter motion. For $\lambda =$

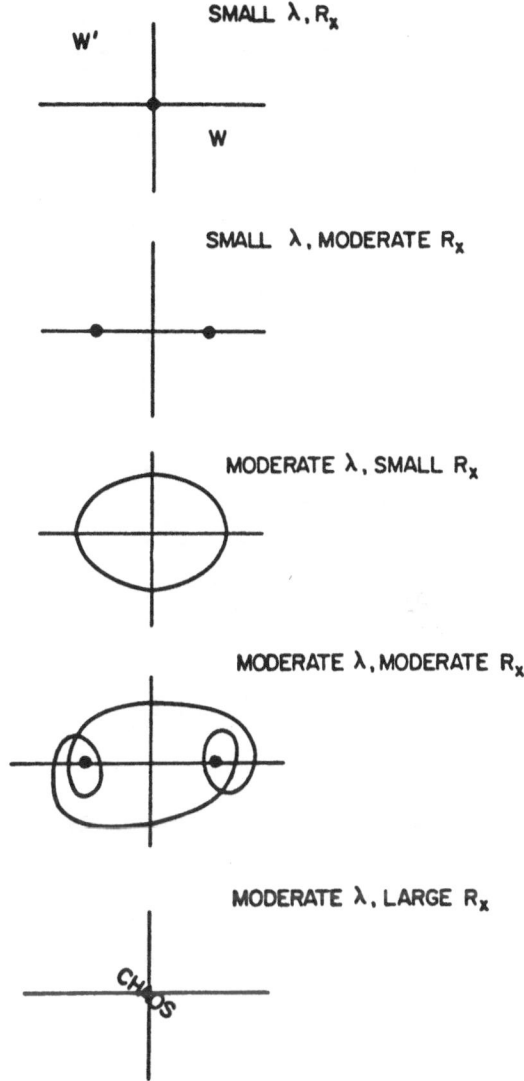

Fig. 4. Sketch of representative phase plane orbits.

200, the first deviations from the ellipse are evident, although the phase plane plot remains a single
closed curve. However at λ = 175, 150, 130, 115 three
closed curves are in evidence.   The largest   of these
derives from the pure flutter motion, while   the two

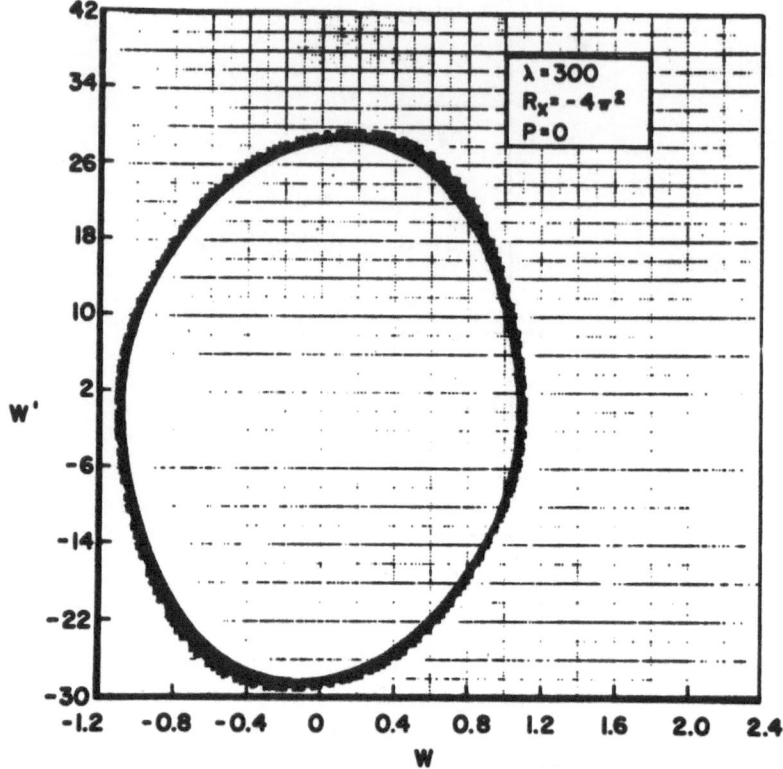

Fig. 5. Phase plane plot: effect of flow velocity.

smaller ones are associated with buckling or divergence. At $\lambda = 175$ the three closed curves are particularly clearly defined and as λ = 150, 130, 115 the
motion becomes progressively more chaotic. At $\lambda = 100$
the phase plane plot is simply a point, representative
of a pure buckling or divergence instability.

Secondly λ was held fixed at 150 and R_x varied
from $-2.5\pi^2$, $-3\pi^2$, $-3.5\pi^2$, $-4\pi^2$, $-5\pi^2$, $-6\pi^2$. At R_x =
$-2.5\pi^2$, $-3\pi^2$ the three closed curves are particularly
clearly defined and as R_x = $-3.5\pi^2$, $-4\pi^2$, $-5\pi^2$, $-6\pi^2$,
the motion becomes progressively more chaotic. See
Figs. 9 and 10. Nevertheless the basic three closed
curves pattern remains albeit in a more obscure form.

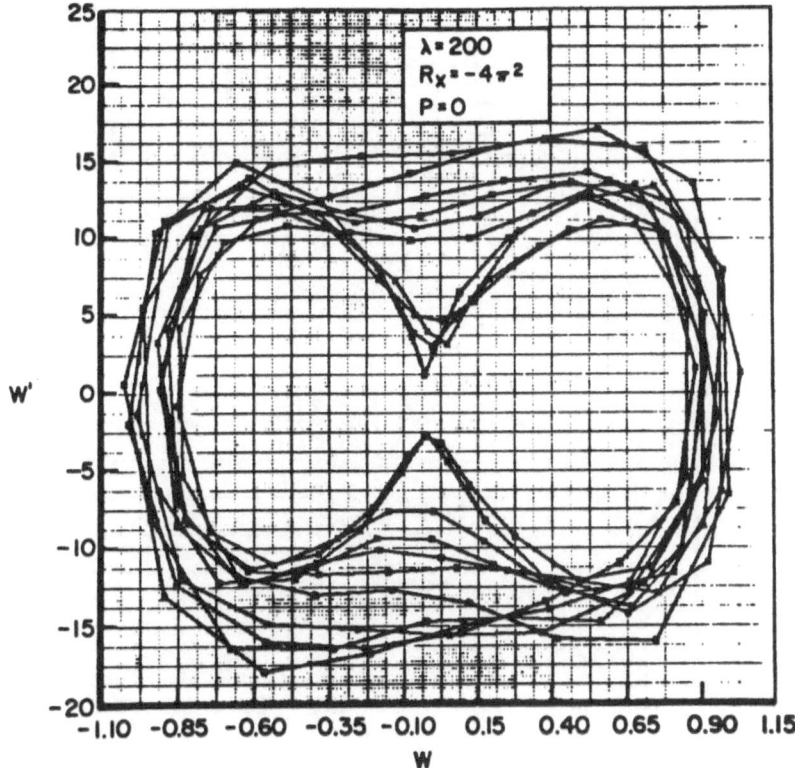

Fig. 6. Phase plane plot: effect of flow velocity.

The chaotic motion that occurs for certain parameter combinations of λ and R_x might well be termed random. However, it is clear that such motion evolves continuously in parameter space from motion that is decidedly deterministic. Moreover, even taken solely as motion at a point in parameter space (some fixed λ and R_x), it is manifestly bounded in the phase plane and can be characterized, for example, by minima and maxima of displacement and velocity, W and W'.

To further characterize the motion, autocorrelations and power spectra of the motion were calculated. Two representative power spectra are shown in Figs. 11 and 12 for $\lambda = 150$ and $R_x/\pi^2 = -3$ and -6, respectively. As can be seen the former results show two discrete frequencies dominating the power spectrum which is the expected result for a periodic motion with two

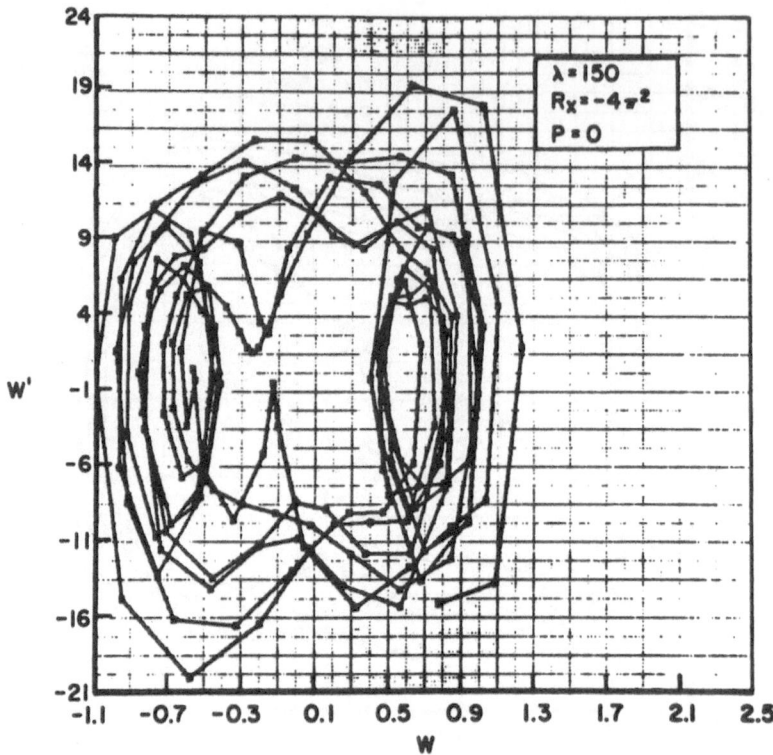

Fig. 7. Phase plane plot: effect of flow velocity.

dominant orbits in the phase plane. Recall Fig. 9. However, in the latter case the power spectrum, although still showing peaks, has a continuous distribution such as that conventionally expected of a random process. f is a nondimensional frequency in cycles per unit time.

Quite aside from its own intrinsic interest, the present results may serve as a paradigm for similar chaotic motions that result from limit cycles associated with partial differential equations. Perhaps the best known of these is the chaotic (turbulent) motion that is the pseudo limit cycle associated with the Navier-Stokes equations. There is at least one distinction, however, between the two. For the present problem there are two parameters, λ and R_x, each of which governs a distinctive instability, flutter and buckling (divergence). It is the interaction of these

177

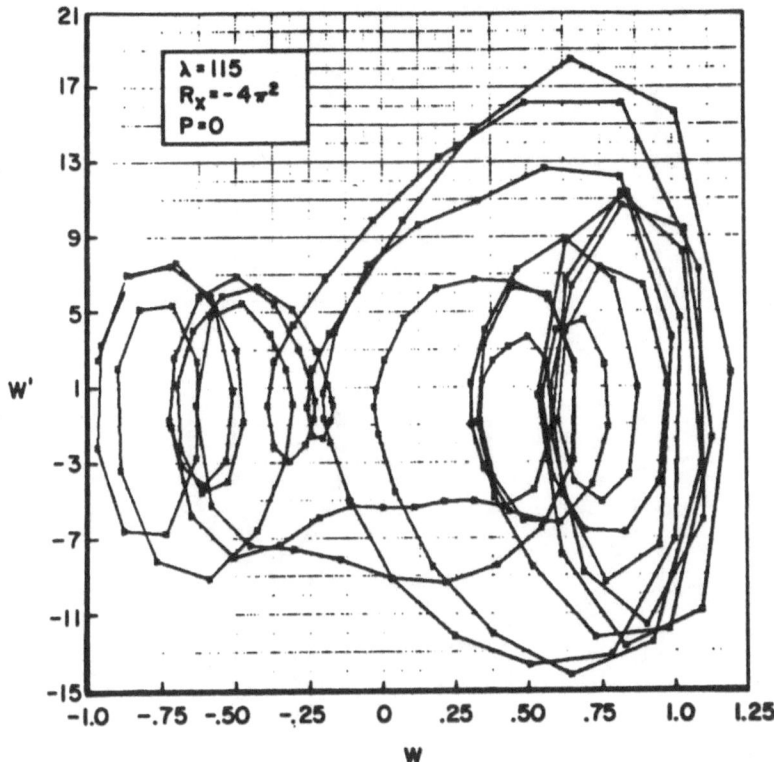

Fig. 8. Phase plane plot: effect of flow velocity.

two essentially deterministic motions for certain combinations of λ and R_x that leads to chaotic motion.
On the other hand, only a single parameter, the Reynolds number $\equiv$ Re, appears in the Navier-Stokes equations (assuming incompressible flow). At this stage,
one can only speculate that Re may play a dual role
(as suggested by Lin's interpretation of <u>linear</u> hydrodynamic stability theory [14]) and/or that more than
one instability mode is governed by Re.

<u>Effect of Initial Conditions.</u> For the base case
of $\lambda = 150$ and $R_x = -4\pi^2$, various initial disturbance
amplitudes were studied. In all cases the initial
deflection of the plate was assumed to be a pure first
natural mode, i.e., $W(x, t = 0) = A_1(t = 0) \sin \pi x/a$,
with various $A_1(t = 0)$ chosen. In addition to the

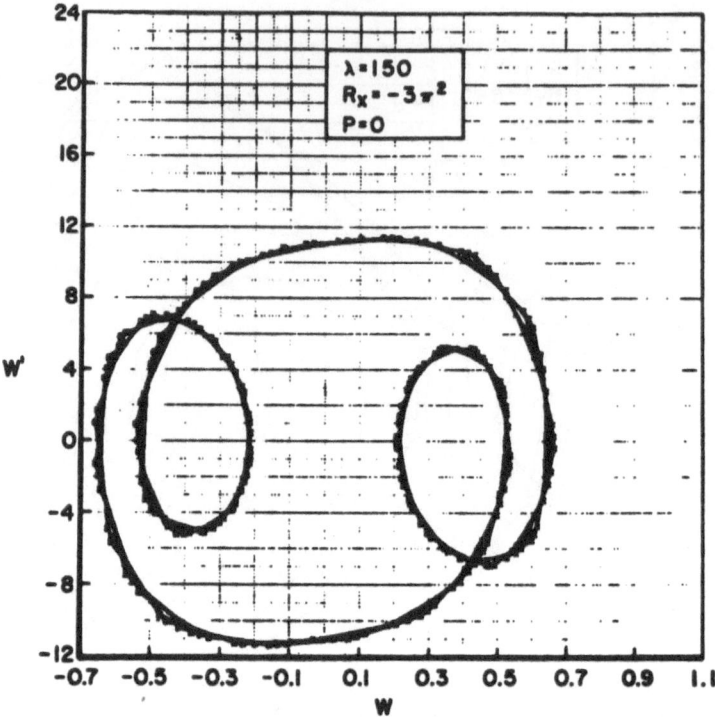

Fig. 9. Phase plane plot: effect of in-plane load.

base case value of $A_1(t = 0)$ = 0.1, values of 0.2,
0.5, 1.0 and 2.0 were chosen. The long time steady
state results were essentially the same, showing no
sensitivity to the initial disturbance amplitude
chosen.   It should be emphasized that this result is
not universal since Ventres and Dowell [15] have
already shown that for plates of finite (large)
length/width ratio initial conditions can play an
important role.

Effect of Static Pressure Differential. Another
parameter of interest is a static pressure differential across the plate. This, of course, destroys the
symmetry of the geometry by giving the plate deflection a preferred direction. Two trend studies were
conducted. First of all, the simpler case of zero-

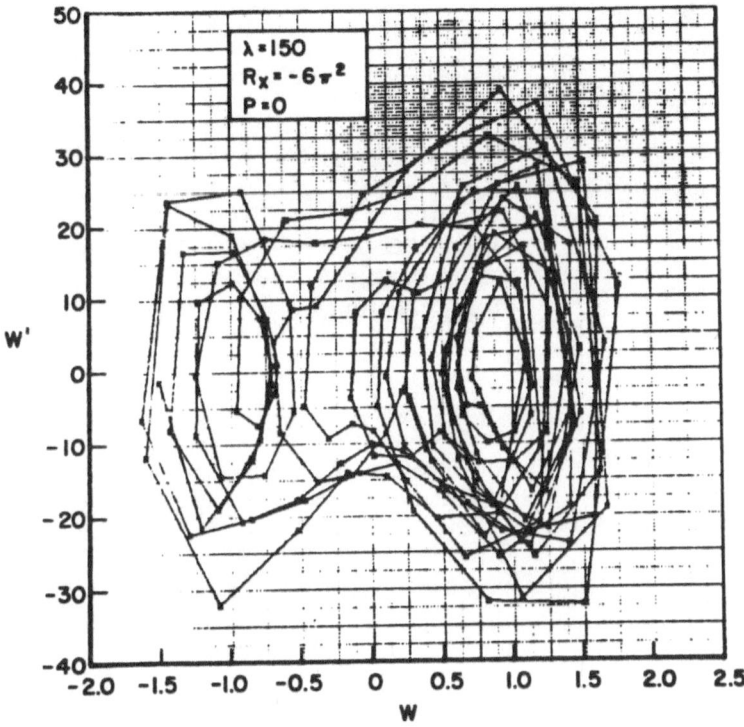

Fig. 10. Phase plane plot: effect of in-plane load.

in-plane load, $R_x = 0$, was studied at $\lambda = 400$ for various pressure differentials, P. At P = 0 the plate flutters in a limit cycle oscillation whose phase plane trajectory is an ellipse centered about the origin, $W = W; = 0$. As P is increased from 0 to 100, 200, 250 the ellipse decreases in size and its center moves to the right. At P = 300 the ellipse has collapsed completely to a point; the static pressure differential having stiffened the plate sufficiently so that flutter is completely suppressed. As expected the motion does not exhibit any tendency to chaos for $R_x = 0$.

Next the base case of $\lambda = 150$, $R_x = -4\pi^2$ was reconsidered for various amounts of static pressure differential ranging over R = 0, 12.5, 25, 37.5, 50,

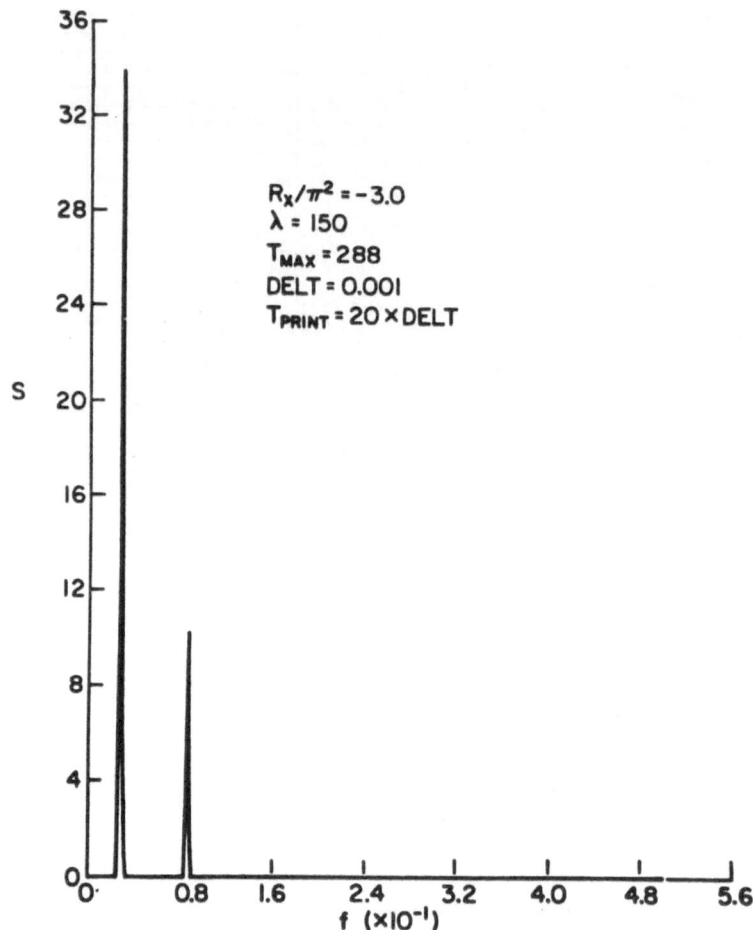

Fig. 11. Power spectra.

56.25, 62.5, 68.75, 75, 100, 200. For P ≥ 68.75, the
static loading was sufficient to suppress all flutter
motion. For small P, say 0, 12.5, 25, 37.5, the
motion was qualitatively similar and exhibited a somewhat chaotic appearance. However, it will be recalled
that for R = $-3\pi^2$ the distinct three closed curves
were present and, with that knowledge, these three
closed curves may still be seen, somewhat dimly to be
sure, at R_x = $-4\pi^2$. The most remarkable result
occurs, however, at P = 50, 56.25. See Fig. 13. Here

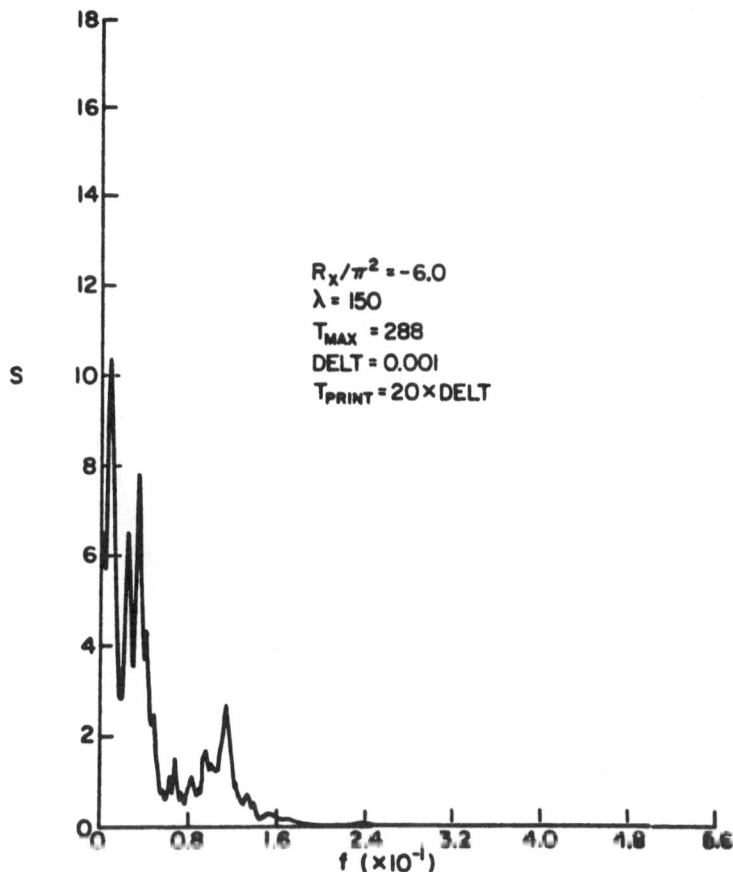

Fig. 12. Power spectra.

the motion has a very distinct, nonchaotic character
with two closed orbits in evidence. That is, the
static pressure differential has suppressed the chaotic character of the motion and one of the three
orbits that appears for smaller P. For P = 62.5 the
motion returns to a chaotic state with no readily dis-
cernible pattern and the P > 68.75, the limit cycle
degenerates to a point and all motion ceases.

Poincaré Plots. To gain further insight into the
nature of the motion, an alternative format for presentation of the data was considered, i.e., a Poincare
plot. For our purposes such a plot is one where the

182

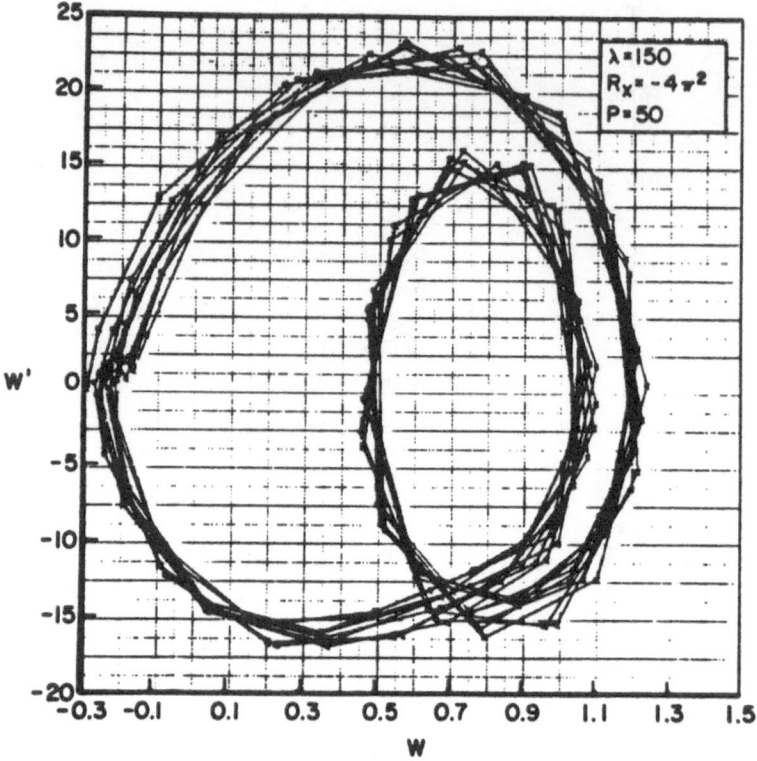

Fig. 13. Phase plane plot: effect of static pressure
         differential.

continuous phase plane diagram is sampled at discrete
time intervals. Of course, since our time histories
are generated by a digital computer and are thus
necessarily discrete, all of the previous phase plane
diagrams were, strictly speaking, Poincaré plots.
However previously, the time interval used was short
compared to any time intervals characteristic of the
motion, so that the diagrams were effectively continuous.   The effect of systematically increasing this
time interval was examined for two example cases, λ =
150 and R_x = $-3\pi^2$ or $-4\pi^2$.   The former, R_x = $-3\pi^2$, is
a periodic motion, and as the sampling time interval
becomes equal to the period of the motion, the Poincaré plot reduces to a single point. The latter, R_x =
$-4\pi^2$, is a chaotic motion where no such reduction is

expected, but other interesting insights may be obtained. A unit time interval was selected to be $\Delta\tau = 0.02$.

Results were obtained for $R_x = -3\pi^2$ and time intervals of 1, 2, 5, 10, 20, 34, and 40 units. For a small number of units the complete character of the phase plane diagram is evident; for 34 units essentially a single point is obtained indicating that this is the period of the periodic motion.

Similar results were obtained for $R_x = -4\pi^2$. Because the motion is now chaotic, the character of the motion becomes very difficult to ascertain, since the sampling time interval is increased. This is emphasized when the sampling points in the phase plane are unconnected. No period is found in which the phase plane diagram is reduced to a single point.

An alternative Poincaré plot may be constructed by specifying one of the dynamical variables and taking slices in variable space of the remaining coordinates. For example, one could specify $A_1' = 0$ and examine, at those times when this is true, the values of A_1, A_2, A_2', A_3, A_3',.... However for even four modes, the remaining variable space has dimension $7 = 4 \times 2 - 1$. Hence it would be difficult to display and interpret such results.

To reduce the dimension of this Poincare plot, one might proceed as follows (which is suggested by physical considerations). Consider a particular physical velocity, W', and the corresponding displacement, W, for each such W' that appears in the phase plane. (Of course, W is a linear combination of the A and W' of the A_n. Hence, they are suitable <u>lower</u> dimensional representations of the individual coordinates, A_n and A'.) Furthermore, the range of W (for each such W'), call it ΔW, is a measure of the system chaos. A strictly periodic system would have a range of zero, i.e., for each such W', there is a (small) finite number of discrete W. Conversely if all W are possible between W_{min} and W_{max} the system motion might be termed completely chaotic. The larger $R \equiv \Delta W/(W_{max} - W_{min})$, the more chaotic the motion.

In Figs. 14 and 15 Poincaré plots are shown for $W' \equiv 0$, $\lambda = 150$ and $R_x/\pi^2 = -3$ and -6, respectively. The former results display the expected repetitive, periodic character for an oscillation with two dominant frequencies. For the latter a pattern appears to be present, but it has a far more complicated character. One might call this pattern, suggestively, large scale turbulent motion. For a more extensive discussion of Poincaré maps, see Appendix A.

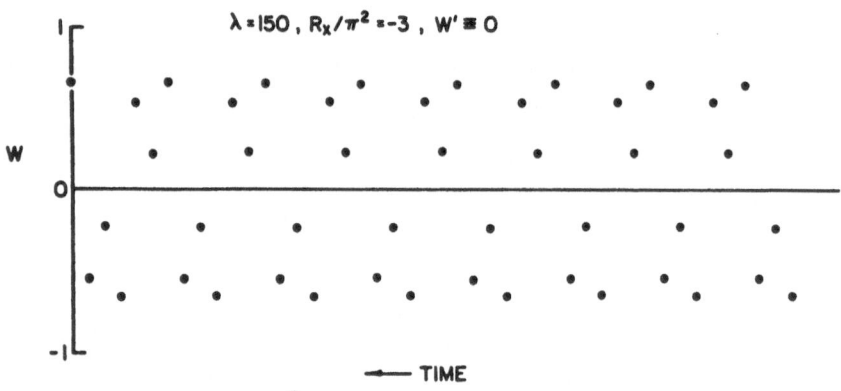

Fig. 14. Pointcare plot.

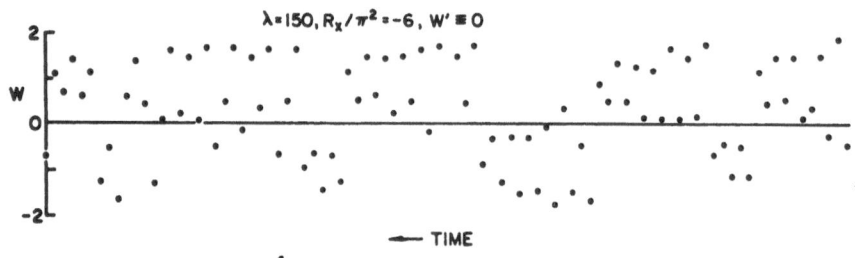

Fig. 15. Poincare plot.

Effect of Number of Modes Retained. Two mode calculations were made which gave qualitatively similar results to those using four modes. Based on the present results and earlier studies [1,4] using up to six modes, it is expected that four mode calculations will give quantitatively accurate results for the parameter combinations studied here.

Conclusions. Chaotic motions occur for certain combinations of in-plane compressive load and fluid flow velocity. These evolve continuously from simpler motions with changes in these two parameters. Phase plane plots effectively display the results; conventional Poincaré plots are less useful, although extensions of these appear fruitful (Appendix A).

Results from the qualitative theory of nonlinear differential equations (differential dynamics) were helpful in motivating the present study and interpreting the results obtained. For further in-depth discussion of the observation and evolution of chaos, see Appendix A.

Finally, there are several questions that frequently are asked about this work. These are dealt with here.

- How many modes are required to give the chaotic motion, how many are required to give essentially converged quantitative results, and what is the largest number that has been used? The answers are, respectively, two, four (for the range of nondimensional parameters studied in this paper), and as many as twenty (the maximum number of modes employed being determined by computing cost). Four modes were normally used in the present calculations.

- Is the time step size used in the numerical simulation small enough to ensure numerical stability of the results? The usual estimates of step size required for numerical stability for the corresponding linear system were made based on the highest frequency mode and also numerical testing was done with various step sizes. These results were in good agreement and showed that for a sufficiently small time step, no numerical instability occurred and the results for the time simulation were closely repeatable.

called Hopf bifurcation) and Euler buckling (sometimes called static bifurcation). In the corresponding linear model, flutter occurs as a result of the coalescence of two eigenfrequencies with increasing flow velocity and buckling occurs upon the vanishing of an eigenfrequency with increasing compressive mechanical in-plane load. Near the point in the parameter space of flow velocity and mechanical in-plane load where the flutter and buckling stability boundaries merge, chaos appears. As the mechanical in-plane load increases, the range of flow velocity for which chaotic motion occurs also increases. See also the further discussion in Chapter VII as to why chaotic motions occur.

5.2. Nonlinear oscillator models in bluff body aeroelasticity

Such models [16-20] have been developed to provide a phenomenological description of observed motion of bluff (nonstreamlined) elastic bodies in a streaming fluid. See in Fig. 16. The dynamics of the elastic body are described by the equations of solid mechanics. These reduce, in their simplest form, to an oscillator with mass, stiffness and damping characteristics driven by a (fluid) force. By analogy, the dynamics of the fluid, as characterized by the fluid force on the body, are assumed to be described by a fluid oscillator. The properties of the fluid oscillator (forms of the terms in the differential equation for the fluid force and their coefficients) are deduced by requiring the coupled solid - fluid oscillator system to have solutions similar to those observed in practice.

In the present approach a more fundamental (but still phenomenological) approach is taken by deducing the fluid oscillator properties from fluid mechanical considerations (experimental and theoretical) alone. It should be noted that Iwan and Blevins [20] made a similar effort largely based on the theoretical field equations of fluid mechanics. The desire, of course,

- Is the result truly random or are the time histories repeatable from one run to the next for a given set of initial conditions? The results <u>are</u> repeatable from one computer run to the next for the same initial conditions. Hence, calling the motion chaotic is perhaps a better choice of term than random.

- Do the results depend on the initial conditions? Apparently not. Initial displacements in the first modal amplitude, A_1, from .1 to 2 were chosen with no perceived difference in the steady state oscillations and their character.

- How do you know that the steady state oscillation has been reached and the transient has decayed? The time for the transient oscillation to decay can be computed for the corresponding linear system, and also numerical experiments were performed for varying time simulation record lengths. Results from these are consistent and show that for a sufficiently long time simulation the transient has decayed. Typically, the maximum time of the simulation was $\tau = 18$ and the (conservative) time at which the steady results were taken to be reached was 9 (corresponding to, typically, 20 cycles of oscillation for the transient to decay and 20 cycles of steady state motion). However, see Appendix A for discussion of some exceptional cases.

- Are similar results obtained when a plate of finite width that bends in two directions is considered? Yes.

- Perhaps the most profound and difficult question is why do chaotic motions occur? Here a partial, intuitive answer is given. The appearance of chaotic motion seems to arise as a consequence of the presence of two parameters, in this case flow velocity and mechanical in-plane load, which govern two distinct types of instability, in this case flutter (sometimes

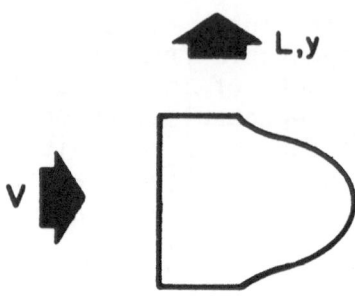

Fig. 16. Sketch of bluff body geometry in a fluid flow.

is to obtain the equations of the solid oscillator and the fluid oscillator by separate, independent means and to use them together to <u>predict</u> the behavior of the combined fluid-elastic system. The present approach is based on a combination of experimental and theoretical fluid mechanical considerations.

For informative critiques of nonlinear oscillator models, References 21 and 22 are recommended. A more extensive discussion of the present approach is contained in References 23 and 24.

<u>Model Development</u>. For simplicity, a single-degree-of-freedom system consisting of a bluff cylindrical structure transverse to an oncoming two-dimensional stream will be considered. Extensions to multidegree of freedoms and three-dimensional flow should be possible [25].

The structural equation of motion is well known

$$m \, [\ddot{y} + 2\zeta_0 \omega_0 \dot{y} + \omega_0^2 y] = q \, D \, C_L \qquad (2.1)$$

where

    m - mass of structure
    y - transverse deflection of the structure
    ζ_0 - critical damping ratio of in vacuo structural mode

ω_0 - natural structural mode frequency

q - $(1/2)\rho V^2$, flow dynamic pressure

ρ - flow density

V - flow velocity

D - reference dimension of aerodynamic cross-section

C_L - aerodynamic lift coefficient

The crux of the matter is what oscillator equation may be used to determine C_L? This oscillator model will be determined by requiring that

(1) at high frequencies the fluid oscillator model give the expected virtual mass relationship between C_L and y, i.e.,

$$C_L = - B_1 \, \rho \, D^2 \, \ddot{y} \text{ as } \omega \to \infty \qquad (2.2)$$

where B_1 is a nondimensional aerodynamic coefficient determined from (potential flow) theory and/or experiment (in a still fluid),

(2) at low frequencies the model give the expected quasisteady relationship between C_L and y, i.e.,

$$C_L = f \, (\dot{y}/V) \text{ as } \omega \to 0 \qquad (2.3)$$

For example, if one were to use the Parkinson galloping model [21,22] to evaluate the function, $f(\dot{y}/V)$, one might ask that

$$C_L = A_1 \, \dot{y}/V - A_3 \, (\dot{y}/V)^3 + \dots \text{ as } \omega \to 0 \qquad (2.4)$$

(3) for small C_L (and no structural motion $y \equiv 0$) the natural frequency of the <u>fluid</u> oscillation be determined by the <u>Strouhal</u> wake frequency, e.g.,

$$\ddot{C}_L + \omega_s^2 \, C_L = 0 \text{ as } C_L, \, y \to 0 \qquad (2.5)$$

where ω_S is the dimensional fluid frequency, and $\omega_S = V/D \, k_S$, where k_S is a Strouhal number which depends on Reynolds number,

(4) for a stationary cylinder the root mean square value of C_L be that observed in experiment. If one uses (2.5) with no additional terms involving C_L, this would require that the initial condition on C_L be

$$C_L \, (t = 0) = 2^{1/2} \, C_{L_{rms}} \qquad (2.6)$$

Alternatively one might add a nonlinear term to equation (2.5), which permits a limit cycle behavior whose amplitude is $C_{L_{rms}}$ (and which is independent of initial conditions). This latter procedure presumably is closer to the physics of the fluid, whose oscillations are those of a fluid limit cycle due to an unstable laminar flow. Previous authors [16-20] have suggested adding a nonlinear term to equation (2.5), which makes it into a van der Pol oscillator, and we shall follow their lead here. The specification of the nonlinear term provides a basis for flexibility, but also arbitrariness, in the model. The next step beyond a van der Pol oscillator might be to introduce a (strange attractor) nonlinearity, which would produce a chaotic motion (turbulence). The problem discussed in Section 5.1 offers a means for doing this. This would appear to model even more closely the fluid physics.

Derivation of Model Equation. There are several ways one may derive the fluid oscillator equation. The following is representative and reasonably straightforward.

From equation (2.2), one may deduce that

$$\ddot{C}_L = - B_1 \frac{D}{V^2} \, \ddddot{y} \quad \text{as } \omega \to \infty \qquad (2.7)$$

From equations (2.7) and (2.5), one infers that

$$\ddot{C}_L + \omega_s^2 \, C_L = - B_1 \frac{D}{V^2} \, \ddot{\dddot{y}} \qquad (2.8)$$

and to satisfy equation (2.3) one modifies equation (2.8) to become

$$\ddot{C}_L + \omega_s^2 \, C_L = - B_1 \frac{D}{V^2} \, \ddot{\dddot{y}} + \omega_s^2 \, f \left(\frac{\dot{y}}{V}\right) \qquad (2.9)$$

or, using equation (2.4) rather than equation (2.3),

$$\ddot{C}_L + \omega_s^2 \, C_L = - B_1 \frac{D}{V^2} \, \dddot{y}$$

$$(2.10)$$

$$+ \omega_s^2 \left[A_1 \frac{\dot{y}}{V} - A_3 \left(\frac{\dot{y}}{V}\right)^3 + A_5 \left(\frac{\dot{y}}{V}\right)^5 - A_7 \left(\frac{\dot{y}}{V}\right)^7 \right]$$

Finally, the nonlinear term may be added to equation (2.10), which for $y \equiv 0$ gives a limit cycle oscillation of magnitude $2^{1/2} C_{L_{rms}} = C_{L_o}$; here a van der Pol oscillator is used.

$$\ddot{C}_L - \epsilon \left[1 - 4 \left(\frac{C_L}{C_{L_o}}\right)^2 \right] \omega_s \, \dot{C}_L + \omega_s^2 \, C_L = - B_1 \frac{D}{V^2} \, \ddot{\dddot{y}}$$

$$(2.11)$$

$$+ \omega_s^2 \left[A_1 \frac{\dot{y}}{V} - A_3 \left(\frac{\dot{y}}{V}\right)^3 + A_5 \left(\frac{\dot{y}}{V}\right)^5 - A_7 \left(\frac{\dot{y}}{V}\right)^7 \right]$$

where ϵ is to be determined from subsequent fluid mechanical information. It is natural to introduce a nondimensional time, $\tau \equiv tV/D$ and a nondimensional displacement, $x \equiv y/D$. Then equation (2.11) becomes

$$C_L'' - \epsilon \left[1 - 4 \left(\frac{C_L}{C_{L_o}}\right)^2 \right] k_s \, C_L' + k_s^2 \, C_L = - B_1 \, x''''$$

$$(2.12)$$

$$+ k_s^2 \left[A_1 \, x' - A_3 \, (x')^3 + A_5 \, (x')^5 - A_7 \, (x')^7 \right]$$

and equation (2.1) becomes

$$x'' + 2 \, \zeta_o k_o \, x' + k_o^2 \, x = \mu \, C_L \qquad (2.13)$$

where $\mu \equiv \rho \, D^2/2m$ is a fluid/structural mass ratio. Thus, the problem is reduced to one with ten parameters which we divide into three classes.

   I. A_1, A_3, A_5, A_7, B_1

   II. k_s, $C_{L_{rms}}$

   III. μ, k_o, ζ_o

   Class I parameters are assumed to be universal, slowly varying functions of Reynolds number for a prescribed body cross-sectional shape; k_s and $C_{L_{rms}}$ are assumed to be weak functions of Reynolds number and are often assigned typical values, $k_s \sim 1$, $C_{L_{rms}} \sim 0.25$. μ, K_o and ζ_o are usually varied to examine their impact on the coupled fluid-oscillator motion [23,24].

Self-Consistency Checks on the Model and Other Remarks. The above derivation of the model is, perhaps, more properly termed a construction. Various analytical and numerical consistency checks have been proposed [23].

Check 1. One requires that $C_L \to - B_1 x''$ as $k \to \infty$. This requires that $x \to 0$ at a certain rate as $k \to \infty$.

Check 2. One requires that $C_L \to A_1 x' + A_3 (x')^3 + \ldots$ as $k \to 0$.

Check 3. When $k \sim k_s \sim 1$, there is no formal guarantee that the present model is accurate. The high and low frequency terms have been combined in a patching process, rather than by a matching procedure [26].

Check 4.   Neglecting all nonlinear terms, a linearized stability analysis may be performed. One may show that the linearized system is always unstable. The question is then, is it reasonable for the coupled solid-fluid oscillator system to be always unstable <u>in a linearized sense</u>? This question would appear as much philosophical as technical. It is perhaps interesting to note that for $A_1 < 0$, the linearized system is stable at large k_0, i.e., small $V/\omega_0 D$.

Check 5.   As $x \to 0$, the model is well behaved.

Check 6.   As $x \to \infty$, the model predicts that C_L will be dominated by its forcing due to structural motion and its own self-induced (turbulent) C_L will be small by comparison, i.e., $C_L \gg C_{L_{rms}}$. This is plausible, but by no means confirmed by experimental data. A key question is whether there is significant <u>inter-action</u> between the lift generated by <u>turbu-lence</u> and that due to motion. If so, the present model would require as a minimum the addition of terms of the form of products of C_L and x and/or their time rates of change.

    <u>Concluding Remarks</u>. Much of the impetus for using nonlinear fluid oscillator models comes from the careful experimental work of Bishop and Hassan [27]. They were fully conscious that their experimental data for lift (and drag) on an oscillating cylinder (with prescribed motion) could be interpreted as describing the behavior of a nonlinear oscillator.

    For further discussion of the uses of such a model including numerical studies, comparisons with other theoretical models and experiments, the reader is referred to References 23 and 24. Also see Appendix B. Here our attention is limited to the construction of the model and its possible improvement based on the insight gained in Part I. Clearly, from the latter, the nonlinear lift oscillator might be gener-

alized to a two degree of freedom system (lift and
drag as the dependent variables?) and chaotic lift
created to model fluid turbulence in the absence of
any body motion. The fundamental question is whether
such a model is generic for a given level of modeling.
That is, can all of the coefficients in such an oscillator model be consistently if not uniquely determined? Further study is required to answer such questions, indeed to understand whether such questions
are meaningful. In the longer term, one may hope that
the qualitative theory of differential equations
(e.g., see Holmes [9] and Guckenheimer [28]) may lend
insight and provide more rigorous methods for establishing generic models for this class of dynamical
systems.

5.3. Flutter of Airfoils at Transonic Mach Numbers

It is well known that the aerodynamic forces on
an airfoil may be nonlinear functions of the airfoil
motion at transonic speeds. Less well appreciated
until recently is the fact that for sufficiently small
motions the aerodynamic forces are linear even at
transonic Mach numbers. Of course, for sufficiently
large motions the aerodynamic forces are nonlinear at
subsonic and supersonic Mach numbers. Nevertheless it
is true that the amplitudes of motion for which the
aerodynamic forces become nonlinear are smaller in the
transonic range than in the subsonic or supersonic
range. Here we summarize two recent studies which
address the questions of

- when may the aerodynamic forces be treated as
  linear in the transonic range, and

- when they must be treated as nonlinear, how
  can a flutter analysis be efficiently conducted at the cost of some approximation?

Linear/Nonlinear Behavior in Unsteady Transonic
Aerodynamics. The accurate calculation of the aerodynamic forces in unsteady transonic flow requires the
solution of the nonlinear flow equations. The aeroelastician, on the other hand, seeks to treat his

problems (flutter, for example) by means of linear equations whenever possible. He may do this, even when the underlying steady flow is nonlinear, if the forces are linear over some (perhaps small) range of unsteady amplitude of motion.

Calculations were made for an NACA 64A006 airfoil oscillating in simple, harmonic motion in pitch over a range of amplitudes, frequencies, and Mach numbers. (Similar results were also obtained for a supercritical airfoil.) The primary aerodynamic method used was the well known LTRAN2 code of Ballhaus and Goorjian [29] that provides a finite-difference solution to the low frequency, small disturbance, two-dimensional potential flow equation.

Figures 17 and 18 present the lift, pitching moment, and shock motion as a function of amplitude of pitching motion for reduced frequencies, k, of 0.0 and 0.2, respectively. The mean angle of attack is zero and the Mach number is 0.86. The reduced frequency is based on chord. In general, the lift is a linear function of oscillation amplitude over a wider range than is the moment. The shock motion departs from linearity at lower amplitudes than either lift or moment. Based on these results and others for several Mach numbers and frequencies, the following criterion for linearity was selected: force and moment are linear functions of amplitude whenever the shock motion is less than five percent of chord during a cycle of oscillation. This choice is conservative in that in many cases the linear range would be larger than indicated by the criterion. Note that the linear range is larger for the unsteady case ($\alpha_1 < 0.5°$ in Fig. 18) than for the steady one ($\alpha_1 < 0.25°$ in Fig. 17). Similar results were obtained for other Mach numbers.

Using the above criterion, the 'boundary between the regions of linear and nonlinear behavior for the NACA 64A006 airfoil are shown in Fig. 19. The calculations were made for steady flow, k = 0 and for k = 0.2. Mach number is given as a function of amplitude of oscillation. The linear range (low amplitude) lies to the left of the curve. In steady flow, the linear

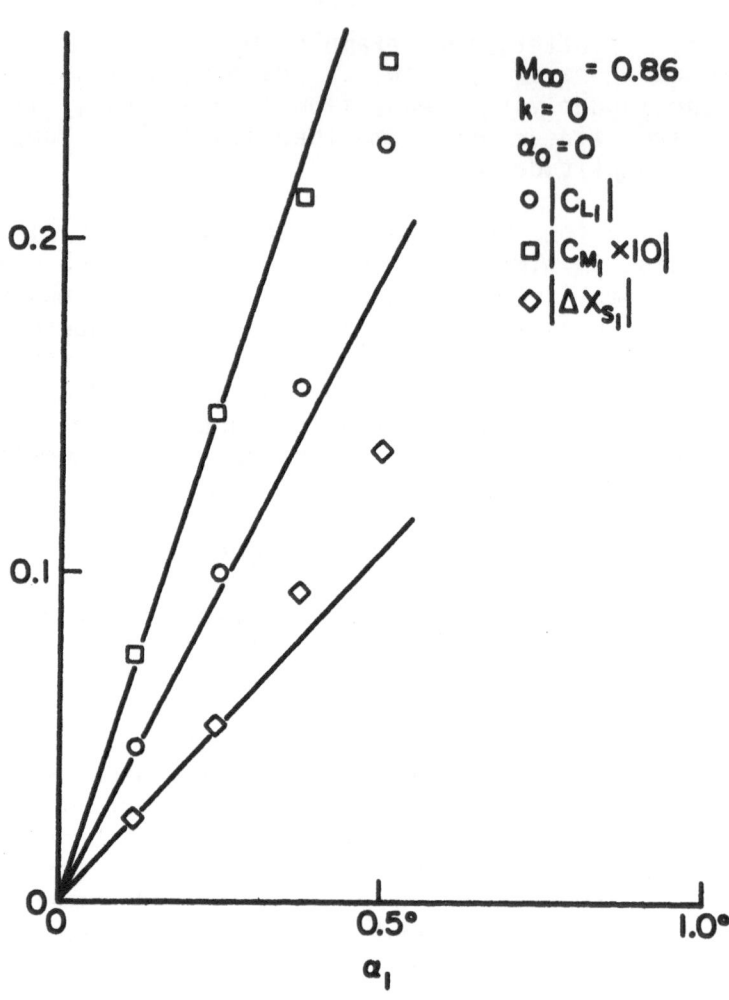

Fig. 17. Effect of dynamic angle of attack on dynamic
          forces and shock motion.

range practically disappears for Mach numbers near
0.89. The beneficial effect of increasing frequency
in increasing the linear range is apparent.   Finally,
the dashed lines, to which the boundaries are asymptotic, may be mentioned.   The lower line represents
the critical Mach number below which the local flow
remains subsonic, i.e., there is no shock.   The upper
line may be described as the upper limit of the

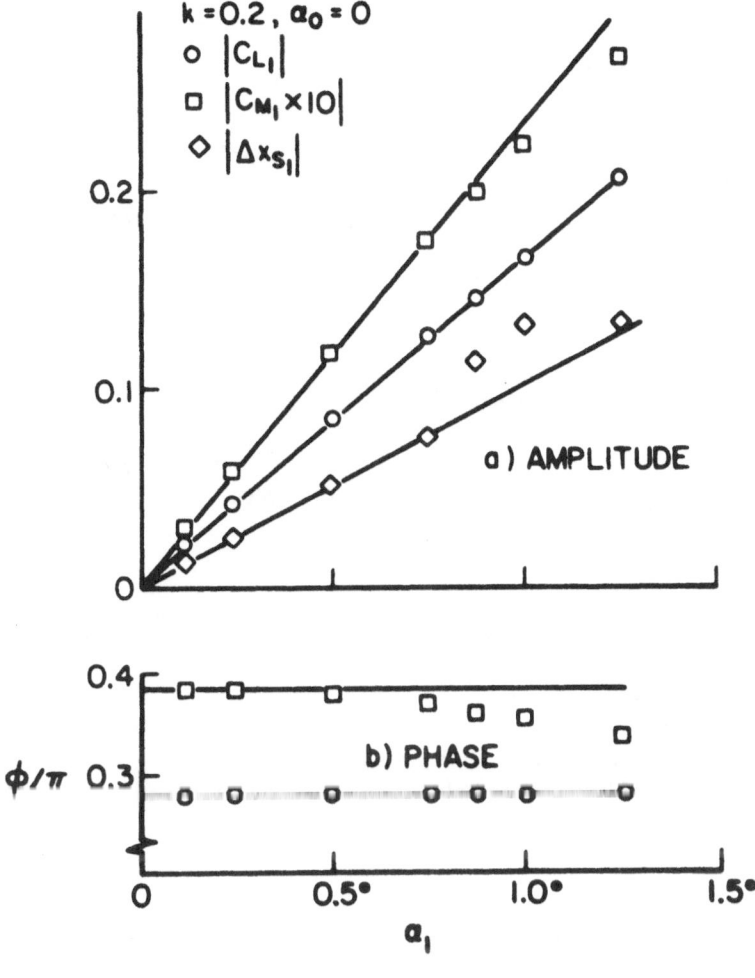

Fig. 18. Effect of dynamic angle of attack on dynamic
forces and shock motion.

transonic range; above the upper line the local flow
is supersonic on both surfaces at the trailing edge
(i.e., upper and lower shocks have reached the trailing edge). Below the lower line the flow is called
subcritical; above the upper line it is called transcritical.

Results for transonic flow over steady and oscillating airfoils have been obtained which indicate that

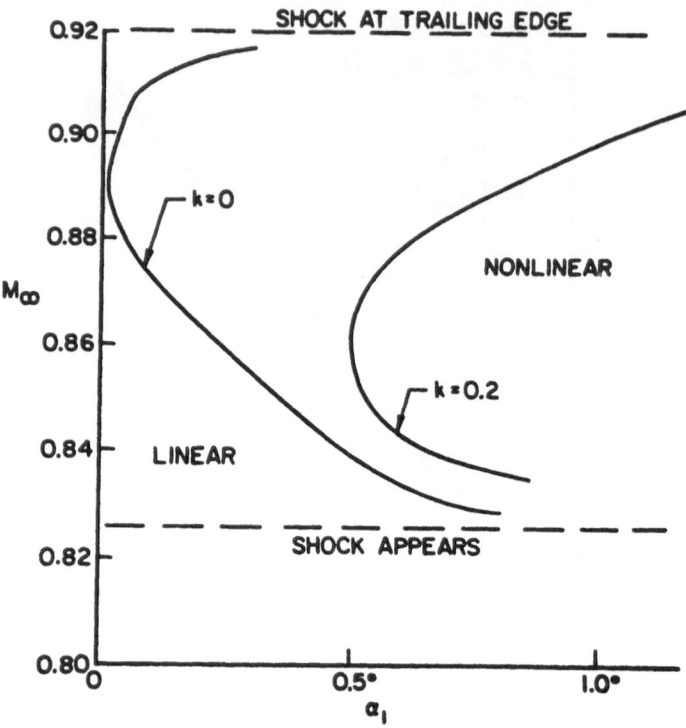

Fig. 19. Effect of frequency on boundary for linear/ nonlinear behavior.

forces are a linear function of oscillation amplitude over a range of parameters. The increased linearity with increasing frequency was demonstrated. It further was shown that the region of linearity also increases as the Mach number either decreases toward subcritical flow or increases toward completely transcritical flow. For additional results and more extensive discussion, the reader may consult References 30 and 31, and Chapter VI.

Flutter Analysis Using Nonlinear Aerodynamic Forces. Recent developments in computational aerodynamics have led to renewed interest in the prediction of flutter boundaries of an airfoil in the transonic flow regime. To date flutter calculations have either assumed the transonic aerodynamic forces could be approximated as <u>linear</u> functions of the airfoil motion so that traditional linear flutter analysis methods could be used or, alternatively, taken a brute force approach by simultaneously numerically integrating in time the structural and aerodynamic equations. The latter method does, of course, fully account for aerodynamic nonlinearities.

Here we study the nonlinear effect of transonic aerodynamic forces on a flutter boundary by utilizing a novel variation of the describing function method that takes into account the first fundamental harmonic of the nonlinear oscillatory motion. By using an aerodynamic describing function, traditional flutter analysis methods may still be used while <u>including</u> (approximately) the effects of aerodynamic nonlinearities.

The method used to calculate the describing functions is briefly this. A step change in angle of attack is specified and the transient aerodynamic force time history (calculated numerically by an appropriate aerodynamic code) is identified as a nonlinear impulse function. The Fourier transform of this impulse function (which in general depends on the step input level or amplitude) is taken as the aerodynamic describing function. Calculations have shown that this describing function agrees very well with the one determined by using a harmonic angle of attack input to the aerodynamic code. The latter is, of course, much more expensive and time consuming to calculate for the range of frequencies needed in flutter analysis.

A representative flutter result is shown in Fig. 20 for a typical section, NACA 64A006 airfoil, where the flutter boundary (flutter speed) as well as reduced frequency, bending/torsion amplitude ratio, and phase, are given in terms of amplitude of motion. The notation is that conventionally used in flutter analysis [1]. A purely linear flutter analysis would give the results corresponding to the flutter amplitude approaching zero. It is seen that the nonlinear aerodynamic effect is appreciable for amplitudes of motion

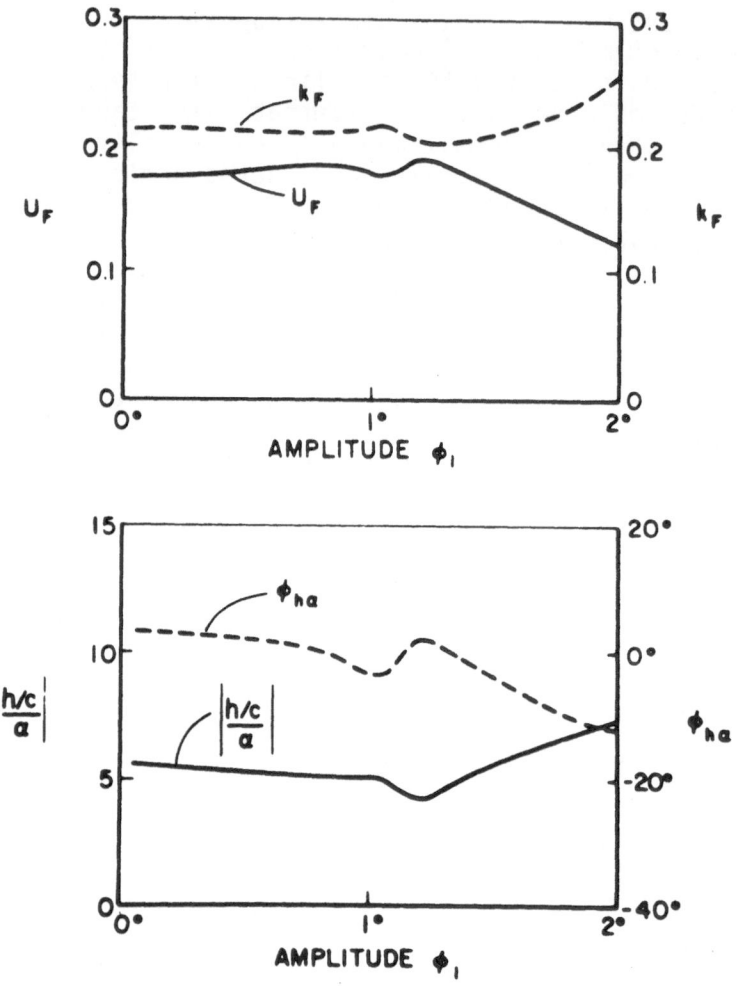

Fig. 20. Flutter boundary and flutter mode vs. amplitude of motion (NACA 64A006, M_∞ = 0.86, a =
-0.3, x_{cg} = 0.2, r_g^2 = 0.24, R = 0.2, μ =
60).

more than 0.5°. A particularly dramatic change occurs
in the range of amplitudes from 1° to 1.25°. For
greater amplitudes of motion, the flutter speed decreases as the amplitude increases. On physical
grounds the interpretation of the flutter speed vs.
amplitude curve is as follows: On the portions of the
curve where flutter speed increases with amplitude, a

nonlinear limit cycle occurs; on those portions where flutter speed decreases with amplitude, the curve represents a dividing boundary between disturbance amplitudes which lead to lower amplitude limit cycles and those which lead to higher amplitude limit cycles. In particular, in the latter case, the nonlinear flutter speed may be lower than that given by linear flutter theory.

A fuller discussion of the method and additional numerical results are contained in Reference 32 and Chapter VI.

REFERENCES

1. E. H. Dowell, H. C. Curtiss, Jr., R. H. Scanlan and F. Sisto, A Modern Course in Aeroelasticity, Sitjhoff-Noordhoff, Leyden, The Netherlands, 1979.

2. E. H. Dowell, Aeroelasticity of Plates and Shells, Noordhoff, Leyden, The Netherlands, 1975.

3. E. Breitbach, Effects of Structural Nonlinearities on Aircraft Vibration and Flutter, AGARD Report R-665, 1977.

4. E. Simiu and R. H. Scanlan, Wind Effects on Structures, John Wiley and Sons, New York, 1978.

5. R. L. Bisplinghoff and H. Ashley, Principles of Aeroelasticity, John Wiley and Sons, New York, 1962.

6. R. L. Bisplinghoff, H. Ashley and R. L. Halfman, Aeroelasticity, Addison-Wesley Publishing Co., Cambridge, Mass., 1955.

7. Y. C. Fung, An Introduction to the Theory of Aeroelasticity, John Wiley and Sons, New York, 1955.

8. R. H. Scanlan and R. Rosenbaum, Aircraft Vibration and Flutter, MacMillan Company, New York, 1951.

9. P. J. Holmes, Bifurcations of Divergence and Flutter in Flow-Induced Oscillations: A Finite Dimensional Analysis, J. Sound and Vibration, 53 (1977), pp. 471-503.

10. P. J. Holmes and J. Marsden, Bifurcations to Divergence and Flutter in Flow-Induced Oscillations: An Infinite Dimensional Analysis, Automatica, 14 (1978), pp. 367-384.

11. E. H. Dowell, Flutter of a Buckled Plate as an Example of Chaotic Motion of a Deterministic Autonomous System, J. Sound and Vibration, 85 (1982), pp. 333-344.

12. E. H. Dowell, Observation and Evolution of Chaos for an Autonomous System, J. Applied Mechanics, 51 (1984), pp. 664-673.

13. E. H. Dowell, Nonlinear Oscillations of a Fluttering Plate I., AIAA Journal, 4 (1966), pp. 1267-1275, Also Part II, AIAA Journal, 5 (1967), pp. 1856-1862.

14. C. C. Lin, The Theory of Hydrodynamic Stability, Cambridge University Press, 1955.

15. C. S. Ventres and E. H. Dowell, Comparison of Theory and Experiment for Nonlinear Flutter of Loaded Plates, AIAA Journal, 8 (1970), pp. 2022-2030.

16. R. T. Hartlen and I. G. Currie, Lift Oscillator Model of Vortex-Induced Vibration, Proc. ASCE, EM, 5 (1970), pp. 577-591.

17. I. G. Currie, R. T. Hartlen and W. W. Martin, The Response of Circular Cylinders to Vortex Shedding, IUTAM-IAHR Symposium on Flow-Induced Vibrations (1974), Edited by E. Naudascher, pp. 128-142.

18. O. M. Griffin, R. A. Skop and G. H. Loopman, The Vortex-Excited Resonant Vibrations of Circular Cylinders, J. Sound and Vibration, 31 (1973), pp. 235-249.

19. R. A. Skop and O. M. Griffin, A Model for the Vortex-Excited Response of Bluff Cylinders, J. Sound and Vibration, 27 (1973), pp. 225-233.

20. W. D. Iwan and R. D. Blevins, A Model for the Vortex-Induced Oscillation of Structures, J. Applied Mechanics, 41 (1974), pp. 581-585.

21. E. Simiu, and R. H. Scanlan, <u>Wind Effects on Structures</u>, John Wiley and Sons, New York (1978), pp. 194-212.

22. G. V. Parkinson, Mathematical Models of Flow-Induced Vibrations of Bluff Bodies, <u>IUTAM-IAHR Symposium on Flow-Induced Vibrations</u> (1974), Edited by E. Naudascher, pp. 128-142.

23. E. H. Dowell, Nonlinear Oscillator Models in Bluff Body Aeroelasticity, <u>J. Sound and Vibration</u>, 75 (1981), pp. 251-264.

24. E. J. Cui and E. H. Dowell, An Approximate Method for Calculating the Vortex-Induced Oscillation of Bluff Bodies in Air and Water, presented at the <u>ASME Winter Annual Meeting</u>, Washington, DC, November 16-20, 1981.

25. R. A. Skop and O. M. Griffin, On a Theory for the Vortex-Excited Oscillations of Flexible Cylindrical Structures, <u>J. Sound and Vibration</u>, 41 (1975), pp. 263-274.

26. M. Van Dyke, <u>Perturbation Methods in Fluid Mechanics</u>, The Parabolic Press, Stanford, California, 1975.

27. R. E. D. Bishop and A. Y. Hassan, The Lift and Drag Forces on an Oscillating Cylinder, <u>Proc. Roy. Soc.</u> A, 277 (1964), pp. 32-75.

28. J. Guckenheimer, Patterns of Bifurcation, in <u>New Approaches to Nonlinear Problems</u>, edited by Philip J. Holmes, Philadelphia, <u>SIAM</u>, 1980.

29. W. Ballhaus and P. Goorjian, Implicit Finite-Difference Computations of Unsteady Transonic Flow About Airfoils, <u>AIAA Journal</u>, 7 (1977), pp. 1728-1735.

30. E. H. Dowell, S. R. Bland and M. H. Williams, Linear/Nonlinear Behavior in Unsteady Transonic Aerodynamics, <u>AIAA Journal</u>, 21, (1983), pp. 38-46.

31. E. H. Dowell, S. R. Bland, and M. H. Williams, Linear/Nonlinear Behavior in Unsteady Transonic Aerodynamics, Princeton University MAE Technical Report No. 1520, May 1981.

32. T. Ueda and E. H. Dowell, Flutter Analysis Using Nonlinear Aerodynamic Forces, J. Aircraft, 21 (1984), pp. 101-109.

CHAPTER VI

UNSTEADY TRANSONIC AERODYNAMICS AND AEROELASTICITY*

Abstract

In recent years substantial progress has been made in the development of an improved understanding of unsteady aerodynamics and aeroelasticity in the transonic flow regime. This flow regime is often the most critical for aeroelastic phenomena yet it has proven the most difficult to master in terms of basic understanding of physical phenomena and the development of predictive mathematical models. The difficulty is primarily a result of the nonlinearities which may be important in transonic flow. The emerging mathematical models have relied principally on finite difference solutions to the governing nonlinear partial differential equations of fluid mechanics. Here are addressed fundamental questions of current interest which will provide the reader with a basis for understanding the recent and current literature in the field.

Four principal questions are discussed.

(1) Under what conditions are the aerodynamic forces essentially linear functions of the airfoil motion?

(2) Are there viable alternative methods to finite difference procedures for solving the relevant fluid dynamical equations?

(3) Under those conditions when the aerodynamic forces are nonlinear functions of the airfoil motion, what is the significance of the multiple (nonunique) solutions that are sometimes observed?

---

*An earlier version of this chapter has appeared in Recent Advances in Aerodynamics, edited by A. Krothapelli and C. A. Smith, Springer-Verlag, 1986.

(4) What are effective, efficient computational pro-
     cedures for using unsteady transonic aerodynamic
     computer codes in aeroelastic (e.g., flutter)
     analyses?

Nomenclature

C_L, C_M lift, moment coefficients

$C_{L_\alpha}, C_{M_\alpha}$ lift, moment curve slopes

C_p pressure coefficient

c airfoil chord

K $= (\gamma + 1)M_\infty^2 \tau / \beta^3$

k $= \omega c / U_\infty$; reduced frequency

M Mach number

s $= (\beta^2 t U_\infty / c) / M_\infty^2$

t time

x, y spatial coordinates in freestream and
         vertical directions

x_S shock location

Δx_S shock displacement normalized by the
            airfoil chord

α_0, α_1 mean angle of attack; dynamic angle of
                     attack in degrees

β $= (1 - M_\infty^2)^{1/2}$

γ ratio of specific heats

ν $\qquad = kM_\infty^2/\beta^2$

$\phi^{(0)},\phi^{(1)}$ velocity potentials of steady flow and unsteady airfoil motion respectively

ϕ $\qquad$ phase angle

τ $\qquad$ thickness ratio of airfoil

ω $\qquad$ frequency

Δ $\qquad$ gradient operator

Subscripts

∞ $\qquad$ freestream

L $\qquad$ local; also lift

M $\qquad$ moment

max $\qquad$ maximum

0,1 $\qquad$ mean, dynamic

TE $\qquad$ trailing edge

Superscripts

c $\qquad$ shock first forms

tc $\qquad$ shock reaches the trailing edge

Section 5

M $\qquad$ number of structural modes

NF $\qquad$ number of reduced frequencies needed for a flutter analysis

NR $\qquad$ number of response levels for a nonlinear flutter analysis

P   number of parameters

T_A computational time for aerodynamic code to reach a steady state lift value for a prescribed airfoil motion

T_F computational time for a simultaneous fluid-structural calculation to complete a transient

T_{AF} computational time for aerodynamic code to determine aerodynamic forces for one reduced frequency

Section 6

$A(\ ), A_L, A_M$ indicial response functions

α distance of elastic axis from mid-chord: percent semichord, positive downstream

b   semichord length

c   full chord length

C_L^N nonlinear lift coefficient

C_M^N nonlinear moment coefficient about mid-chord

C_{Me}^N nonlinear moment coefficient about elastic axis

$\bar{D}_L, \bar{D}_M$ components of describing function

F   output of describing function

G   structural transfer function

H   nonlinear aerodynamic transfer function

$\hat{H}$ aerodynamic describing function

h   plunging displacement of elastic axis (positive down)

h_c plunging displacement of mid-chord (positive down)

I_α moment of inertia per unit span about elastic axis

k $= \dfrac{c\omega}{u}$, reduced frequency

L   lift force

M   moment force about mid-chord (positive nose-up)

m   mass per unit span

M_∞ Mach number of uniform airflow

R $= \dfrac{\omega_h}{\omega_\alpha}$, uncoupled frequency ratio

r_α dimensionless radius of gyration about elastic axis (based on semichord); $r_\alpha^2 = r_{cg}^2 + (x_{cg} - \alpha)^2$.

r_{cg} dimensionless radius of gyration about center of gravity (based on semichord)

S_α static unbalance

s   dimensionless variable of Laplace operator; s = ik for harmonic oscillation

t   time

U   dimensionless airspeed $\dfrac{u}{c\omega_\alpha \sqrt{\mu}}$.

u   dimensional airspeed

x_{cg} distance of center of gravity from
                     mid-chord; percent semichord, positive
                     downstream

α pitching displacement

ϕ effective induced angle-of-attack; see
                     equation (6.1)

ϕ_1 amplitude of ϕ oscillation

μ mass ratio $\dfrac{m}{\pi \rho b^2}$

ω_h, ω_α uncoupled circular frequency of the airfoil
                     in plunging and in pitch, respectively

ρ air density

τ dimensionless time $\dfrac{ut}{c}$

Superscripts

T   transpose of matrix

^   quantity associated with describing
                     function

' $= \dfrac{d}{dt}$

−   quantity in the subsidiary domain of
                     Laplace Operator

6.1. Introduction

The four questions cited in the summary are chosen to provide the framework of this chapter. This selection was made for several reasons.

- They are fundamental questions that are expected to be of lasting significance.

- Answers to these questions have important consequences for aeroelastic applications of unsteady transonic aerodynamics.

- Recent work has led to at least partial answers.

The four questions are addressed in Sections 6.2, 6.3, 6.4, and 6.5-6.6, respectively. Each section may be read relatively independently of the others, and the reader may wish to take advantage of that option.

6.2. Linear/Nonlinear Behavior in Unsteady Transonic Aerodynamics

6.2.1 Motivation and general background

The aeroelastician uses linear dynamic system theory for most aeroelastic analyses. The motivation for doing so is clear. Extensive experience, understanding, and effective computational/experimental procedures have been developed for linear systems. By contrast, although nonlinear methods of analysis and experimentation are available, the results are far more expensive to obtain and also more difficult to interpret. Hence linear models, where applicable, are very powerful, relatively simple, and extremely valuable. Thus, it is highly important to determine the domain of validity of any linear model.

Here our concern is with possible aerodynamic nonlinearities in transonic flow. Of course, aerodynamic nonlinearities may arise in other flow regimes; however, it is in transonic flow where they tend to be most important. Indeed, it is often observed that the transonic flow regime is inherently nonlinear in the governing field equations. However, at any Mach number for any airfoil, if the angle of attack is sufficiently small, the aerodynamic forces and shock motion will be linear in the angle of attack. Moreover, as the frequency of the angle of attack motion increases, the range of angle of attack over which linear behavior persists increases. It is our purpose here to study when linear or nonlinear behavior occurs using as our principal analytical method the low frequency, transonic small disturbance (LTRAN2) procedure of Ballhaus and Goorjian (1977, 1978). Any other present or future nonlinear aerodynamic method could (and should) be used for similar purposes.

It will be helpful to discuss first the shock and its motion, which is sometimes a source of confusion. A consequence of any consistent linearization of steady transonic small disturbance aerodynamic theory in the dynamic angle of attack is that a concentrated force or pressure pulse (sometimes called a shock doublet) will appear at the location of the steady state shock (Williams, 1979a, 1979b). The strength of the pressure pulse is equal to the steady state shock pressure jump and its width is proportional to the dynamic angle of attack. By contrast, elsewhere on the airfoil chord (away from the shock doublet whose center is at the steady state shock location), the pressure magnitudes (in a transonic linear theory) are proportional to the dynamic angle of attack and become smaller in proportion as the dynamic angle of attack is smaller. Of course this latter behavior is also true in classical theory. The most important (although not the only) distinction between classical, linear theory and transonic, linear theory is the presence of the shock (and its motion) in the latter that creates the concentrated shock force doublet. LTRAN2 and some other transonic computer codes include both the shock and its motion while classical aerodynamic theory includes neither. Some inconsistent transonic methods include the shock's presence, but not its motion.

The behavior described above is seen in a non-linear dynamic theory as well, when the dynamic angle of attack becomes small. Consider Fig. 1 which was obtained using LTRAN2. It shows the chordwise differential (lower surface minus upper) pressure distribution for an NACA 64A006 airfoil at $M_\infty = 0.86$ for several angles of attack. Here, for simplicity, the reduced frequency is set to zero, so there is no distinction (numerically) between steady and dynamic angle of attack. As may be seen for small angles of attack, say $\alpha = 0.125$ deg, 0.25 deg, the pressure distribution has a shock doublet centered at the mean (zero angle of attack) shock location, $x_S/c = 0.584$. The width of the shock doublet is indicated by the vertical lines, the forward one is at the lower surface shock location and the rearward one at the upper surface location. The shock doublet width is proportional to α for the smaller α; however, as α increases to 1 deg, the lower surface shock disappears while the upper surface shock moves to the trailing edge and remains there.

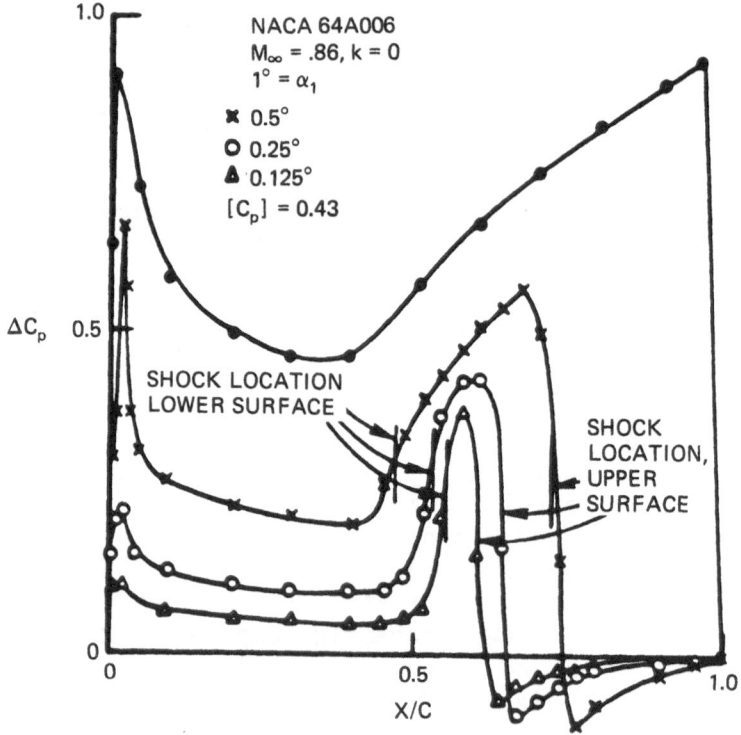

Fig. 1. Differential pressure distribution.

Also, for the smaller α the shock doublet magnitude is essentially equal to the pressure jump through the shock at $\alpha = 0$ deg, i.e., 0.43. Away from the shock doublet, the pressures are proportional to α for small α.

Finally, note a matter of practical importance. For small α as the shock doublet width narrows, any finite difference scheme nonlinear in α will have a resolution problem as $\alpha \rightarrow 0$. By contrast a method a priori linearized in α avoids this difficulty as it computes the shock motion explicitly, e.g., see Williams (1979a, 1979b) and Fung et al. (1978). Also see the discussion of Tijdeman (1977) and Tijdeman and Seebass (1980) for a critical assessment of theory and experiment. The experimental study of Davis and Malcolm (1979) is particularly relevant here as it provides confirmation of the above in broad terms.

6.2.2. NACA 64A006 airfoil at $M_\infty = 0.86$ pitching about its Leading Edge

The following principal issues were studied (Dowell et al., 1983): effect of dynamic angle of attack at various reduced frequencies on dynamic forces and shock motion; boundary for linear/nonlinear behavior; effect of reduced frequency and dynamic amplitude on aerodynamic transfer functions; effect of dynamic angle of attack on steady state forces and shock displacement; and effect of steady-state angle of attack on dynamic forces and shock motion. For the sake of brevity, only the first two issues will be considered here.

6.2.2.1. Effect of dynamic angle of attack on dynamic forces and shock motion

It is desirable to assess at what dynamic amplitude nonlinear effects become important in order to determine the relative linear vs nonlinear behavior of lift, pitching moment, and shock motion. Note that the total lift (moment, shock motion) is characterized by $C_L = C_{L_0} + C_{L_1}$, where C_{L_0} is defined to be the lift due to the mean angle of attack, α_0, and C_{L_1} that due to the dynamic angle of attack, α_1 for given α_0. In classical linear theory (but not transonic linear theory), C_{L_1} is independent of α_0.

In Figure 2 lift, pitching moment, and shock displacement amplitudes are shown as a function of dynamic amplitude, α_1, for a reduced frequency of k = 0.2. Lift and moment coefficient have their usual definitions and the moment is about the midchord. The shock displacement is normalized by the airfoil chord. Phases are also presented for lift and pitching moment. The shock motion phase was also computed; however, it tended to be less accurately determined. Since it is not needed for our present purposes, it is not shown.

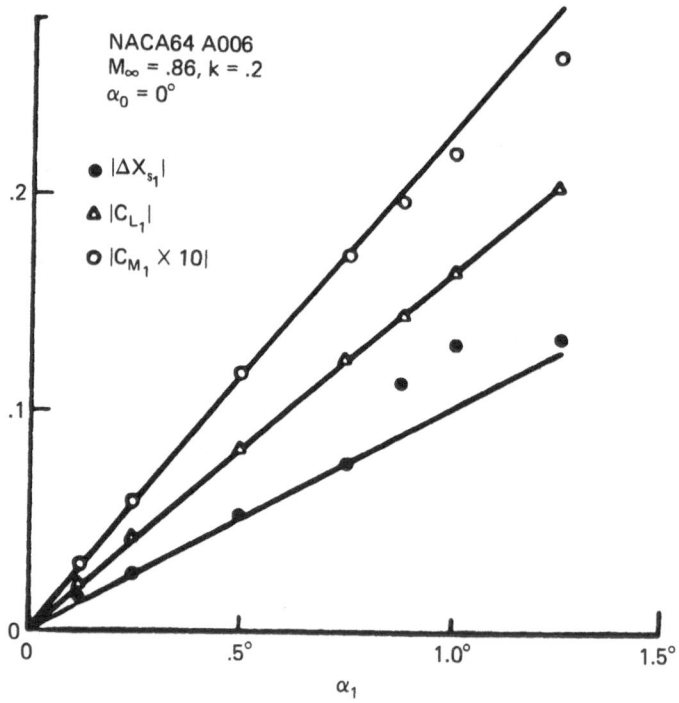

Fig. 2a. Effect of dynamic angle of attack on dynamic forces and shock motion: amplitudes.

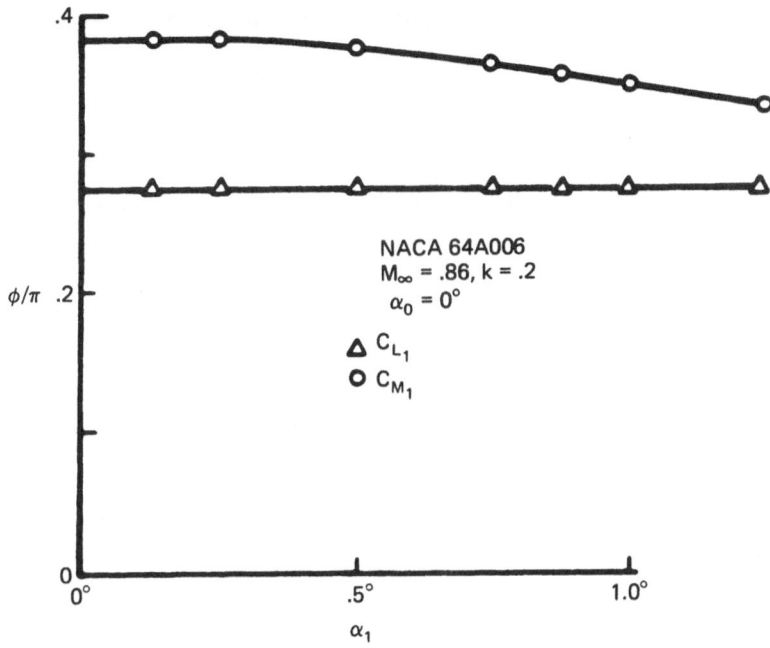

Fig. 2b. Effect of dynamic angle of attack on dynamic
forces and shock motion: phases.

It is seen that lift tends to remain linear to higher dynamic amplitudes than moment, which, in turn, tends to remain linear to higher amplitudes than shock motion. Moreover, as will be seen, the larger the reduced frequency, the greater the range of linear behavior. Phase information generally, although not universally, is a more sensitive indicator of departure from linearity than lift, moment, or shock amplitude information. In a strictly linear theory, of course, the phase is independent of the dynamic angle of attack.

It is noted that no measurable higher harmonic content was found in any of the numerical results. The results were virtually sinusoidal signals for lift, moment, and shock motion; hence, determination of magnitude and phase was readily done by any one of several conventional methods. The exception was shock motion phase, which is difficult to determine accurately by any method because of the relatively coarse finite difference mesh resolution of the shock.

218

6.2.2.2. Boundary for linear/nonlinear behavior

It is highly desirable to provide a criterion by which the aeroelastician may assess when a linear dynamical theory may be used.

Figure 3 has been constructed from Figure 2 and other similar results by identifying the k, α_1 combinations for which the pitching moment deviates by 5% in amplitude or phase from linearity. As expected, at higher k the pitching moment remains linear to larger α_1.

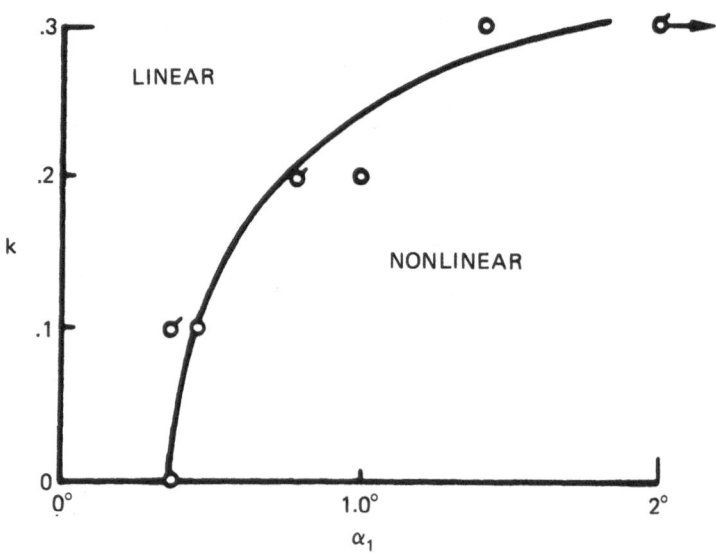

Fig. 3. Boundary for linear/nonlinear behavior in terms of reduced frequency and dynamic angle of attack.

Although Fig. 3 provides very useful information, it requires a nonlinear dynamical theory to construct it. A question thus arises: Is there a similar, but perhaps more conservative, criterion that may be used

219

with a linear dynamical theory? The answer is provided by the shock motion. In Fig. 4 a similar boundary to that shown in Fig. 3 is constructed (again
from information such as that provided by Fig. 2)
based on shock motion rather than pitching moment. It
is observed in Fig. 2 that for shock displacement
amplitudes of less than 5% the shock motion (as well
as lift and pitching moment) behave in a linear
fashion. Hence, a 5% shock motion boundary is shown
in Fig. 4. Note that this boundary could be constructed from a linear dynamical theory. A second
boundary (less conservative) based on the first
detectable deviation of shock motion from linearity is
also shown. Finally, the boundary from Fig. 3 is
shown for reference. These results are consistent
with those of Ballhaus and Goorjian (1978) who also
suggested that shock motions of less than 5% chord
correspond to linear behavior.

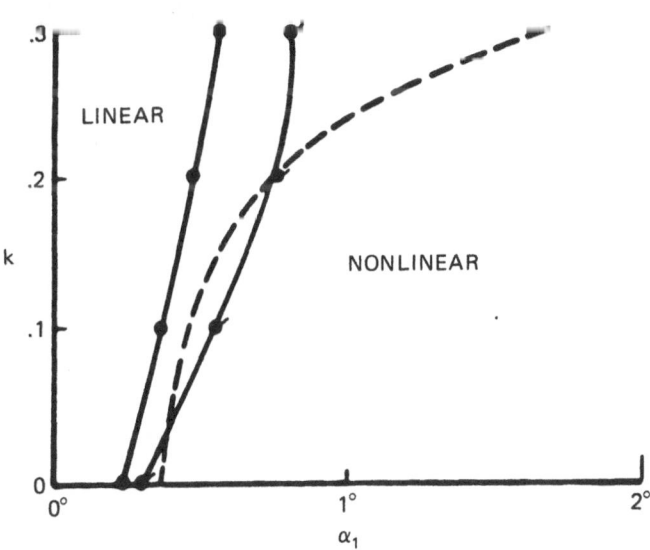

Fig. 4. Conservative boundary for linear/nonlinear
behavior based on shock motion amplitude.

Thus, it is concluded that a simple criterion for departure from nonlinearity based on shock motion may be used. It can be evaluated by a linear dynamical theory in principle (which enhances its practical utility), although the present results were obtained using a nonlinear, dynamical theory.

A brief digression is in order to explain why the shock motion criterion is extremely useful to the aeroelastician. After the flutter mode is determined from a conventional flutter analysis using linearized but transonic aerodynamics, one may compute the amplitude of the flutter motion which will correspond to a 5% shock motion using the linear transonic aerodynamic model employed in the flutter analysis. This will give the aeroelastician the limit on amplitude for which the linear, flutter calculation is valid. This is very useful information.

6.2.3. Mach number trends

6.2.3.1. Similarity law

Here the effects of Mach number are studied systematically for the NACA 64A006 airfoil. We note that a similarity rule holds for low frequency, transonic flow, which gives the following results for any family airfoils:

$$C_p = \frac{\tau}{\beta} \bar{C}_p (x/c, s; K, \nu, \alpha/\tau) \tag{2.1}$$

where $\bar{C}_p$ is a universal function of its arguments and

$$\beta \equiv (1 - M_\infty^2)^{1/2}, \quad \nu \equiv k M_\infty^2 / \beta^2$$

$$K \equiv \frac{(\gamma + 1) M_\infty^2}{\beta^3}, \quad s \equiv \frac{\beta^2 t U_\infty / c}{M_\infty^2}$$

$$\tau \equiv \text{thickness ratio of airfoil}$$

$$\alpha \equiv \text{angle of attack}$$

Equation (2.1) may be further specialized for the case $\alpha \to 0$, by expanding in a Taylor series, i.e.,

$$C_p = \frac{\tau}{\beta} \bar{C}_p(x/c,s;K) + \frac{\alpha}{\beta} \text{Re}\, \{e^{i\nu s}\, \bar{C}_{p1}(x/c;K,\nu)\} \quad (2.2)$$

This is the similarity law for dynamic linearization in α, i.e., $\alpha = \alpha_1$. Zero mean angle of attack is assumed for simplicity, $\alpha_0 = 0$, although the result is readily extended. From equation (2.2) it is seen that similarity for the harmonic component requires only that K and ν be the same for two different flows.

Finally, it is noted that since the shock is simply a discontinuity surface of ϕ_x, it satisfies a similarity law expressed by

$$x_S = x_S(\beta y/c,s;K,\alpha/\tau,\nu) \quad (2.3)$$

For the limit, $\alpha \to 0$

$$x_S = x_{S_0}(\beta y/c;K) + \alpha/\tau \text{Re}\, \{e^{i\nu s} x_{S_1}(\beta y/c;K,\nu)\} \quad (2.4)$$

The similarity law given by equation (2.1) was known to Miles (1959). Equations (2.2)-(2.4) are extensions of his results.

Using the similarity rules, the results for the 64A006 airfoil may be used to obtain results for any other airfoil of the same family, in particular, the 64A010.

From equation (2.4) it may be inferred that the 5% shock motion criterion has the functional form (for a given family of airfoils)

$$\alpha/\tau = F(\nu,K) \quad (2.5)$$

It is interesting to note that Fung et al. (1978) proposed a criterion for the validity of linearization of the form

$$\frac{\alpha}{\tau} / K \ll 0.1 \qquad (2.6)$$

Equation (2.6) is clearly a special case of equation (2.5). Using equation (2.5), the data of Figs. 3, 4 (and 5, subsequently) may be reinterpreted in terms of similarity variables, and thereby generalized.

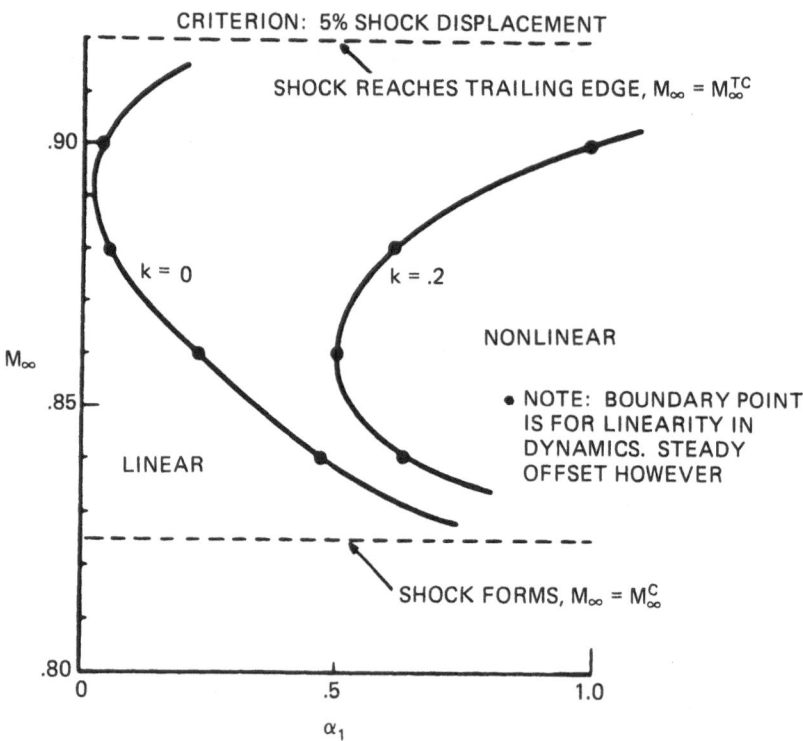

Fig. 5.   Effect of reduced frequency on boundary for linear/nonlinear behavior: Mach number vs. dynamic angle of attack.

6.2.3.2. Boundary for linear/nonlinear behavior

Using results such as those shown in Fig. 2 and invoking the 5% shock displacement criterion, a linear/nonlinear boundary may be constructed in terms of Mach number vs amplitude of airfoil oscillation. Of course, as the shock reaches very near the trailing edge, the 5% criterion would need to be modified. Results are shown in Fig. 5 for $k = 0$ and 0.2. Note that for steady flow ($k \equiv 0$) the angle of attack must be very small when $M_\infty = 0.88$ and 0.9 for linear behavior to occur. However, as we have seen before, the 5% shock displacement criterion is conservative. That is, lift and moment tend to remain linear in α to higher α than this criterion would suggest. Nevertheless, the trend should not change using any other reasonable criterion. By contrast, for $k = 0.2$ the linear region is much enlarged.

For $M_\infty < M_\infty^c$ or $M_\infty > M_\infty^{tc}$ the linear region is for all practical purposes unbounded. In practice, in this region other physical effects, e.g., viscosity, are likely to come into play before inviscid, small disturbance, transonic theory nonlinearities become important. M_∞^c is the Mach number when the shock first forms and M_∞^{tc} that when it reaches the airfoil trailing edge.

6.2.3.3. Aerodynamic transfer functions

In the linear region it is of interest to display aerodynamic transfer functions vs Mach number. Perhaps the most familiar of these is lift curve slope, C_{L_1}/α_1. Its amplitude is shown in Fig. 6a from LTRAN2 for $k = 0$. Also shown are results from full potential theory, classical subsonic theory, and local linearization. The latter is shown for $M_\infty > M_\infty^{tc}$ i.e., the shock is at the trailing edge. It uses the local trailing edge supersonic Mach number in classical (supersonic) theory. One concludes that for $M_\infty < M_\infty^c$, classical theory gives reasonable results, and for $M_\infty > M_\infty^{tc}$ local linearization gives reasonable results.

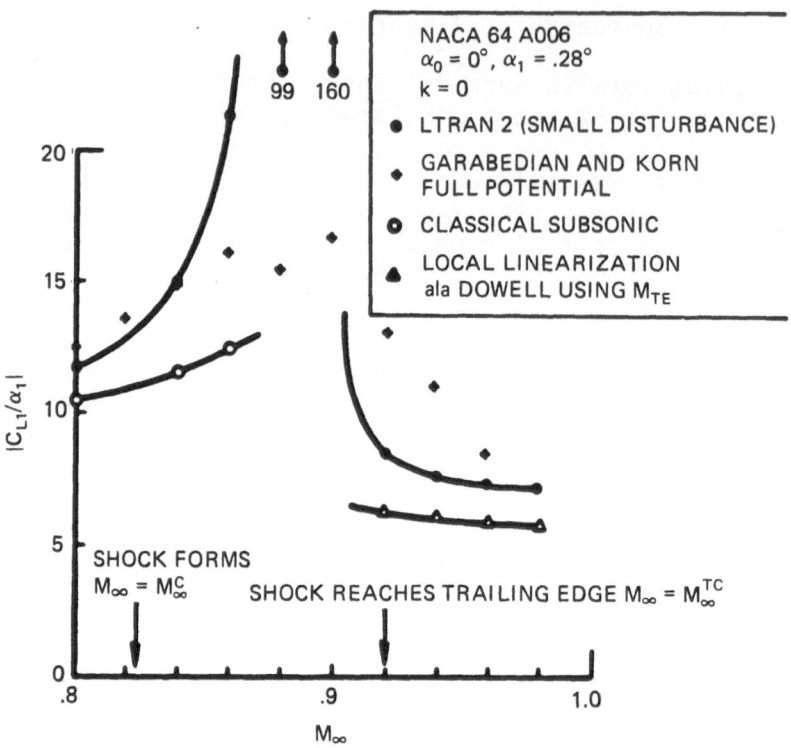

Fig. 6a.   Effect of Mach number on lift curve slope.

For $M_\infty^C < M_\infty < M_\infty^{tc}$, LTRAN2 gives markedly different results although it likely fails for $M_\infty = 0.88, 0.90$. Note the difference between transonic small disturbance theory (LTRAN2), which falls well off scale at $M_\infty = 0.88$ and 0.9, and full potential theory (Bauer et al., 1972).

It should be noted that the full potential results shown in Fig. 6a were obtained using a nonconservative finite difference scheme. Full potential results obtained using a quasiconservative finite difference scheme (for technical reasons results were only obtained for $M_\infty < 0.87$) are essentially identical to those of transonic small disturbance theory using a conservative finite difference scheme (LTRAN2). Hence, the difference shown in Fig. 6 should be attributed to the distinction between conservative and nonconservative finite differences and not to the distinction between small disturbance and full potential theory. To the extent that the nonconservative finite difference method may be said to have some form of numerical (as opposed to physical) viscosity, the differences may be attributed to the qualitative distinction between inviscid and viscous flow.

In Fig. 6b results are shown for $k = 0.2$. For reference, the LTRAN2 results for $k = 0$ are also shown. Again it is seen that the classical subsonic theory and local linearization theory give reasonable results (better than for $k = 0$) for $M_\infty < M_\infty^C$ and $M_\infty > M_\infty^{TC}$, respectively. Moreover, LTRAN2 appears to give reasonable results over the entire Mach number range, although there is no better theory to validate it. Note that from $M_\infty = 0.9$ to 0.92 there is a somewhat abrupt change.

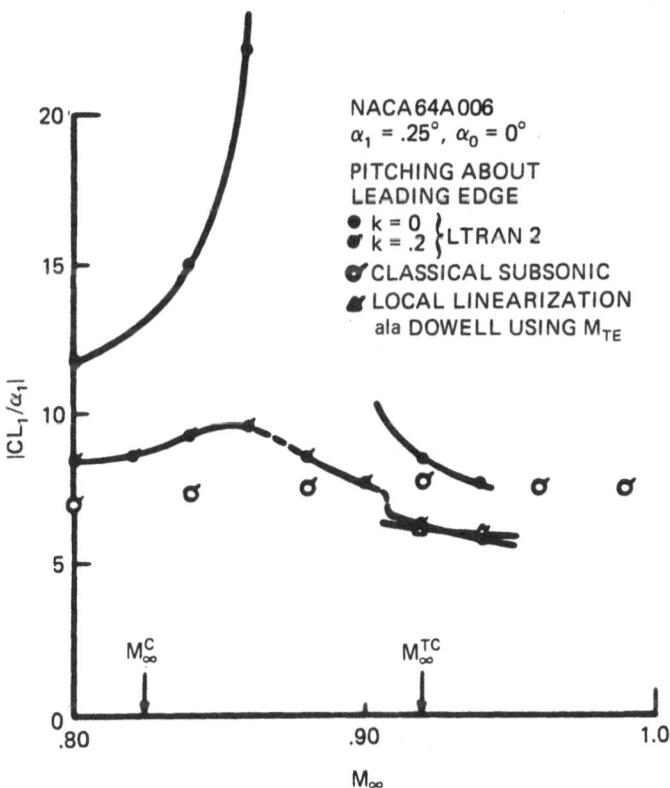

Fig. 6b. Effect of Mach number on lift curve slope.

6.2.4. Conclusions

For $M_\infty < M_\infty^c$, where no shock exists, the aerodynamic forces are linear over a substantial range of angle of attack. This is also true for $M_\infty > M_\infty^{tc}$, i.e., where the shock has moved to the trailing edge. For $M_\infty^c < M_\infty < M_\infty^{tc}$ a boundary of linear/nonlinear behavior may be constructed, which shows the angle of attack must be quite small for linear behavior to occur for steady flow. However, the region of linear behavior increases substantially for unsteady flow.

In the range $M_\infty^c < M_\infty < M_\infty^{tc}$, transonic small disturbance theory (LTRAN2) and full potential theory appear to fail for steady flow for some narrow band of M_∞ where they substantially overestimate the shock displacement and, hence, the aerodynamic forces. This is tentatively attributed to the absence of viscosity in the theories.

Classical subsonic theory and local linearization are useful approximate tools for unsteady flow provided their limitations are recognized.

Aerodynamic transfer functions are expected to retain their utility even when nonlinear dynamic effects are important. This is for several reasons, including:

(1) Nonlinear effects diminish with increasing frequency.

(2) At high frequencies, classical linear theory is expected to be reasonably accurate and indeed most inviscid theories will approach classical theory as the frequency becomes larger (Williams, 1979a, 1979b).

(3) The preceding suggests that several theories may be used to provide a composite aerodynamic representation in the frequency domain. For example, one might use BGK for $k = 0$, LTRAN2 for $k = 0.05-0.2$, Williams for $k = 0.2-1.0$, and classical theory (which Williams' theory smoothly approaches) for $k > 1.0$.

A similarity law for low frequency transonic small disturbance theory is available that reduces the number of aerodynamic computations required and generalizes results for one airfoil to an entire family.

Although two-dimensional flows have been discussed here, the general concepts and approach should be useful for three-dimensional flows. In particular, one expects the effect of three-dimensionality to increase the region of linear behavior for transonic flows. For example, the accuracies of transonic small disturbance theory, local linearization, and classical theory should be enhanced by three-dimensional effects.

No transonic method of aerodynamic analysis can be expected to give useful information to the aeroelastician unless the mean steady flow it predicts and uses is accurate. Hence, it is highly desirable to be able to input directly the best steady flow information which is available including that from experiment. The latter would include implicitly viscosity effects on the mean steady flow; in particular it would place the mean shock in the correct position.

The reader may wish to consult the lucid survey article by Tijdeman and Seebass (1980) which provides a context in which to evaluate the present results and conclusions. Also Nixon and colleagues have discussed extensively how the transonic, linear theory may be used in aeroelastic analyses. For example, see Nixon and Kerlick (1980) and Nixon (1981). Finally see the subsequent discussion in Section 5.

6.3. Viable alternative solution procedures to finite difference methods

Although continuing advances in computer technology will lead to diminishing costs, economics alone will dictate for the next decade a substantial effort to improve the efficiency of finite difference methods and/or consider less expensive alternative solution techniques. Here, the latter is discussed drawing largely on the recent work of Hounjet (1981b) and Cockey (1983). Both of these authors have used integral equation methods (IEM), although from rather different points of view. Prior work by Hounjet (1981a) was based on the Williams-Eckhaus model (Williams, 1979a, 1979b), which also is the point of departure for Cockey. The motivation for considering IEM and a concise description of earlier work is well covered by Hounjet (1981b), Morino (1974), Morino and Tseng (1978), Albano and Rodden (1969), Nixon (1978), Voss (1981), Williams (1978), and Liu (1978).

Both Hounjet and Cockey adopt a transonic small disturbance equation approximation and the associated velocity potential is divided into steady, $\phi^{(0)}$, and unsteady (due to airfoil motion), $\phi^{(1)}$, parts. By assuming (infinitesimally) small airfoil (harmonic) motion, the governing field equation for $\phi^{(1)}$ is linear with variable coefficients that depend on $\phi^{(0)}$, viz.,

$$(1-M_\infty^2)\phi_{xx}^{(1)} + \phi_{yy}^{(1)} - 2ikM_\infty^2\phi_x^{(1)} + k^2M_\infty^2\phi^{(1)}$$

$$= [(M_L^2 - M_\infty^2)\phi_x^{(1)}]_x \tag{3.1}$$

where M_∞ is the freestream Mach number, k is the reduced frequency, $k = \omega c/2u$, in which c denotes the airfoil chord. c is used to make lengths dimensionless. M_L is the local Mach number:

$$M_L^2 = M_\infty^2 + [3 - (2 - \gamma)M_\infty^2]M_\infty^2\phi_x^{(0)} \tag{3.2}$$

Subscripts on $\phi^{(1)}$ denote spatial derivatives.

From this point the approaches of Hounjet and Cockey follow different paths. Note that by setting the right hand side of equation (3.1) to zero, one retrieves classical aerodynamic theory.

6.3.1. Hounjet

By using the Green's function of classical aerodynamic theory [the LHS of equation (3.3)], one obtains an integral equation for the unknown $\phi^{(1)}$. It is

$$\phi^{(1)}(x,y) = \int_0^\infty \Delta\phi^{(1)}(u) \frac{\partial}{\partial y} [E(x - u,y;k,M)]du$$

(3.3)

$$+ \int_{-\infty}^\infty \int_{-\infty}^\infty m(u,v)E(x - u,y - v;k,M)dudv$$

On the right hand side of equation (3.3), the first term is a integral along the airfoil chord (and wake), while the second integral is over the <u>entire</u> (but see below) flow field. E represents an elementary source solution that satisfies the radiation condition and the following equation:

$$(1 - M_\infty^2)E_{xx} + E_{yy} - 2ikM_\infty^2E_x + k^2M_\infty^2E$$

$$= e^{i\bar{k}x}\delta(x)\delta(y)$$

(3.4)

where

$$\bar{k} \equiv kM_\infty^2/\beta^2$$

m is given by

$$m(x,y) = [3 - (2 - \gamma)M_\infty^2]M_\infty^2[\Phi_x^{(0)}\Phi_x^{(1)}]$$

(3.5)

In classical (integral equation) aerodynamic theory, of course, the second integral of equation 2.3 is not present because $\Phi_x^{(0)}$ is zero. In transonic IEM the second term may be neglected everywhere in the flow field where $\Phi_x^{(0)} = 0$, i.e., the steady flow field is sensibly uniform. Hence, only a relatively small part of the total flow field will contribute to the second integral term and this simplifies the subsequent calculation very substantially. This is the key point which allows the possibility of an efficient computer code. Then the numerical solution of equation (3.3) proceeds as described by Hounjet (1981b).

Numerical results obtained by Hounjet show two principal features:

(1) The accuracy (for a linearization in the dynamic airfoil motion) is the same as that obtained by finite difference procedures.

(2) The computer time is approximately one quarter of that of LTRAN2 (a popular finite difference code).

Recently, Hounjet has extended this approach to three-dimensional flow fields.

6.3.2. Cockey

First rewrite equation (2.1) as

$$(1 - M_\infty^2)\phi_{xx}^{(1)} + \phi_{yy}^{(1)} - 2ikM_\infty^2\phi_x^{(1)} + k^2M_\infty^2\phi^{(1)}$$

$$- (M_L^2 - M_\infty^2)\phi_{xx}^{(1)} - 2M_L\frac{dM_L}{dx}\phi_x^{(1)} = 0$$

(3.6)

If one could determind the Green's function for the LHS of equation (3.6), then an integral equation for $\phi^{(1)}$ will only have an integral along the airfoil chord, shock, and wake and none in the flow field per se (except possibly along the shock). Unfortunately, obtaining this Green's function is difficult because the last two forms of the LHS of equation (3.6) have variable coefficients.

Thus, Cockey modifies equation (3.6) by a local linearization approximation to M_L and $\frac{dM_L}{dx}$ for the purposes of obtaining the Green's function (approximately). The subsequent calculation follows standard techniques except that integrals extend along the shock as well as the airfoil chord (and wake).

Numerical results obtained by Cockey show two principal features:

(1) The accuracy is substantially less than that obtained by finite difference procedures, even though the shock and its movement is taken into account.

(2) However, the computer cost is no greater than that associated with classical aerodynamic theory.

(1) is, of course, a disappointing result. However, the successful incorporation of the shock into the Cockey model and Hounjet's substantial success suggest a possible way of advantageously combining the features of methods of Hounjet and Cockey.

6.3.3. A possible synthesis of the Hounjet and Cockey methods

Consider again equation (3.1). Recognizing that in the Cockey method, M_L, $\dfrac{dM_L}{dx}$ are approximated for the purpose of obtaining a Green's function equation (3.1) is rewritten as follows:

$$(1 - M_\infty^2)\phi_{xx}^{(1)} + \phi_{yy}^{(1)} - 2ikM_\infty^2\phi_x^{(1)} + k^2M_\infty^2\phi^{(1)}$$

$$- (M_{LA}^2 - M_\infty^2)\phi_{xx}^{(1)} - 2M_{LA}\frac{dM_{LA}}{dx}\phi_x^{(1)} \qquad (3.7)$$

$$= (- M_{LA}^2 + M_L^2)\phi_{xx}^{(1)} + 2 (M_L\frac{dM_L}{dx} - M_{LA}\frac{dM_{LA}}{dx})\phi_x^{(1)}$$

Setting the RHS of equation (3.7) to zero, we retrieve the Cockey model. However, if the RHS is retained, then Hounjet's method could be used to solve the resulting integral equation where now Cockey's approximate Green's function is used corresponding to the LHS of equation (3.7) rather than the classical Green's function (as Hounjet has used) corresponding to the LHS of equation (3.1). Presumably the advantage of using the hybrid approach [i.e., equation (3.7) rather than equation (3.1)] is that the RHS of equation (3.7) is usually smaller than the RHS of equation (3.1) and thus the region in the flow field which is included in Hounjet's approach may be smaller and thus the resulting computer code will be more efficient.

We note finally that this approach (Hounjet's original method or the hybrid approach suggested here) is reminiscent of Lighthill's theory of jet noise (Goldstein, 1976) except that here, of course, the RHS in any of its several possible forms is known exactly.

6.4. Nonuniqueness, transient decay times, and mean values for unsteady oscillations in transonic flow

6.4.1. Early Work

Kerlick and Nixon (1981) have made the important point that, when using a finite difference, time marching computer code to investigate the lift on an oscillating airfoil in transonic flow, it is necessary to carry the solution sufficiently far forward in time that an essentially steady state solution is obtained. Moreover, they offer a method for estimating the transient time before the steady state is reached. They note that, if one stops the time marching solution before the transient is complete and the steady state is reached, then one may reach the incorrect conclusion that a change in the mean lift has occurred due to the oscillating motion of the airfoil when in fact no such change has occurred.

Nevertheless, what is perhaps surprising is that for a narrow Mach number range the time for the transient to decay and a steady state to be reached is extraordinarily long. Moreover, for a very narrow range of Mach number a non-zero mean value of lift can occur for an airfoil of symmetrical profile oscillating about a zero angle of attack.

6.4.2. Recent Work

In Kerlick and Nixon (1981) a NACA 64A006 airfoil was studied using the LTRAN2 computer code for a freestream Mach number of .875 with the airfoil oscillating at a peak angle of attack of .25° and with various reduced frequencies, k. It was shown that typically up to six cycles of airfoil oscillation must be considered for the mean lift to be less than 1% of the corresponding oscillatory lift peak value. Dowell et al. (1983) (also using the LTRAN2 computer code) examined various airfoils, 64A006, 64A010, MBB-A3, at various Mach numbers and reduced frequencies. The calculations in Dowell et al. (1983) were carried forward in time through six cycles of airfoil oscillation. As was found by Kerlick and Nixon, the results of Dowell et al. (1983) for the (symmetric) 64A006 airfoil showed the mean or average lift with the airfoil oscillating to be essentially unchanged from its value for no airfoil oscillation (i.e., zero) for most Mach numbers studied after six cycles of oscillation. However, at $M_\infty = .88$ and particularly at $M_\infty = .9$ this was not the case. Hence, the present author incorrectly concluded that a change in mean or average lift had occurred. The correct conclusion is that at $M_\infty = .9$, .88, many cycles of oscillation (> 40) are required for the mean lift to decay to essentially zero.

The results presented in Fig. 7 are for $M_\infty = .9$, $k = .2$ and various peak oscillatory angle of attack (Dowell et al., 1983). They are for the lift coefficient; similar results (not shown for brevity) were obtained for moment (about mid-chord) coefficient and shock displacement (normalized by airfoil chord).

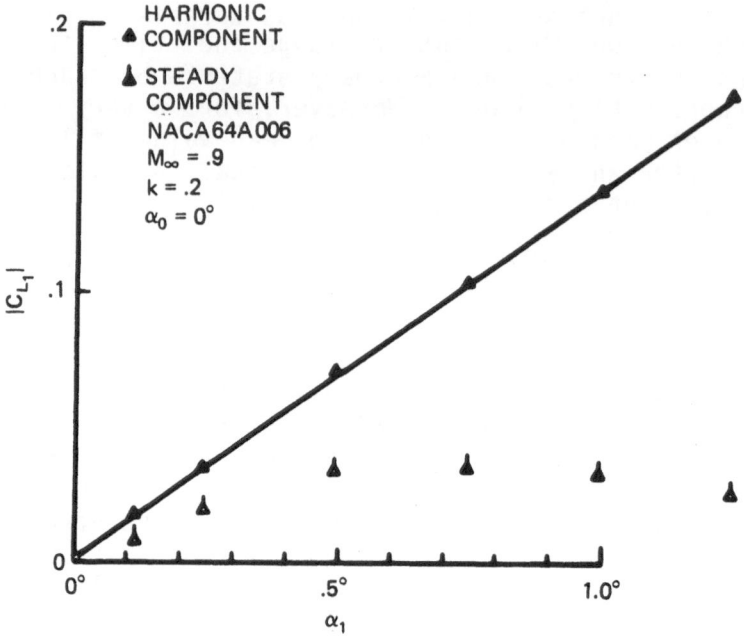

Fig. 7. Steady state and first harmonic lift components vs. dynamic angle of attack, NACA
64A006, M_∞ = .9, k = .2, α_0 = 0°.

Both mean steady values and first harmonic are shown
after six cycles of oscillation. As can be seen the
change in mean lift at this Mach number can be as much
as 50% of the peak oscillatory lift. For M = 0.88 the
mean or average components are never more than 10% of
the first harmonic oscillatory components, and, hence,
these results are not displayed.

Inspired by the note of Kerlick and Nixon, the
case of k = .2 and a peak oscillating angle of .5° was
carried forward in time through many more cycles of
airfoil oscillationn for several M_∞. A typical result
is shown in Fig. 8 for M_∞ = .9 in terms of lift vs.
angle of attack through 20 cycles of oscillation.
Time varies along the (hysteresis) curve. As may be
seen after twenty cycles, the average lift is still
distinctly different from zero and, moreover, is a
substantial fraction of the oscillatory lift. The
average lift is defined as

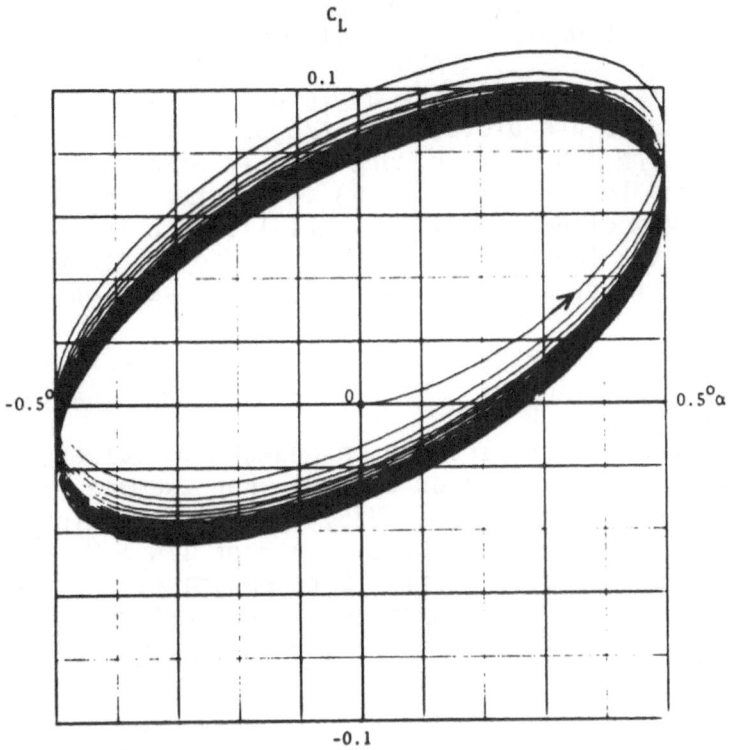

Fig. 8.   NACA 64A006 airfoil, M_∞ = 0.9, k = 0.2. Hysteresis curve of lift vs. dynamic angle of attack.

$$C_{L_{AVG}} = \frac{C_L^+ + C_L^-}{2}$$

and the oscillatory lift as

$$C_{L_{OSC}} = \frac{C_L^+ - C_L^-}{2}$$

where C_L^+, C_L^- are peak values, positive and negative respectively, of any two adjacent peaks.

The essential results can be summarized compactly in Fig. 9 where $C_{L_{AVG}}$ is plotted vs. 1/N, where N is the cycle number of the angle of attack oscillation. The results are plotted vs. 1/N so that the limit as N → ∞ or 1/N → 0 may be more readily examined. As may be seen for M_∞ = .86 and .875, the average lift rapidly declines and is essentially zero for N > 10 or 1/N < .1. (However it should be noted that the oscillatory lift, $C_{L_{OSC}}$, converges much more rapidly than the than the average lift. For the sake of brevity, $C_{L_{OSC}}$ is not shown.) By contrast for M_∞ = .9 and .885 the average lift persists in measurable values through a much larger number of cycles. For example, for M_∞ = .9 at N = 40 or 1/N = .025, there is a clear trend toward zero average lift as N → ∞ or 1/N → 0, but at N = 6 this would be difficult to perceive. Of course, these results have important practical consequences. A run for N = 40 takes 1672 sec of CPU time on an IBM 3033.

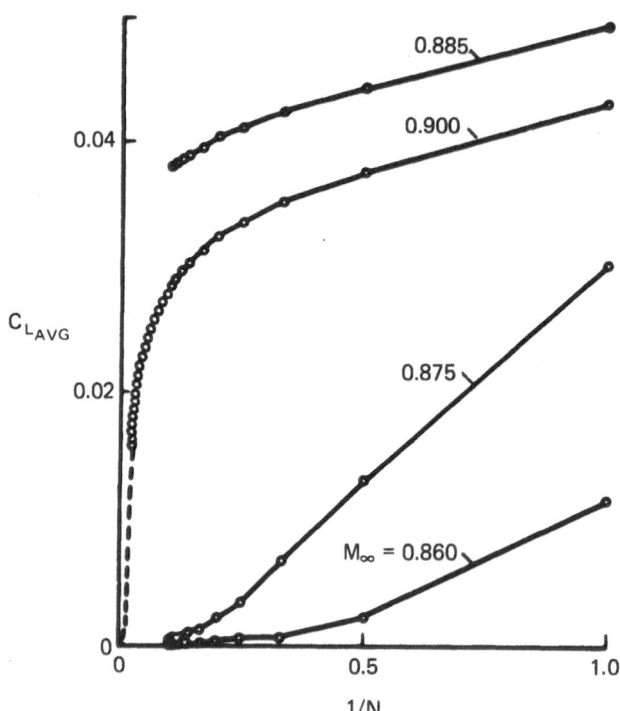

Fig. 9. Average lift vs. the inverse of the number of cycles of oscillations.

Hence, at this point the present author concluded that while a non-zero average lift is mathematically possible for a nonlinear aerodynamic system responding to an oscillating angle of attack, no such lift was observed using LTRAN2 for the range of parameters studied. However, at some Mach numbers the time for the average lift to decay to zero is extraordinarily long.

Next the present author greatly benefitted from a discussion with Dr. Peter Goorjian [19]. He had carried out calculations (Ballhaus and Goorjian, 1977) at M_∞ = .89 which indicated a non-zero average lift did occur. Dr. Goorjian's results are more fully discussed in Dowell et al. (1983), where an extended version of the present account is given. Hence, the present author and his colleague, Dr. Ueda, also carried out calculations at this M_∞ and the results are shown in Fig. 10 (analogous to Fig. 8) and Fig. 11

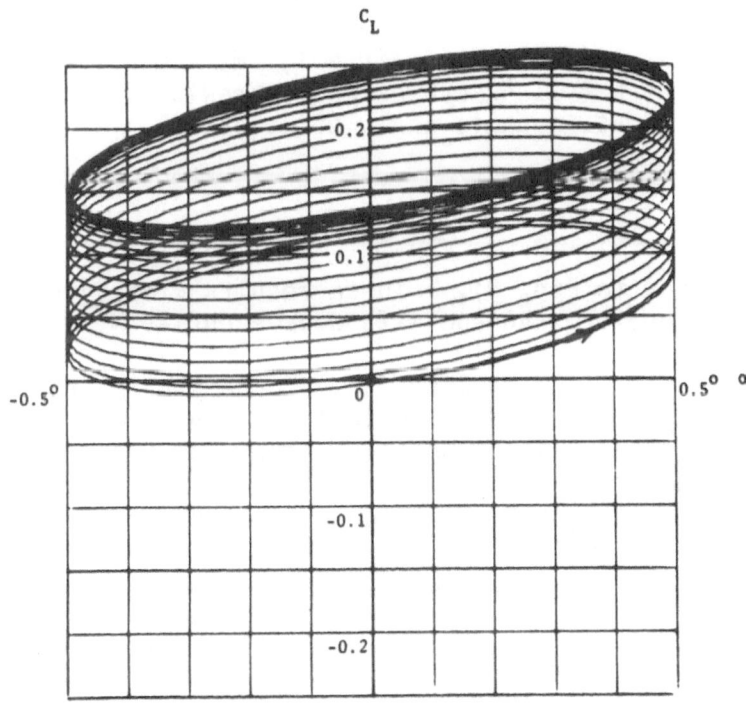

Fig. 10a.   NACA 64A006 airfoil, M_∞ = 0.89, k = 0.2, N = 1-20 cycles.   Hysteresis curve of lift vs. dynamic angle of attack.

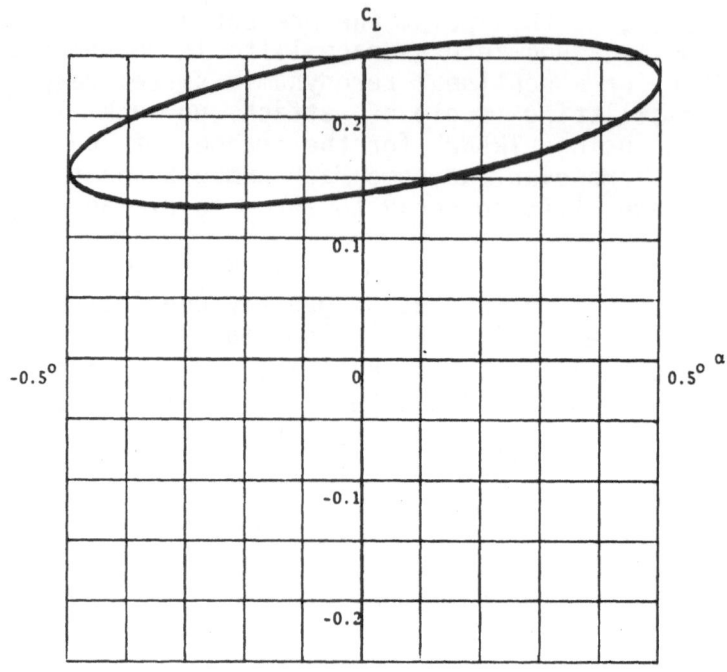

Fig. 10b.   NACA 64A006 airfoil, M_∞ = 0.89, k = 0.2, N = 21-40 cycles. Hysteresis curve of lift vs. dynamic angle of attack.

(analogous to Fig. 9).   These results clearly suggest that a non-zero average lift does occur at M_∞ = .89.

As explained by Goorjian, if one starts the airfoil oscillation with an initial negative angle of attack rather than a positive one, the mean lift is correspondingly negative (with the same magnitude) rather than positive. Indeed, the entire hysteresis curve is a double mirror image with C_L and α both undergoing sign inversions. Moreover, if the oscillating airfoil is now rendered motionless at exactly zero angle of attack, the mean lift persists at some finite, nonzero value. This implies that three solutions for lift occur at this single (zero) angle of attack. This result is fully consistent with that of Steinhoff and Jameson (1981) who obtain nonunique (multivalued) steady flow solutions by direct calculation for the full potential equations.

More recently, Salas (1983) has studied the full potential equations and the Euler equations for steady flow. For similar conditions he has found nonunique solutions for the full potential equations, but not for the Euler equations. However, it is very difficult to prove by numerical calculations, the absence of nonunique solutions for all possible conditions of interest.

Questions still remain of course.

- Why does this nonzero average lift occur only over a narrow range of Mach number? Note that the Mach number, $M_\infty \simeq .89$, at which nonunique solutions are observed corresponds to the Mach number for which the range of linear behavior is smallest as $k \to 0$. See Fig. 11.

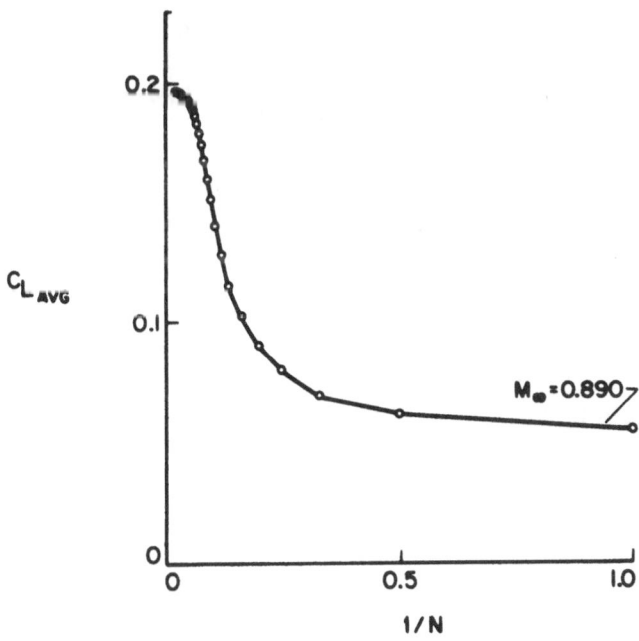

Fig. 11. Average lift vs. the inverse of the number of cycles of oscillations. $M_\infty = .89$.

- At what level of mathematical modeling, if any, do nonunique solutions no longer occur?

- Is the result physically significant? In particular, what would be the counterpart, if any, for a viscous fluid model?

6.4.3. Studies of Williams et al. and Salas et al.

Other studies, which have provided at least partial answers to these questions and added to our knowledge concerning the nonuniqueness of the potential flow solutions, include those of Williams, Bland, and Edwards (1985) and Salas and Gumbert (1985). Williams et al. have studied the stability (instability) of the multiple static equilibrium solutions predicted by potential flow theory (zero lift, positive and negative lift) and shown that, as expected, the zero lift solution is unstable. Furthermore, they have investigated the long transient decay times associated with the approach to the two stable, non-zero lift, static equilibria. The long transient times are not totally unexpected, since, even in the classical fully linear aerodynamic theory, the decay times are long as the freestream Mach number nears one. Here it is the average local Mach number that is near one.

They also have combined a potential flow model with a simplified (lag-entrainment) boundary layer theory to see if viscous effects remove the multiple equilibria and restore the stability of the zero lift solution. Apparently they do not.

Finally, they have noted that the small-disturbance potential flow equations are a consistent physical model and should represent a proper limit of the Euler equations as the thickness of the airfoil tends toward zero. Unfortunately present finite difference codes have difficulty solving the Euler equations in this limit.

Salas and Gumbert (1985) have computed steady flow results from potential flow theory for several (relatively thick) airfoils over a range of Mach num-

ber. They show that the predicted nonuniqueness is
common if the angle of attack is sufficiently large.
However, the nonuniqueness only occurs over a limited
range of Mach number at zero angle of attack.

For one airfoil, NACA 0012, at two Mach numbers
M_∞ = .67 and .75, Euler solutions are also provided
by Salas and Gumbert and compared to those from potential flow theory. See Figs. 6 and 7 of their paper.
The linearity of lift with angle of attack as predicted by the Euler code is typical for thick airfoils
and is remarkable. It is argued by Salas and Gumbert
that the deviation of the results of potential theory
from those of the Euler code is an indication of the
failure of the potential theory. This may well be;
however, such an explanation still does not address
the predictions of nonuniqueness by the potential flow
model for thin airfoils at nearly zero angle of
attack.

Fortunately (or otherwise depending on one's
point of view), most airfoils of practical interest
are of a certain thickness where the Euler codes
apparently give reliable answers and where nonuniqueness has not been observed. Moreover, and perhaps
more importantly, recently published results by
Williams (1986) show that for three-dimensional flows
over wings of moderate to small aspect ratio, the nonuniqueness observed in the two-dimensional flows over
airfoils is absent.

6.4.4. Aileron buzz (flutter) and the Steger-Bailey
calculation

No discussion of unsteady transonic aerodynamics
and aeroelasticity would be complete without a summary
of this pioneering study by Steger and Bailey (1980).
They used an implicit finite difference computer code
to simulate the fluid dynamic behavior of a two-layer
algebraic eddy viscosity model of the Navier-Stokes
equations for the flow about a NACA 65-213 airfoil
with an oscillating control surface (aileron). Using
a simple (one-degree-of-freedom with inertia only)
dynamic model of the control surface itself, they

determined the stability (flutter) boundary of the control surface and also limit cycle behavior amplitude and frequency during the flutter oscillation.

In earlier experimental work, Erickson and Stephenson (1947) had concluded that aileron buzz is a one-degree-of-freedom flutter in which shock-wave motion causes a phase shift in the response of the control surface hinge moment to the aileron motion. They observed oscillatory aileron motion for various combinations of Mach number, geometry and mechanical damping.

For the specifics of the fluid dynamic modeling and computational method, Steger and Bailey (1980) should be consulted. Here we focus on the aeroelastic results.

Representative results are shown in Fig. 7 and 10 of their paper. In Fig. 7 the transient oscillation of the aileron is shown for M_∞ = .82, a mean airfoil angle of attack of α = -1.0°, and an initial control surface angle of δ_a (t = 0) = 4°. As can be seen, after several cycles of oscillation the aileron approaches an apparent steady state limit cycle oscillation. On the other hand, for δ_a (t = 0) $\cong$ 0°, the aileron oscillation decayed to zero after a transient motion and no steady state oscillation was observed. As the Mach number was increased to M_∞ = .83, however, a steady state oscillation was observed even for δ_a (t = 0) $\cong$ 0°. At lower Mach numbers, M_∞ = .76 and .79, no steady state oscillation was observed even for δ_a (t = 0) = 4°.

Such results were used to construct a stability boundary as shown in Fig. 10 of their paper which was compared to that found experimentally. The comparison is reasonably good. Moreover, the (limit cycle) amplitude, and frequency of the flutter oscillation predicted by the calculation was in reasonable agreement with that measured, as well. Of course, the demonstrated sensitivity of the limit cycle motion to initial conditions gives some reason for caution in comparing calculated and experimental results.

Steger and Bailey (1980) also carried out inviscid calculations. For α = -1.0° they found flutter occurred at M_∞ = .84, but not at .83. Recall that with viscous effects included flutter occurred at M_∞ = .83 and .82 (but not at .79). Hence, there is some difference between the two calculated results, inviscid and viscous, with respect to the predicted onset of flutter (the stability boundary). More important, however, the inviscid calculation failed to predict a steady state oscillation, instead only showing an exponentially, diverging or decayig oscillation. Thus, the dominant nonlinear effects leading to a finite limit cycle amplitude appear to be due to viscosity. As Steger and Bailey concluded, "From these data we find that while inviscid unsteady shock wave motion is the driving force of transonic aileron buzz, the viscosity is nevertheless crucial and can both sustain and moderate the flap motion."

They continued, "Finally for the aileron held fixed at a higher Mach number, M_∞ = .85, we find that the viscous flow [itself] does not reach a steady state [here they really mean, static equilibrium] but buffets [reaches an oscillatory steady (or chaotic?) state]...if the aileron is then released, it no longer oscillates in a simple sinusoidal motion. Viscous effects appear to be much more dominant Lat this Mach number] and change the frequency and amplitude of aileron motion. The motion appears to repeat every fourth oscillation [the motion is periodic with two dominant frequencies in the ratio of 4 to 1]...".

Altogether these results are most fascinating and one can only hope that more such will be forthcoming.

6.5. Effective, efficient computational approaches for determining aeroelastic response using unsteady transonic aerodynamic codes

6.5.1. Various computational approaches and their relative merits

The basic issue is whether

244

- one should run an aerodynamic code for prescribed airfoil motions and store the resultant aerodynamic data prior to carrying out a
  flutter analysis (Option I)

or whether

- one should run an aerodynamic code in conjunction with a structural dynamics code to determine simultaneously the time history of the
  airfoil and aerodynamic forces (Option II)?

There are, of course, two types of aerodynamic
computer codes available, (1) those which calculate in
the time domain (Ballhaus and Goorjian, 1977, 1978;
Houwink and van der Vooren, 1980; Hessenius and Goorjian, 1982; Borland and Rizzetta, 1982a, 1982b;
Borland et al., 1982; Rizzetta and Borland, 1983), and
(2) those which calculate in the frequency domain
(Williams, 1979a, 1979b; Hounjet, 1981b; Ehlers and
Weatherill, 1982). The use of a code of the type (2),
of course, precludes the pursuit of option II. However, the use of a computer code of type (1) permits
the use of option I or II.

Below, each type of computer code will be considered in terms of how it may be used most effectively,
and for a type (1) code the relative merits of options
I and II will be discussed. Until otherwise noted, it
will be assumed that we are anticipating a linear
flutter analysis that only seeks to determine the
conditions for the onset of flutter. A type (2) aerodynamic code implicitly assumes this to be the case.
Again, as will be discussed later, use of a type (1)
aerodynamic code will permit either a linear or a nonlinear flutter analysis to be conducted.

6.5.2. Type (1) Aerodynamic Code-Time Domain

Assume that flutter solutions are to be found for
P parameter combinations (e.g., dynamic pressure
values, tip tank masses, control surface stiffnesses)
and M structural modes. Moreover, assume that T_A is
the computational time it takes for the aerodynamic
code to reach a steady state lift value for a pre-

scribed airfoil motion and T_F is the computational time it takes for a simultaneous fluid - structural time marching calculation to complete a transient. The relative sizes of T_A and T_F depend on airfoil profile and Mach number (for T_A) and structural damping, (T_F). Generally, $T_F > T_A$, but there are exceptions. For simplicity, think of the Mach number as fixed, since calculations at several M_∞ will simply increase all computations by the same factor.

Consider now the relative merits of options I and II.

Option I: Generate and Store Aerodynamic Data Prior to Aeroelastic Calculation

The total computational time to generate the aerodynamic forces is

$$M * T_A \qquad (5.1)$$

which is independent of P. There is some additional time required for flutter solutions per se, but it is assumed this is negligible compared to the time required to generate the aerodynamic forces.

Option II: Generate Aerodynamic Data and Structural Data Simultaneously

The total computational time will be

$$P * T_F \qquad (5.2)$$

which, of course, is independent of M. Clearly, which of the two options is most attractive depends on whether $M * T_A \lessgtr P * T_F$. For option II to be more attractive, the number of modes, M, should be somewhat larger than the number of parameters, P. Thus option II will be more attractive in a design verification study while option I will tend to be more attractive in a preliminary design phase.

6.5.3. Type (2) Aerodynamic Code-Frequency Domain

Here only option I has been used to date; however, see the discussion below. Let NF be the number of reduced frequencies needed for the flutter analysis. Let T_{AF} be the time for the aerodynamic code to determine the aerodynamic forces for one frequency. Assume that an aerodynamic influence coefficient approach is used so that the number of modes does not influence the computational time.*

Option I

The computational time to generate the aerodynamic forces is

$$NF * T_{AF} \qquad\qquad (5.3)$$

Compare equation (5.3) to the computational time associated with time domain aerodynamics [see previous discussion and equation (5.1)]:

$$NF * T_{AF} \overset{>}{\underset{<}{}} M * T_A$$

If T_{AF} and T_A are comparable (one might expect competition would tend to make them so), then the method of choice as between equations (5.1) and (5.3) will depend on the number of frequencies, NF, compared to the number of modes, M, needed in the aeroelastic calculations.

Option IA

Although this approach has not been pursued to date, one could take the frequency domain aerodynamic forces, curve fit them with Padé approximants or comparable representation (Dowell, 1980), use these to deduce a differential equation aerodynamic force representation (Tran and Petot, 1980; McIntosh [31]),

---

*If a relaxation scheme is used rather than direct inversion of the aerodynamic matrix, then another set of issues arises.

and then do a time marching flutter solution. The computational time would still be approximately

$$NF * T_{AF} \qquad (5.3a)$$

The comparable computational time using a direct time-marching aerodynamic code was (see previous discussion)

$$P * T_F \qquad (5.2a)$$

Assuming $T_{AF} = T_F$, one concludes that, if the number of parameters is large compared to the number of reduced frequencies needed, then the frequency domain aerodynamic method tends to be the method of choice over the time domain method which uses option II.

6.5.4. Summary Comparison

By comparing the estimates, equations (5.1), (5.2), (5.3), (5.3a), one may make an initial judgement as to the method of choice in a given situation.

6.5.5. Nonlinear Flutter Analysis

6.5.5.1. General Considerations

If a nonlinear flutter analysis is needed, then only the time domain aerodynamic method is available, type (1).

Of course, the aeroelastic calculation may be done in either the time or frequency domain. However, there is still a trade-off between options I and II. Now, for option I, a further multiplicative factor must be used which is the number of airfoil response levels, NR, which are of interest. For linear flutter analysis only one response level is of interest (strictly speaking an infinitesimal response level which approaches zero). Thus the computational times to compare are

Option I: NR * M * T_A

Option II: P * T_F.

It should be noted, moreover, that for nonlinear flutter analysis the number of parameters, P, will tend to be somewhat larger than for linear flutter analysis.

The relative attractiveness of the two options is as before but with a bias shift toward option II, be-cause of the factor NR appearing in option I. The use of option I in a nonlinear flutter analysis does in fact lead to a flutter analysis in the frequency domain, and the methodology by which that is done is described below in Section 6. Of course, this method-ology reduces to the classical frequency domain flut-ter solution method when the airfoil response levels are small.   The solution procedures for option II as currently practiced are straightforward and will not be elaborated upon further here. It is also worthy of note that

- a frequency domain nonlinear flutter analysis usually introduces approximations beyond those of a time domain analysis, but

- fortunately, linear flutter analysis will often suffice and hence the whole set of ques-tions is frequently moot.

6.6. Nonlinear flutter analysis in the frequency do-main and comparison with time marching solutions

The nonlinear effects of transonic aerodynamic forces on the flutter boundary of a typical section airfoil are discussed.   The amplitude dependence on flow velocity is obtained by utilizing a novel varia-tion of the describing function method that takes into account the first fundamental harmonic of the non-linear oscillatory motion.   By using an aerodynamic describing function, traditional frequency domain flutter analysis methods may still be used while including (approximately) the effects of aerodynamic nonlinearities. Results from such a flutter analysis are compared with those of brute force and periodic shooting time marching solutions. The aerodynamic forces are computed by the LTRAN2 aerodynamic code for a NACA 64A006 airfoil at $M_\infty = 0.86$.

6.6.1. Motivation and background

Recent developments in computational aerodynamics have led to renewed interest in the prediction of flutter boundaries of an airfoil in the transonic flow regime (Ballhaus and Goorjian, 1978; Yang et al., 1979; Isogai, 1980). Until recently, flutter calculations have either assumed the transonic aerodynamic forces could be approximated as linear functions of the airfoil motion so that traditional linear flutter analysis methods could be used or, alternatively, taken a brute force approach by simultaneously numerically integrating in time the structural and aerodynamic equations. The latter method does, of course, fully account for aerodynamic nonlinearities.

Ballhaus and Goorjian (1978) calculated the aeroelastic response of a NACA 64A006 airfoil with a single-degree-of-freedom control surface by simultaneously integrating numerically in time the structural equation of motion and also the aerodynamic equations. They used their own code, LTRAN2, for unsteady transonic flow. The indicial method, whereby an aerodynamic impulse function is first calculated by the aerodynamic code and then used in the flutter calculation via a convolution integral, was also studied. The indicial method assumes linearity of the aerodynamic forces with respect to airfoil motion. The flutter of the same airfoil but with two-degrees-of-freedom was analyzed by Yang et al. (1979) with aerodynamic forces obtained by three different methods. These forces were obtained by the time integration method, the indicial method (both of these employed the LTRAN2 code), and the harmonic analysis method in the frequency domain using the ULTRANS2 code. The latter method also assumes linearity of the aerodynamic forces. In general, all three methods agree well for the range of parameters studied by Yang. After the flutter boundary was obtained, the response was confirmed near the flutter boundary by simultaneous time integration of the governing structural and aerodynamic equations. Isogai (1980) studied the transonic behavior of the NACA 64A010 airfoil by using his own USTS transonic aerodynamic code, which can be applied to supercritical Mach numbers for reduced frequencies, $0 < k < 1.0$. (By contrast, the aerodynamic methods used by Yang were limited to small k.) Isogai used the time integration method for evaluating the aerodynamic forces, but then converted them to linearized harmonic forces for the flutter calculations. See Yang et al. (1981) for further discussion of previous work on flutter calculations including that of other investigators who have used brute force, simultaneous numerical integration of the structural and aerodynamic equations.

A discussion of when the aerodynamic forces may be treated as linear in the airfoil motion is given in Dowell et al. (1983) and previously in the present chapter. The analysis of Yang et al. (1979) and Isogai (1980) described above assumed linear characteristics for the aerodynamic forces in the flutter calculations. Linearity can be ensured if the amplitude of the airfoil oscillation is sufficiently small (Dowell et al., 1983), even though the governing fluid equations are inherently nonlinear for transonic flow fields. Yang et al. (1979) fixed an amplitude of pitching motion at 0.01 radian (0.574°), whereas Isogai (1980) used 0.1 in degrees for the computation of the aerodynamic forces. Dowell et al. (1983) pointed out that increasing the value of the reduced frequency increases the range of amplitude of oscillation for which linear behavior exists in transonic flow. However, the aerodynamic forces often begin to deviate from linear behavior for amplitudes of relatively small value such as 1.0° in pitching motion. Such amplitudes may be attained due to the disturbances an aircraft wing encounters during its flight. It is of importance, therefore, to clarify the aerodynamic nonlinear effect on a flutter boundary, especially when the nonlinear effect may create an aeroelastic softening system, i.e., the flutter speed decreases as the amplitude of oscillation increases. Such softening behavior may cause a dangerous unconservative estimation of a flutter boundary by linear analysis.

Here we study the nonlinear effect of transonic aerodynamic forces on a flutter boundary by utilizing a novel variation of the describing function method (Hsu and Meyer, 1968), which takes into account the first fundamental harmonic of the nonlinear oscillatory motion. By using an aerodynamic describing function, traditional frequency domain flutter analysis methods may still be used while including (approximately) the effects of aerodynamic nonlinearities. Brute force time marching calculations are also presented for comparison purposes.

The method used to calculate the describing functions is briefly this. A step change in angle of attack is specified and the transient aerodynamic force time history (calculated numerically by an appropriate aerodynamic code) is identified as a nonlinear impulse function. The Fourier transform of this impulse function (which in general depends on the step input level or amplitude) is taken as the aerodynamic describing function (nonlinear transfer function). Calculations have shown that this describing function agrees very well with the one determined by using a harmonic angle of attack input to the aerodynamic code. The latter method of calculation is, of course, much more expensive and time consuming for the range of frequencies needed in flutter analysis. The LTRAN2 computer code is used for determining the aerodynamic forces. However, any other nonlinear code could be used in a similar fashion.

6.6.2. Typical airfoil section

A typical airfoil section subjected to transonic flow is considered as shown in Figure 12. Since it can be assumed that the structural deformation is linearly dependent on the aerodynamic load for wings of ordinary modern aircraft during its normal flight, a linear structural transfer function is used. The

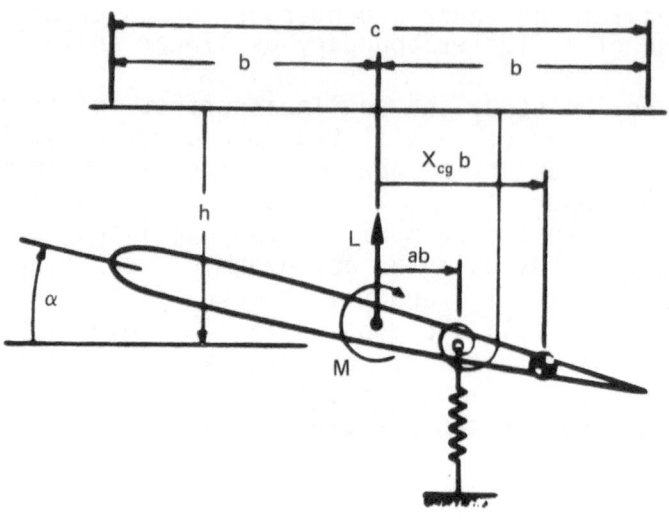

Fig. 12. Typical section airfoil.

aerodynamic force, however, may depend in a nonlinear
manner on the structural deformation in the transonic
flow range (Dowell et al., 1983). In order to include
the nonlinear effect of large (r) amplitudes of motion
on the aerodynamic forces and, hence, on a flutter
boundary, we use a nonlinear aerodynamic transfer
function by employing the describing function method.

6.6.3. Aerodynamic describing function

Here we give a summary of the relevant standard
describing function results and place the present
method in context.

If we assume that the frequency of motion is
relatively low, the aerodynamic forces due to the airfoil motion can be approximated as a function of the
effective induced angle-of-attack, which is given by

$$\phi = \alpha + \frac{\dot{h}}{u} c \qquad (6.1)$$

This quasi-steady approximation is compatible with the low frequency assumption in the LTRAN2 transonic unsteady aerodynamic code which we use in the present flutter calculation. Taking into account the nonlinear effects of the amplitudes of motion, we assume the aerodynamic forces take the form:

$$L = \frac{1}{2} \rho u^2 c C_L^N(\phi, \dot{\phi}), \qquad (6.2)$$

$$M = \frac{1}{2} \rho u^2 c^2 C_M^N(\phi, \dot{\phi}), \qquad (6.3)$$

where C_L^N, C_M^N are functionals of ϕ, $\dot{\phi}$, i.e., they may, in principle, include the complete time history of ϕ and $\dot{\phi}$.

For general periodic time dependent motion, the effective angle-of-attack ϕ can be expanded in a Fourier series as

$$\phi = \frac{\phi_0}{2} + \sum_n^N [\phi_{I,n} \cos(nk\tau) + \phi_{R,n} \sin(nk\tau)] \qquad (6.4)$$

According to the describing function method, only the first harmonic of ϕ is taken as an input to the aerodynamic force transfer function, i.e.,

$$\phi = \phi_1 \sin k\tau \qquad (6.5)$$

This input motion generates aerodynamic forces through the nonlinear fluid element, call it H, which, in general, includes higher harmonics. Thus,

$$C_L^N(\phi, \dot{\phi}) = \frac{C_{L_{1,0}}(\phi_1)}{2}$$

$$+ \sum_n^N [C_{L_{I,n}}(\phi_1)\cos(nk\tau) + C_{L_{R,n}}(\phi_1)\sin(nk\tau)], \qquad (6.6)$$

$$C_M^N(\phi,\dot\phi) = \frac{C_{M_{1,0}}(\phi_1)}{2}$$

$$+ \sum_n^N [C_{M_{I,n}}(\phi_1)\cos(nk\tau) + C_{M_{r,n}}(\phi_1)\sin(nk\tau)] \tag{6.7}$$

The describing function method, however, replaces the nonlinear element H by another nonlinear element H with the property that it operates on any sinusoidal input, equation (6.5), by passing its fundamental frequency in exactly the same manner as H; however, whatever the input frequency k, H generates no higher harmonics. This replacement allows us to write

$$\hat{C}_L^N(\phi,\dot\phi) = C_{L_{I,1}}(\phi_1)\cos(k\tau) + C_{L_{R,1}}(\phi_1)\sin(k\tau)$$

$$= D_{L_R}(\phi_1)\phi + D_{L_I}(\phi_1)\dot\phi/k, \tag{6.8}$$

$$\hat{C}_M^N(\phi,\dot\phi) = C_{M_{I,1}}(\phi_1)\cos(k\tau) + C_{M_{R,1}}(\phi_1)\sin(k\tau)$$

$$= D_{M_R}(\phi_1)\phi + D_{M_1}(\phi_1)\dot\phi/k \tag{6.9}$$

where

$$D_{L_R} = \frac{1}{\pi\phi_1} \int_0^{2\pi} C_L^N(\phi,\dot\phi)\sin(k\tau)d(k\tau), \tag{6.10}$$

$$D_{L_I} = \frac{1}{\pi\phi_1} \int_0^{2\pi} C_L^N(\phi,\dot\phi)\cos(k\tau)d(k\tau), \tag{6.11}$$

$$D_{M_R} = \frac{1}{\pi\phi_1} \int_0^{2\pi} C_M^N(\phi,\dot\phi)\sin(k\tau)d(k\tau), \tag{6.12}$$

$$D_{M_I} = \frac{1}{\pi \phi_1} \int_0^{2\pi} C_M^N(\phi, \dot{\phi})\cos(k\tau)d(k\tau) \qquad (6.13)$$

Using complex notation for equations (6.8) and (6.9) yields a more compact result, i.e.,

$$\hat{C}_L^N(\phi, \dot{\phi}) = D_{L_1}(\phi_1)\phi, \qquad (6.14)$$

$$\hat{C}_M^N(\phi, \dot{\phi}) = D_{M_1}(\phi_1)\phi \qquad (6.15)$$

where

$$D_{L_1}(\phi_1) = D_{L_R} + iD_{L_I} \qquad (6.16)$$

$$D_{M_1}(\phi_1) = D_{M_R} + iD_{M_I} \qquad (6.17)$$

In equations (6.14) and (6.15), the coefficients, $\hat{C}_L^N$ and $\hat{C}_M^N$, also have complex values whose real parts correspond to equations (6.8) and (6.9), respectively.

If the amplitude ϕ_1 is fixed, the equations (6.14) and (6.15) take a form identical to that for a linear system. This implies the applicability of the same stability analysis as that for a linear system.

The coefficients defined in equations (6.10)-(6.13) to construct the describing function can be computed by a time integration code for transonic flow. It is also possible to evaluate them from wind-tunnel experimental data measured on a harmonically (or impulsively) excited airfoil. In the present study, we utilize an extended nonlinear version of the indicial method (Ballhaus and Goorjian, 1978) to calculate the aerodynamic coefficients.

Since the describing function assumes the same form as a linear transfer function when the amplitude is fixed, we can regard a typical such element, H_ϕ, which relates any representative aerodynamic forces, F, to airfoil motion, ϕ, as a linear system with respect to variations in frequency.

$$F(\tau) = \hat{H}_\phi(ik, \phi_1)\phi_1 e^{ik\tau} \qquad (6.18)$$

This relation corresponds in the subsidiary (effectively frequency) domain of the Laplace operator to the following:

$$\bar{F}(s) = \hat{H}(s, \phi_1)\frac{\phi_1}{s - ik} \qquad (6.19)$$

If we put k = 0, then equation (6.19) represents an indicial response relationship.

$$\bar{\hat{A}}(s, \phi_1) = \hat{H}_\phi(s, \phi_1)\frac{\phi_1}{s} \qquad (6.20)$$

By using equation (6.20) we can obtain the describing function, $H_\phi(ik, \phi_1)$, from the indicial response to a step input with amplitude ϕ_1. Furthermore, if we neglect the effect of higher harmonics, an assumption already made in the describing function method, then $\hat{H}_\phi(ik, \phi_1)$ can be approximated by using the indicial response $\bar{A}(s, \phi_1)$ of the element H as

$$\hat{H}_\phi(ik, \phi_1) = \bar{A}(ik, \phi_1)ik/\phi_1 \qquad (6.21)$$

From a linear system, starting from (6.18) one may proceed through (6.19), (6.20), to (6.21) and vice versa by standard mathematical methods. However, as the careful reader will note this is not strictly possible for a nonlinear system, i.e., (6.19) and (6.20) follow from (6.18) only by analogy to linear system results. Indeed we may take (6.18) and (6.21) [or (6.20)] as two independent definitions of $\hat{H}_\phi(ik, \phi_1)$, the describing function that will be used in the flutter analysis. However, by numerical example we will show that, in fact, the two definitions lead to similar results. This is fortunate, because the less obvious definition, (6.21), is far easier to use in practice for generating aerodynamic forces to employ in flutter calculations.

6.6.4. Working form of the aeroelastic system equations

The governing structural equations of the system are given in nondimensional form by

$$\frac{\pi\mu}{2}\left(\frac{h}{c}\right)'' + \frac{\pi\mu}{2}\left(\frac{S_\alpha}{mc}\right)\alpha'' + \frac{\pi\mu}{2}\left(\frac{c\omega_h}{u}\right)^2\left(\frac{h}{c}\right) = -C_L^N, \quad (6.22)$$

$$\frac{\pi\mu}{2}\left(\frac{S_\alpha}{mc}\right)\left(\frac{h}{c}\right)'' + \frac{\pi\mu}{2}\left(\frac{I_\alpha}{mc^2}\right)\alpha'' + \frac{\pi\mu}{2}\left(\frac{I_\alpha}{mc^2}\right)\left(\frac{c\omega_\alpha}{u}\right)^2\alpha = C_{Me}^N$$

$$(6.23)$$

From (6.22) and (6.23) the structural transfer function for the state vector, $[(h/c)\alpha]^T$, is

$$G^{-1}(s,U) = \begin{matrix} \frac{\pi}{2}\{\mu s^2 + R^2/U^2\} & \frac{\pi\mu}{4}(x_{cg} - a)s^2 \\ \\ \frac{\pi\mu}{4}(x_{cg} - a)s^2 & \frac{\pi}{8}r_\alpha^2\{\mu s^2 + 1/U^2\} \end{matrix} \quad (6.24)$$

As to the aerodynamic describing function, we first assume the indicial response $A(r,\phi 1)$ to a step change in ϕ in (6.21) can be expressed in the following form for the lift and moment forces;

$$A_L(\tau,\phi_1) = a_0^L(\phi_1) + \sum_{i=1}^N a_i^L(\phi_1)e^{b_i^L\tau}, \quad (6.25)$$

$$A_M(\tau,\phi_1) = a_0^M(\phi_1) + \sum_{i=1}^N a_i^M(\phi_1)e^{b_i^M\tau} \quad (6.26)$$

where a_0^L and a_0^M are chosen to be identical to the steady state values for $\phi = \phi_1$, since every b_i is chosen to be a negative real number. The coefficients, a_i^L, a_i^M are determined by the least square method for fixed values of the b_i's. The b_i are selected to be in the vicinity of the negative of the k values for which the imaginary parts of the aerodynamic transfer function have extrema. This procedure for selecting the b_i is discussed in detail in Dowell (1980).

After determining the coefficients in equations (6.25) and (6.26), the indicial response functions can be written in the subsidiary domain of the Laplace operator (frequency domain) as,

$$\bar{A}_L(s,\phi_1) = \frac{a_0^L}{s} + \sum_{i=1}^{N} \frac{a_i^L}{s - b_i^L}, \tag{6.27}$$

$$\bar{A}_M(s,\phi_1) = \frac{a_0^M}{s} + \sum_{i=1}^{N} \frac{a_i^M}{s - b_i^M} \tag{6.28}$$

Then from equation (6.21), the aerodynamic describing function is obtained for the state variable ϕ as

$$\hat{H}_\phi(ik,\phi_1) = [\bar{D}_L(ik,\phi_1)D_M(ik,\phi_1)]^T \tag{6.29}$$

where

$$\bar{D}_L(ik,\phi_1) = ik\bar{A}_L(ik,\phi_1)/\phi_1, \tag{6.30}$$

$$\bar{D}_M(ik,\phi_1) = ik\bar{A}_M(ik,\phi_1)/\phi_1 \tag{6.31}$$

In order to construct the aerodynamic describing functions so that they are compatible with the structural transfer function, we must transform ϕ and C_M^N to those variables used in the structural equations of motion. The relationships for the state vectors and the moment coefficients are as follows:

$$\overline{\phi}(ik) = (ik \ 1 - \frac{a}{2} \ ik) \quad \frac{\overline{h/c}}{\overline{\alpha}} \ , \qquad (6.32)$$

$$c_{Me}^{N} = c_{M}^{N} + \frac{a}{2} \ c_{L}^{N} \qquad (6.33)$$

As the aerodynamic describing function $\hat{H}(ik, \phi_1)$ is defined by

$$\begin{array}{cc} - \overline{C}_{L}^{N}(ik, \phi_1) & \\ & = \hat{H}(ik, \phi_1) \quad \frac{\overline{h/c}}{\overline{\alpha}} \qquad (6.34) \\ - \overline{C}_{Me}^{N}(ik, \phi_1) & \end{array}$$

it becomes using (6.30)-(6.34)

$$\hat{H}(ik, \phi_1) = \begin{array}{cc} A_{11} & A_{12} \\ A_{21} & A_{22} \end{array} \qquad (6.35)$$

where

$$A_{11} = - \overline{D}_{L_1}(ik, \phi_1) \cdot ik$$

$$A_{12} = - \overline{D}_{L_1}(ik, \phi_1)(1 - \frac{a}{2} \ ik),$$

$$A_{21} = - \overline{D}_{L_1}(ik, \phi_1) \frac{a}{2} \ ik + \overline{D}_{M_1}(ik, \phi_1) \cdot ik,$$

$$A_{22} = \overline{D}_{L_1}(ik, \phi_1) \frac{a}{2} (1 - \frac{a}{2} \ ik) + \overline{D}_{M_1}(ik, \phi_1)(1 - \frac{a}{2} \ ik)$$

Using the structural transfer function equation (6.24) and the aerodynamic describing function equation (6.35), a self-sustained oscillation of the system shown in Fig. 2 is characterized by the equation

$$|G^{-1}(ik,U) - \hat{H}(ik,\phi_1)| = 0 \qquad (6.36)$$

Equation (6.36) corresponds to the so-called flutter determinant if the system is linear. For the present nonlinear system, (6.36) allows one to determine the amplitude of the flutter motion as a function of some system parameter, say airspeed, U.

6.6.5. Extension of the describing function

In the earlier discussion of aerodynamic describing functions, we assumed the aerodynamic forces can be given as functions of a single variable, ϕ. More rigorously, however, the upwash of the mean camber line of an airfoil is given by

$$-\frac{w}{u} = \phi + \frac{c\dot{\alpha}}{u}(x/c - 0.5) \qquad (6.37)$$

where the airfoil is located on $0 < x < c$. The second term in equation (6.37) includes the effect of the angular velocity of the airfoil motion, $\dot{\alpha}$. If we take into account this aerodynamic effect in the flutter analysis, the aerodynamic describing function must be determined separately for h- and α-motion, or alternatively for ϕ and α. Moreover, this procedure makes the describing function method less obvious as to its theoretical basis. Neglecting the effect of $\dot{\alpha}$, however, we encountered fictitious instabilities at high frequencies as well as a decrease of flutter speeds. To eliminate this artifact, an improvement was made to the aerodynamic describing function by adding the component that is derived from the second term of equation (6.37). This is based on the assumption that the $\dot{\alpha}$ effect compared to ϕ is generally secondary for the aerodynamic forces at low reduced frequencies, k, where nonlinear transonic aerodynamic effects are most significant. Thus, the components of the describing function, equation (6.35), are redefined by

$$A_{12} = A_{12} - \bar{D}_{L_2}(ik) \cdot ik$$

$$A_{22} = A_{22} + \bar{D}_{L_2}(ik) \frac{a}{2} \cdot ik + \bar{D}_{M_2}(ik) \cdot ik \qquad (6.38)$$

where A_{12} and A_{22} on the right hand side of equation (6.38) are those of equation (6.35). $\bar{D}_{L_2}$ and $\bar{D}_{M_2}$ are the components obtained from an indicial response of the second term of equation (6.37). For brevity, details are omitted.

6.6.6. Results and discussions

6.6.6.1. Aerodynamic results

All the original aerodynamic data in the follow-ing were computed by the LTRAN2 (Ballhaus and Goorjian, 1977) code for a NACA 64A006 airfoil with Mach number set to $M_\infty = 0.86$ where, characteristical-ly, transonic effects can be observed. The zero angle of attack, steady state shock stands roughly at mid-chord (Dowell et al., 1983). Results were also calcu-lated for a NACA 64A010 airfoil, but these are omitted here.

In this calculation a 113 x 97 finite difference mesh was employed. The non-dimensional time incre-ment, $\Delta\tau$, for integration was selected at $\pi/12$ to ob-tain indicial responses. The lift and moment forces at every five time increments were used to evaluate the coefficients in equations (6.25) and (6.26). In order to compare the aerodynamic forces with those ob-tained by the present extended nonlinear indicial method, the time integration for the airfoil undergo-ing harmonic excitation of pitching motion about mid-chord axis was also performed with 120 time steps per cycle at various reduced frequencies. The latter re-sults [using (6.18)] were compared to those obtained by the indicial method using the Laplace transform versions of equations (6.25) and (6.26), which derive from equation (6.21). In general, good agreement was obtained.

The indicial response for lift to step functions of different amplitudes, ϕ_1, are shown in Fig. 13. Generally, in this type of indicial response, a spike should appear near $\tau = 0$. However, the low frequency theory of LTRAN2 can not follow a piston-wave-type pressure change because of its infinite propagation rate (Ballhaus and Goorjian, 1978). Hence, in the present calculations the lift coefficient increases gradually from zero at $\tau = 0$.

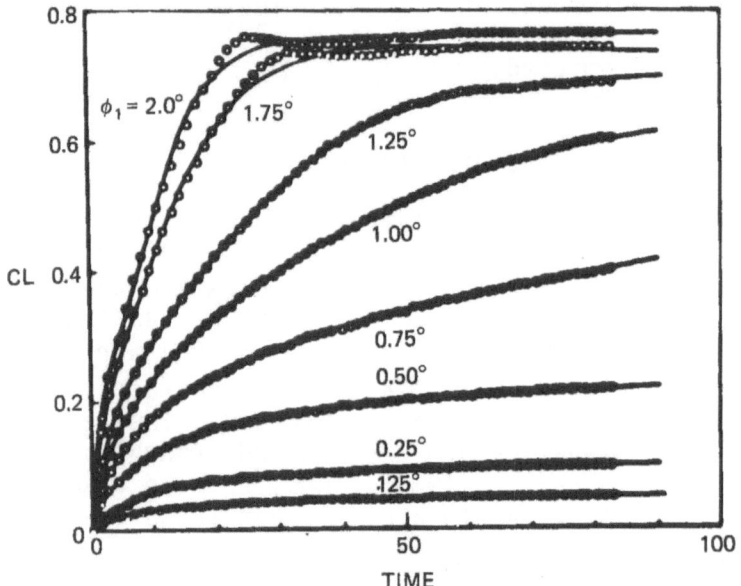

Fig. 13. Indicial responses computed by LTRAN2 and exponential curve fit: lift.

In the results of Figure 13, the curves pre-scribed by equation (6.25) are shown after the coefficients were determined by the least square method using 64 time data points for the indicial response at each amplitude. The b_i^L's and b_i^M's, are selected at six values (N = 6), which are -0.01875, -0.0375, -0.075, -0.15, -0.3, and -0.6. These results obtained from equation (6.25) are in excellent agreement with the indicial responses computed by the LTRAN2 code, especially for the lower amplitude values of step in-puts. Using the coefficients a_i^L, a_i^M, and equation

(6.21) we can obtain the elements of the aerodynamic functions, $\bar{D}_L$ (ik,ϕ_1) and $\bar{D}_M$ (ik,ϕ_1). The real parts of the former are shown in Fig. 14. They are plotted for reduced frequencies up to 0.3. Although the describing function for higher frequencies can be calculated by equations (6.30) and (6.31), they are no longer meaningful at those frequencies because of the low frequency limitation (Ballhaus and Goorjian, 1977) in LTRAN2.

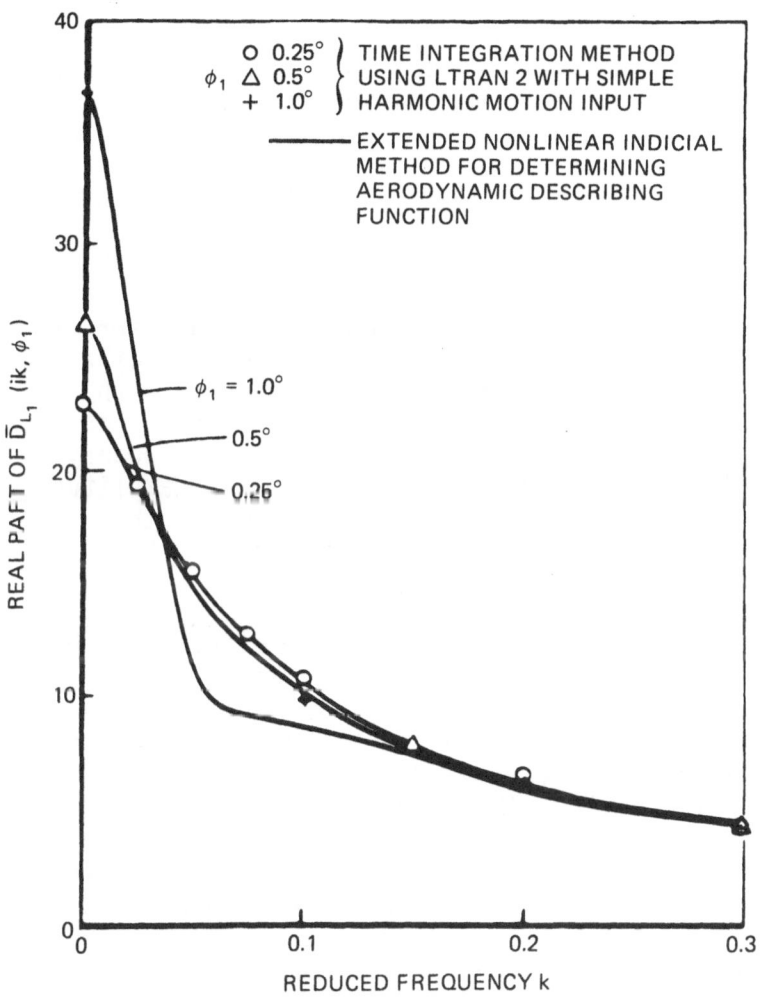

Fig. 14. Comparison of extended nonlinear indicial method with time integration method (NACA 64A006, M_∞ = 0.86).

In Fig. 14, the describing functions thus obtained are also compared with the results of the time integration method for simple harmonic motion inputs that use equations (6.10)-(6.13) [see also (6.18)]. The agreement between the two methods is generally satisfactory. However, it was seen that the agreement is better for smaller amplitude than larger ones, for lift than for moment, and for real part than for imaginary part (Ueda and Dowell, 1982, 1983). It should be emphasized that the extended nonlinear indicial method has substantially greater simplicity and efficiency in determining the aerodynamic describing function, as compared to the time integrationmethod for simple harmonic motion inputs.

6.6.6.2. Flutter results

Some flutter calculations have been done using these aerodynamic describing functions for typical section airfoils. First, the parameters of a typical section airfoil were chosen to compare with the results by Yang et al. (1980). A comparison is made in Ueda and Dowell (1982). The flutter boundary calculated by the present method for an amplitude, ϕ_1, between 0.5° and 1.0° agrees well with that obtained from the linear indicial method by Yang et al. (1980).

To investigate the amplitude effect on the flutter boundary, a typical section airfoil corresponding to case B in Isogai (1980) was considered next, although the results cannot be compared directly with those in Isogai (1980) due to the use of a different airfoil profile. The results for the flutter speed as well as for the reduced frequency, bending/torsion amplitude ratio and phase, are shown in Fig. 15. Those without the $\dot\alpha$ effect are depicted by dashed curves. In this case, the effect of angular velocity on flutter boundaries was very small (Ueda and Dowell, 1982; 1983).

As the aerodynamic describing function method invokes several assumptions, a fully nonlinear time marching solution was computed to verify the above results. The numerical integration scheme adopted for structural equations is the state transition matrix method which Edwards et al. (1982) recommended after examining seven different integrators for a similar calculation (I2 in Edwards et al., 1982). Three time marching calculations with different speed parameters have been carried out for the case shown in Figure 15.

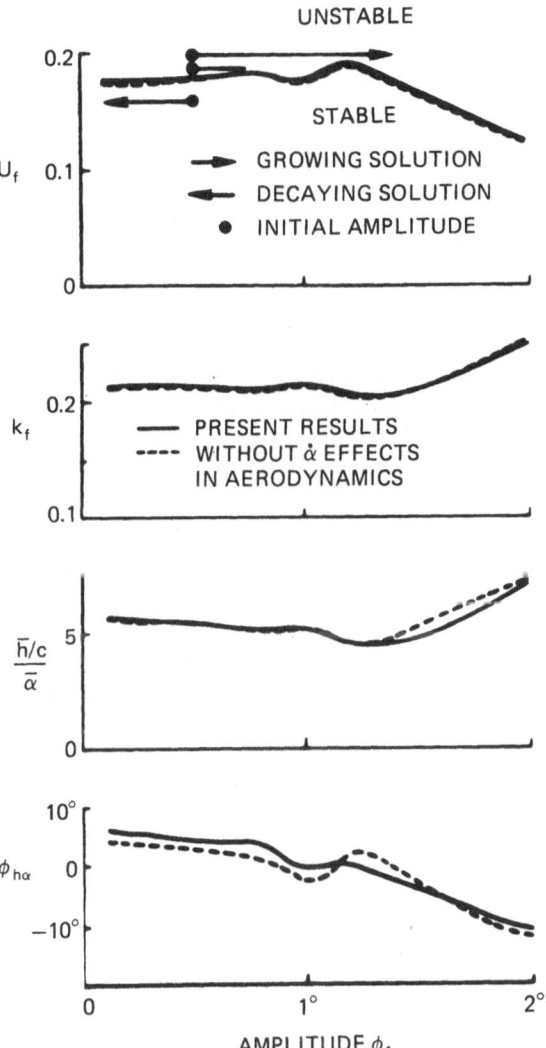

Fig. 15. Flutter parameters vs. amplitudes (NACA 64A006, M_∞ = 0.86, a = -0.3, x_{cg} = 0.24, r^2_{cg} = 0.24, R = 0.2, μ = 60).

The initial state vector was determined from the flutter solution of the describing function method with $\phi_1 = 0.5°$, namely, $U_F = 0.1798$, $k_F = 0.2126$, $|\bar{\frac{h}{c}}/\bar{\alpha}| = 5.446$, and $\phi_{ha} = 4.24°$. If we choose the initial time, $\tau = 0$, as the instant when $\alpha = 0$, the flutter solution gives the initial state vector for the time marching as $\underset{\sim}{x} = (0.03164, 0.00583, -0.00049,0)^T$. The time increment for integration was selected as $\Delta\tau = 0.25$, which, considering the flutter reduced frequency, corresponds roughly to 120 steps per cycle. Although the initial state vector is determined from the describing function flutter motion, the time marching is started from a steady-state initial condition of the airfoil at a static angle of attack for the aerodynamic calculations. For example, the initial effective induced angle of attack becomes 0.305 degrees for this case. It should be noted that the second term of equation (6.37) vanishes at the initial upwash since the starting time is set at the instant when the angular velocity becomes zero. The time marching is continued up to $\tau = 250$, which contains one thousand time steps. The variations of the amplitude ϕ_1 of these solutions are shown in Fig. 15 and time histories in Fig. 16. At $U = 0.16$, the airfoil shows decaying motion, whereas the oscillation is growing at $U = 0.2$. At $U = 0.19$, the oscillation is almost neutrally stable although it is slightly growing. The changes of the peak values in the effective angle of attack of these oscillations are illustrated in Fig. 15. The solid line shows the flutter boundary (limit cycle curve) obtained by the describing function method. In this case, the flutter boundary is nearly horizontal at small amplitudes. As can be seen from the figure, the results from the describing function method agree well with those of the time marching solution. Furthermore, the last cycle of the time marching solution at $U = 0.19$ gives the values of $\phi_1 = 0.723°$, $k = 0.212$, $|\bar{\frac{h}{c}}/\bar{\alpha}| = 5.18$, and $\phi_{ha} < 5°$. The agreement of these values with the results in Fig. 15 is excellent. It should be noted that the damping in the time marching solutions is attributed to the aerodynamic forces, since we use no structural damping nor artificial damping due to

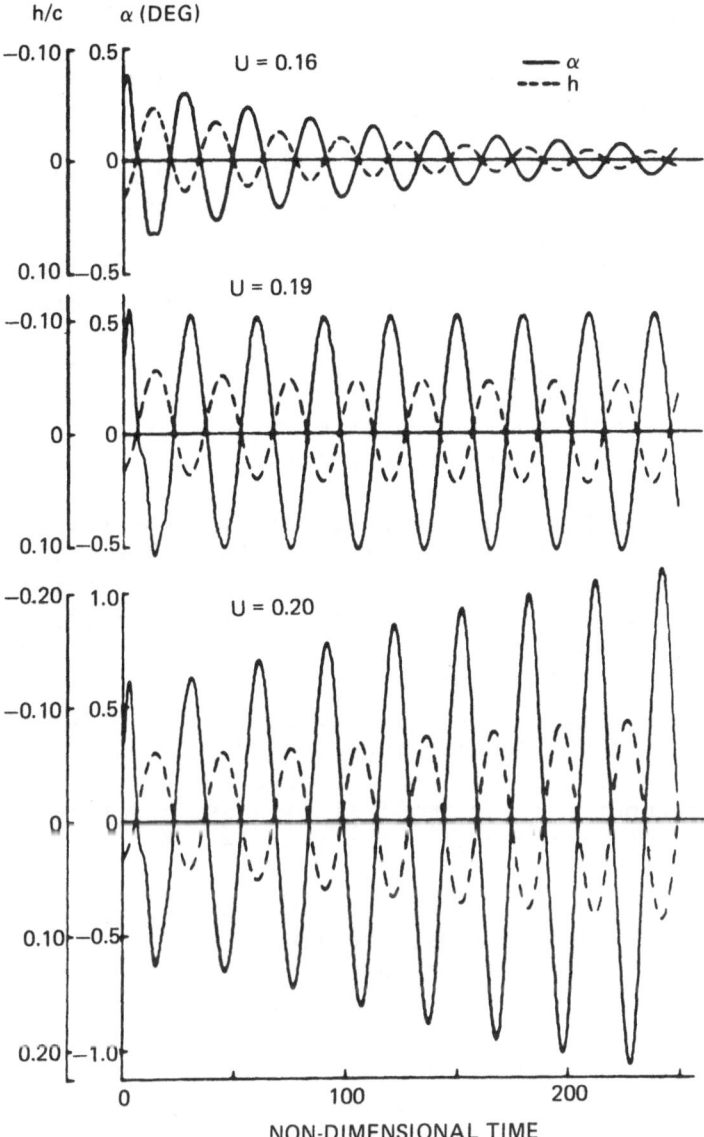

Fig. 16. Time history of time marching solutions.

numerical integration schemes. It is known that the transition matrix integrator gives exactly neutral solutions for free structural vibration irrespective of its time step size.

Since the nonlinear effect is most important at relatively low reduced frequencies (see Fig. 14), the center of gravity was next placed at x_{cg} = -0.25 and the frequency ratio at R = 0.1 in order to obtain a distinctly nonlinear effect. The results are shown in Fig. 17. On those portions of the curve where amplitude increases with airspeed, a stable nonlinear limit cycle is predicted. On those portions of the curve where amplitude increases with decreases in airspeed, an unstable limit cycle occurs.

Further time marching calculations have been performed to confirm the limit cycle. The initial state vector to start time integrations is varied proportionally to that of the flutter solution of the describing function method with the amplitude, ϕ_1 = -.25°, which gives x = (0.04353, 0.00562, -0.005726,0)^T for ϕ_1 = 0.25°. This state vector yields the effective induced angle of attack, -0.0063° at τ = 0. Since the reduced frequencies of flutter are about 0.1 for small amplitudes, the time step size of integration was chosen as $\Delta\tau$ = 0.5.

As the solid curve of the describing function in Fig. 17 anticipates, the limit cycle flutter is also shown by the time marching solutions for small amplitudes. The amplitudes, reduced frequencies, amplitude ratio, and phase angles of these limit cycle oscillations were also calculated from the time history of the solutions and are depicted with open circles in the figure. Connvergence to the limit cycle is determined by changing the initial amplitude. For example, the time history with two different initial amplitudes at U = 0.6 is shown in Fig. 18. Both oscillations converge to the same limit cycle with the amplitude ϕ_1 = 0.355° beyond τ = 400. However, the average displacement for the h-motion at least up to τ = 500 is different for the two initial conditions. This slow convergence for the average displacement to the neutral position may be attributed to the small frequency

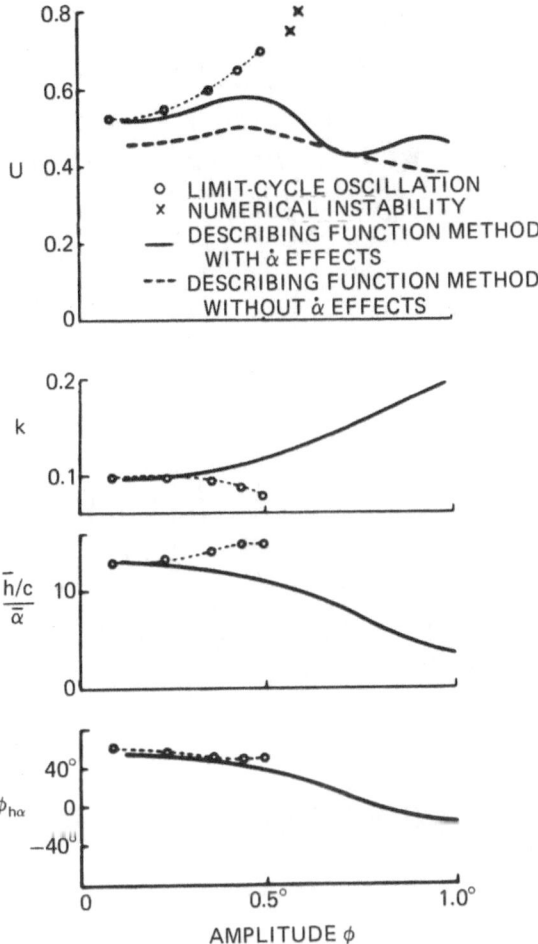

Fig. 17. Limit cycle oscillations (NACA 64A006, M_∞ = 0.86, a = -0.3, x_{cg} = 0.2, r^2_{cg} = .24, R = 0.2, μ = 60).

ratio, R, which implies weak stiffness of the structure against the h-motion. It should be noted that the average translational displacement of an airfoil has no effect on the aerodynamic forces.

The results of the time marching solution can be compared with those obtained by the describing function method in Fig. 17. For small amplitudes, the

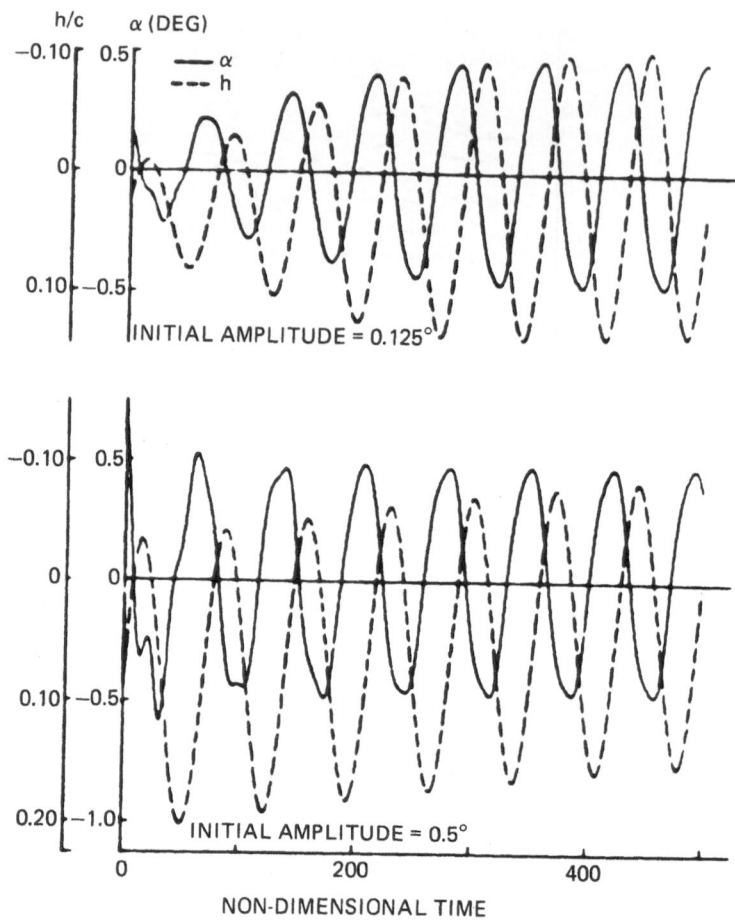

Fig. 18. Time history of limit cycle flutter, (U =
0.6, ϕ_1 = 0.355°, k_F = 0.089, $|\bar{h}/c/\bar{\alpha}|$ =
14.4, $\phi_{h\alpha}$ = 49.7).

agreement of the results is satisfactory. The reduced
frequency of the limit cycle by the time marching
solution, however, decreases as the flutter amplitude
increases, while the describing function method predicts monotonically increasing reduced frequencies.
Furthermore, the time marching solution shows stable
limit cycle flutter up to the speed of U = 0.7, whereas the describing function method predicts no stable
limit cycle solution above U = 0.58. At U = 0.75 and
0.8, the time marching solutions with the initial

amplitude of ϕ_1 = 0.5° are terminated by a numerical instability of the aerodynamic calculations. The peak values of ϕ just before the occurrence of the instability are plotted with the symbol x. This kind of difficulty was frequently encountered when an initial amplitude of more than 0.5 degrees was used in the time marching calculations. Possibly for this reason we failed to detect the divergent unstable limit cycle flutter with larger amplitudes which is predicted by the describing function method.

6.6.6.3. Study by Peters and Ventura

The work of Peters and Ventura (1985) has added very substantially to our understanding of the apparent differences in the results from the time-marching and describing function methods. They have used yet a different flutter solution method, the periodic shooting technique. Briefly it is this. In the periodic shooting method, one seeks a periodic solution by choosing initial conditions and guessing the period of the motion. A time marching solution is then carried out over the assumed period. If at the end of the assumed period, the system conditions are equal to the assumed initial conditions, then a periodic solution has been found. Usually, this will not be the case and an iteration procedure must be used to converge to a set of initial conditions and period that correspond to a steady state motion. This method has two advantages over the other approaches.

- Relative to a conventional time marching solution, which can only determine stable periodic motions, periodic shooting determines both stable and unstable periodic motions.

- Relative to a describing function approach, there is no assumption of simple harmonic motion, and the solutions are essentially exact.

The unexpected result from the periodic shooting solutions was that this method essentially reproduced the results previously obtained by the describing function method and differed in some respects from those results found by a conventional time marching solution. Consider again Fig. 17.

In the plot of U vs ϕ_1 the solid line is the result obtained by <u>both</u> the describing function and periodic shooting solutions. They both show that above a certain flow velocity, U = .58, no periodic motions are possible.

How then does one explain the apparent periodic motions found by the conventional time marching solutions for some U > .58? The answer is now known from the work of Peters and Ventura (1985). As the time marching solution is calculated from some initial conditions for, say, U = .6, the motion is oscillatory and appears to reach a periodic solution for ϕ_1 = .35 after a hundred pseudo-cycles of oscillation or so. At this U and ϕ_1, one is near a true periodic motion which occurs for a slightly smaller U (see Fig. 17) and the growth rate of the oscillation is small. However, if the calculation is continued over several hundred pseudo-cycles, ϕ_1 does increase and as it moves well beyond .35 the rate of growth becomes more rapid and it is clear that for U = .6, in fact, the motion becomes unbounded. This result is important for

- the caution it provides with respect to carrying out and interpreting conventional time marching solutions

and

- its possible implications for conducting and interpreting wind tunnel experiments or flight tests.

6.6.7. Conclusions

An extended nonlinear indicial approach to modeling nonlinear aerodynamic forces for aeroelastic analyses has been developed. The basic approach is based upon describing function ideas.

The flutter boundaries obtained by the describing function method are generally verified by periodic shooting and time marching solutions for sufficiently small amplitude flutter motion. Hence, the former, less costly method is useful for determining the significance of initial departures from linear behavior. More specific conclusions are listed below.

- Generally the accuracy of the describing function method decreases as the amplitude of the motion increases. The describing function method, however, is a powerful tool to predict the characteristics of transonic flutter, since it generally requires a very small amount of computational time for the aerodynamic forces compared to time marching solutions, particularly if a parameter study is to be undertaken.

- The stable nonlinear limit cycle flutter predicted by the describing function method, is also observed in the time marching solutions.

- The periodic shooting method has confirmed the results for the stable and unstable flutter limit cycle oscillations found by the describing function method. This work has also emphasized the care that must be taken when performing conventional time integration flutter solutions.

For an alternative suggestion for achieving the same goals, the reader is directed to Taylor et al. (1980). Also, the recent work of Bland and Edwards (1983) should be cited. They deal with the important (static nonlinear aerodynamic) effects of airfoil shape and thickness (though not nonlinear dynamic effects per se) in a manner similar to that of the present paper. Also see the work of Batina (1985) and Williams (1983). The latter is in the context of panel flutter rather than airfoil flutter. In general the effects of airfoil shape and thickness (static nonlinearities) are more important for flutter than nonlinear dynamics effects per se (Dowell et al., 1983; Bland and Edwards, 1983). Hence, dynamically linear aeroelastic analyses will continue to play a dominant role even in the transonic flow regime.

6.7. Concluding Remarks

6.7.1. Some present answers

The chapter began with four questions. The answers to these will undoubtedly be refined in future years. However, based on present knowledge, partial answers may be formulated as follows:

(1) For sufficiently small airfoil motions (leading to sufficiently small shock motions, ≈ 5% of airfoil chord or less), the aerodynamic forces will be linear functions of the airfoil motion (Tijdeman, 1977; Tijdeman and Seebass, 1980; Seebass, 1984; McCroskey, 1982; Ballhaus and Goorjian, 1977; Fung et al., 1978; Dowell et al., 1983).

(2) At least one viable alternative solution technique to finite difference methods is available, the field panel method of Hounjet (1981b). Further refinements to that approach are expected, possibly including the work of Cockey (1983) as discussed herein and also including the extension to three-dimensional flow fields. The relative merits of finite difference vs. finite element vs. field panel methods remains a subject for future study and, no doubt, vigorous debate.

(3) The nonuniqueness that has been observed under some conditions in transonic small disturbance and full potential flow solutions is inherent in the governing field equations themselves and not a numerical artifact of some solution method (Goorjian [22]; Dowell et al. [15]; Steinhoff and Jameson, 1981; Salas, 1983). To date, corresponding nonuniqueness in solutions to the Euler equations has not been observed. However, it is not clear at this time whether or not such nonuniqueness in the potential flow equation solutions may have its less (or equally) dramatic counterpart in the solutions to the Euler equations under some conditions. This remains an important

topic for future study. Moreover, the physical significance, if any, of such solutions remains to be
clarified.

(4) While the option of simultaneous time integration of the fluid dynamical and structural dynamical equations of motion to determine aeroelastic
response will be attractive for some applications,
flutter analysis in the frequency domain will continue
to be an important, and at times more attractive,
option as well. Methods for generating the frequency
domain aerodynamic forces are now available from

- aerodynamic methods that presuppose infinites-
  imal, harmonic dynamic airfoil motions
  (Williams, 19791, 1979b; Hounjet, 1981b;
  Ehlers and Weatherill, 1982),

- impulse-transfer function ideas (Nixon and
  Kerlick, 1980; Nixon, 1981; Ballhaus and Goorjian, 1978; Ueda and Dowell, 1982, 1983) which
  allow the generation of frequency domain data
  from a single time history record determined
  by a time marching aerodynamic code. This
  approach can be extended (approximately) to
  large airfoil motions where a nonlinear relationship exists between airfoil motions and
  aerodynamic forces by using ideas based on the
  describing function method (Ueda and Dowell,
  1982; 1983).

(5) Simple order of magnitude estimates of the
relative computational times for aeroelastic analyses
can be made and these have been discussed in the text.
Dynamically linear flutter analysis will continue to
play a dominant role even in the transonic flow regime
(Ueda and Dowell, 1982, 1983; Yang et al., 1980;
Edwards et al., 1982; Bland and Edwards, 1983).

6.7.2. Future work

A long list of worthy research topics could be
given. Here we focus on a few that have as their

common theme improved physical modeling and understanding of the fluid dynamic and aeroelastic phenomena of interest.

(1) There is a clear need to understand better
the apparent qualitative difference between the solutions of the potential flow equations and those of the
Euler equations under those parameter conditions where
the former exhibit nonunique solutions. Until this
difference is both understood and resolved, it throws
into question the whole approach of combining potential flow solution with boundary layer corrections to
correct for viscosity under those conditions where
nonunique solutions are observed.

(2) A complementary issue is how can more effective (efficient) solution methods be devised for the
Euler equations for unsteady flows. Subsequent to the
pioneering work of Magnus and Yoshihara (1975), little
has been done.

One possible approach which has several prospec-
tive advantages is to assume that the flow may be
treated as a nonlinear mean steady flow plus a small
(infinitesimal) linear dynamic perturbation. This
technique has been exploited effectively by Williams
(1979a, 1979b), Fung and Seebass (1978), Hounjet
(1981b), Ehlers and Weatherill (1982), and others for
the potential flow model. In one of its limiting
forms this approach has been used by Lighthill (1953),
Williams et al. (1977), and others for modeling boundary layer effects in panel flutter (non-lifting) and
lifting surface aerodynamics using the Euler equations.

Such an approach will lead to a set of time dependent, linear, partial differential equations with
variable (spatially dependent) coefficients. These
will depend in turn on the solution of the time-
independent, nonlinear, partial differential, Euler
equations for the mean steady flow. The expected
advantages of this approach include the following:

(3) Existing and future steady flow codes may be
exploited to maximum advantage to provide input data

for the dynamic perturbation equations and their solution. (Also the finite difference grid developed for
the mean steady flow may be retained for the dynamic
perturbation flow if a finite difference solution
method is used for the latter.)

(4) Solution methods for linear equations with
variable coefficients may be brought to bear, although
these will require extensions and generalizations.
See the work of Hounjet (1981b) and Williams et al.
(1977).

(5) Solving the linear, dynamic equations will

- in all likelihood meet the requirements of the
  aeroelastician (within the context of the
  Euler equations).

- permit inclusion of viscous effects in the
  mean steady flow modeling and thus indirectly
  and partially (but not directly and completely) in the dynamic perturbation equations
  (Lighthill, 1953; Williams et al., 1977).

- allow one to examine the dynamic stability of
  the mean, steady flow itself and thus contribute to a better understanding of the prospect for nonunique solutions of the Euler
  equations.

References

1. Albano, E. and Rodden, W. P., "A Doublet Lattice
   Method for Calculating Lift Distribution on
   Oscillating Wings in Subsonic Flows," AIAA
   Journal 7 (1969), 279-85.

2. Ballhaus, W. F. and Goorjian, P. M., "Implicit
   Finite Difference Computations of Unsteady Transonic Flows About Airfoils," AIAA Journal 15
   (1977), 1728-35.

3. Ballhaus, W. F. and Goorjian, P. M., "Efficient Solution of Unsteady Transonic Flows About Airfoils," Paper 14, AGARD Conference Proceedings No. 226, Unsteady Airload in Separated and Transonic Flows, 1978a.

4. Ballhaus, W. F. and Goorjian, P. M., "Computation of Unsteady Transonic Flows by the Indicial Method," AIAA Journal 16, No. 2 (1978b), 117-24.

5. Batina, J. T., "Effects of Airfoil Shape, Thickness, Camber, and Angle of Attack on Calculated Transonic Unsteady Airloads," NASA TM 86320 (1985).

6. Bauer, R., Garabedian, P., and Korn, D., "Supercritical Wing Sections," Lecture Notes in Economics and Mathematical Systems, 66, Springer-Verlag, 1972.

7. Bland, S. R. and Edwards, J. W., "Airfoil Shape and Thickness Effects on Transonic Airloads and Flutter," AIAA SDM Conference in Lake Tahoe, CA, May 1983, AIAA Paper No. 83-0959.

8. Borland, C. J. and Rizzetta, D. P., "Transonic Unsteady Aerodynamics for Aeroelastic Applications, I: Technical Development Summary," AFWAL TR 80-3107, I, June 1982a.

9. Borland, C. J. and Rizzetta, D. P., "Nonlinear Transonic Flutter Analysis," AIAA Journal, 20, No. 11 (1982b), 1606-15.

10. Borland, C. J., Rizzetta, D. P., and Yoshihara, H., "Numerical Solution of Three-Dimensional Unsteady Transonic Flow Over Swept Wings," AIAA Journal, 20, No. 3 (1982), 340-37.

11. Cockey, W. D., "Panel Method for Perturbations of Transonic Flows with Finite Shocks," Ph.D. thesis, Princeton University, June 1983.

12. Davis, S. S. and Malcolm, G., "Experiment in Unsteady Transonic Flows," _Proceedings of the AIAA/ASME/ASCE 20th Structures, Structural Dynamics and Materials Conference_, St. Louis, MO, April 1979.

13. Dowell, E. H., "A Simple Method for Converting Frequency-Domain Aerodynamics to the Time Domain," _NASA TM 81844_, 1980.

14. Dowell, E. H., Bland, S. R., and Williams, M. H., "Linear/Nonlinear Behavior in Unsteady Transonic Aerodynamics," _AIAA Journal, 21_ (1983), 38-46.

15. Dowell, E. H., Ueda, T., and Goorjian, P. M., "Transient Decay Times and Mean Values of Unsteady Oscillations in Transonic Flow," _AIAA Journal, 21, No. 12_, (1983), 1762-64.

16. Edwards, J. W., et al., "Time-Marching Transonic Flutter Solutions Including Angle-of-Attack Effects," _AIAA Paper 82-0685_, presented at the 23rd SDM Conference, New Orleans, LA, May 1982.

17. Ehlers, F. E. and Weatherill, W. H., "A Harmonic Analysis Method for Unsteady Transonic Flow and its Application to the Flutter of Airfoils," _NASA CR-3537_, 1982.

18. Erickson, A. L. and Stephenson, J. D., "A Suggested Method of Analyzing Transonic Flutter of Control Surfaces Based on Available Experimental Evidence," NACA RM A7F30 (1947).

19. Fung, K. Y., Yu, N. J., and Seebass, R., "Small Unsteady Perturbations in Transonic Flows," _AIAA Journal, 16_ (1978), 815-22.

20. Gibbons, M., Whitlow, W., Jr., and Williams, M. H., "Nonisentropic Unsteady Three Dimensional Small Disturbance Potential Theory," AIAA Paper 86-0863, May 1986.

21. Goldstein, M. E., _Aeroacoustics_, New York: McGraw-Hill, 1976.

22. Goorjian, P. M., Private communication, NASA Ames Research Center.

23. Hessenius, K. A. and Goorjian, P. M., "Validation of LTRAN2-HI by Comparison with Unsteady Transonic Experiment," AIAA Journal, 20, No. 5 (1982), 731-32.

24. Hounjet, M. H. L., "A Transonic Panel Method to Determine Loads on Oscillating Airfoils with Shocks," AIAA Journal, 19 (1981a), 559-66.

25. Hounjet, M. H. L., "A Field Panel Method for the Calculation of Inviscid Transonic Flow About Thin Oscillating Airfoils with Shocks," NLR MP 81043 U, National Aerospace Laboratory, Netherlands. Presented at the International Symposium on Aeroelasticity, Nuremburg, October 5-7, 1981b, Germany.

26. Houwink, R. and van der Vooren, J., "Improved Version of LTRAN2 for Unsteady Transonic Flow Computations," AIAA Journal, 18, No. 8 (1980), 1008-10.

27. Hsu, J. C. and Meyer, A. U., Modern Control Principles and Applications, New York: McGraw-Hill, 1968.

28. Isogai, K., "Numerical Study of Transonic Flutter of a Two-Dimensional Airfoil," Technical Report of National Aerospace Laboratory, Japan, NAL-TR-617T, 1980.

29. Kerlick, G. D. and Nixon, D., "Mean Values of Unsteady Oscillations in Transonic Flow Calculations," AIAA Journal, 19, No. 11 (1981), 1496-98.

30. Lighthill, M. J., "On Boundary Layers and Upstream Influence, II; Supersonic Flows Without Separation," Proceedings of the Royal Society, A217 (1953), 478-507.

31. Liu, D. D., "A Lifting Surface Theory Based on an Unsteady Linearized Transonic Flow Model," AIAA Paper 78-501, 1978.

32. Magnus, R. J. and Yoshihara, H., "Calculation of Transonic Flow Over an Oscillating Airfoil," AIAA Paper 75-98, Jan. 1975.

33. McCroskey, W. J., "Unsteady Airfoils," Annual Review of Fluid Mechanics, 14 (1982), 285-311.

34. McIntosh, S., private communication.

35. Miles, J. W., The Potential Theory of Unsteady Supersonic Flow, Cambridge: Cambridge University Press, 1959, 4-13.

36. Morino, L., "A General Theory of Unsteady Compressible Potential Aerodynamics," NASA CR-2464, December 1974.

37. Morino, L. and K. Tseng, "Time-Domain Green's Function Method for Three-Dimensional Nonlinear Subsonic Flows," AIAA Paper 78-1204, 1978.

38. Murman, E. M., "Analysis of Embedded Shock Waves Calculated by Relaxation Methods," Proceedings of the AIAA CFD Conference, July 1973, 27-40.

39. Nixon, D., "Calculation of Unsteady Transonic Flows Using the Integral Equation Method," AIAA Paper 78-13, January 1978.

40. Nixon, D., "On the Derivation of Universal Indicial Functions," AIAA Paper 81-0328, Jan. 1981.

41. Nixon, D. and Kerlick, G. D., "Calculation of Unsteady Transonic Pressure Distributions by Indicial Methods," Nielsen Engineering and Research Paper 117, 1980.

42. Peters, D. A. and Ventura, L., "Applications of Various Solutions Techniques to the Calculation of Transonic Flutter Boundaries," in Fluid-Structure Interaction and Aerodynamic Damping, Edited by E. H. Dowell and M. K. Au-Yang, ASME, (1985), 29-49.

43. Rizzetta, D. P. and Borland, V. J., "Unsteady Transonic Flow Over Wings Including Inviscid/Viscous Interactions," AIAA Journal, 21, No. 3 (1983), 363-71.

44. Salas, M. D., Jameson, A., and Melnik, R. E., "A Comparative Study of the Nonuniqueness Problem of the Potential Equation," presented at the 6th AIAA CFD Conference, Danvers, Mass., July 13-15, 1983.

45. Salas, M. D., and Gumbert, C. R., "Breakdown of the Conservative Potential Equation," AIAA Paper 85-0367, Jan. 1985.

46. Seebass, R., "Advances in the Understanding and Computation of Unsteady Transonic Flows," Recent Advances in Aerodynamics, ed. A. Krothapalli and C. Smith, 1984.

47. Steger, J. L., and Bailey, H. E., "Calculations of Transonic Buzz," AIAA Journal, 18, (1980), 249-255.

48. Steinhoff, J. and Jameson, A., "Multiple Solutions of the Transonic Potential Flow Equations," AIAA Paper No. 81-1019, AIAA Computational Fluid Dynamics Conference, Palo Alto, CA, June 1981.

49. Taylor, R. F., Bogner, F. K., and Stanley, E. C., "A Stability Prediction Method for Nonlinear Aeroelasticity," AIAA Paper 80-0797, presented at the AIAA/ASME/AHS/ASCE Conference, Seattle, WA, May 12-14, 1980.

50. Tijdeman, H., "Investigation of the Transonic Flow Around Oscillating Airfoils," Ph.D. thesis, Delft University, 1977.

51. Tijdeman, H. and Seebass, R., "Transonic Flow
     Past Oscillating Airfoils," Annual Review of
     Fluid Mechanics, 12 (1980), 181-222.

52. Tran, C. T. and Petot, M., "Semi-empirical Model
     for the Dynamic Stall of Airfoils in View of the
     Application of Responses of a Helicopter Blade in
     Forward Flight," Paper 48, Proceedings, 6th Euro-
     pean Rotorcraft and Powered-lift Aircraft Forum,
     Bristol, England, 1980.

53. Ueda, T. and Dowell, E. H., "Flutter Analysis
     Using Nonlinear Aerodynamic Forces," AIAA Paper
     82-0728, 1982. Also see J. Aircraft, 21 (1984),
     101-109.

54. Ueda, T. and Dowell, E. H., "Describing Function
     Flutter Analysis for Transonic Flow: Extension
     and Comparison with Time Marching Analysis," AIAA
     Paper 83-0958, 1983.

55. Voss, R., "Time-Linearized Calculation of Two-
     Dimensional Unsteady Transonic Flow at Small Dis-
     turbances," DFVLR FB-81-01, 1981.

56. Williams, M. H., "The Effect of a Normal Shock on
     the Aeroelastic Stability of a Panel," J. Applied
     Mechanics, 50, (1983), 275-282.

57. Williams, M. H., Bland, S. R., and Edwards, J.
     W., "Flow Instabilities in Transonic Small-
     Disturbance Theory," AIAA Journal, 23, (1985),
     1491-1496.

58. Williams, M. H., "Unsteady Thin Airfoil Theory
     for Transonic Flows with Embedded Shocks," De-
     partment of Mechanical and Aerospace Engineering,
     Report No. 1376, Princeton University, May 1978;
     also AIAA Journal, 18 (1980), 615-24; also see
     "Unsteady Airloads in Supercritical Transonic
     Flows," Proceedings of the AIAA/ASME/ASCE 20th
     Structures, Structural Dynamics and Materials
     Conference, St. Louis, MO, April 1979b.

59. Williams, M. H., "The Linearization of Transonic Flows Containing Shocks," AIAA Journal, 17 (1979a), 394-97.

60. Williams, M. H., et al., "Aerodynamic Effects of Inviscid Parallel Shear Flows," AIAA Journal, 15, No. 8 (1977), 1159-66.

61. Yang, T. Y., Guruswamy, P., and Striz, A. G., "Aeroelastic Response Analysis of Two-Dimensional, Single and Two Degree of Freedom Airfoils in Low Frequency, Small-Disturbance Unsteady Transonic Flow," AFFDL-TR-79-3077, 1979.

62. Yang, T. Y., Guruswamy, P., and Striz, A. G., "Application of Transonic Codes to Flutter Analysis of Conventional and Supercritical Airfoils," AIAA SDM Conference in Atlanta, GA, AIAA Paper 81-0609, 1981.

63. Yang, T. Y., et al., "Flutter Analysis of a NACA 64A006 Airfoil in Small Disturbance Transonic Flow," J. Aircraft, 17, No. 4 (1980), 225-32.

CHAPTER VII

CHAOTIC OSCILLATIONS IN
MECHANICAL SYSTEMS

Summary

Chaotic oscillations have now been observed in nonlinear mechanical systems by analytical, numerical, and experimental methods. Nevertheless, a more fundamental understanding of why and when such oscillations occur is of great importance. This goal will be pursued here by considering the relationship between chaos induced by forced oscillations vs self-excited oscillations, the relationship of indeterminancy of the final equilibrium state in the initial value problem to chaos in the sustained oscillation problem, comparison of theory to physical experiment, necessary and sufficient conditions for chaos to occur, and the question of convergence of systems of modal ordinary differential equations that derive from partial differential equations.

Introduction

The recent discovery (or rediscovery some would say, giving Poincaré his due) of random-like, broad band oscillations in the response of simple mechanical systems whose parameters are entirely deterministic has led to one of the most exciting periods in the development of nonlinear dynamics since the days of Poincaré. There are some who contend that Poincaré understood chaos (and perhaps he did), but it is only with the advent of modern computational power that our knowledge of chaos has deepened and its presence discovered to be so pervasive in many dynamical systems of physical interest. Classical analytical methods which assume the system motion to be periodic obviously fail and even more modern geometric approaches only hint at the possibility of chaos. For example, see the text by Guckenheimer and Holmes [A].

Here modern geometrical concepts, substantial digital simulation of the dynamic response of non-linear systems, and graphical presentation of data are used to describe the new results in nonlinear dynamics. Two prototypical models are considered. The first is a buckled beam, which has been studied both experimentally and theoretically. The theoretical model in its simplest form leads to a Duffing equation and in its more elaborate form to a set of Duffing equations. The second model considered is entirely theoretical and due to Lorenz, the well known Lorenz equations, which are of some historical importance and for which there is substantial literature.

The present discussion is quite different from Guckenheimer and Holmes [A], who tried valiantly to prove or discuss as many relevant theorems and lemmas as might be useful. The approach is also distinct from the forthcoming book by F. C. Moon [B], which the present author has had the opportunity to read in manuscript. Moon's book emphasizes the physical phenomena and basic conceptual framework, including an authoritative discussion of laboratory experiments. In spirit the present approach is perhaps most similar to the book of Thompson and Stewart [C], however here only two systems exhibiting chaos are considered, but in somewhat greater depth. The three volumes of Guckenheimer and Holmes, Thompson and Stewart, and Moon and this chapter thus span a spectrum of approaches to the subject and are complementary to one another. The present author would recommend all three of these books to the reader with the suggestion that the present chapter or the book by Thompson and Stewart or Moon might be the best place to start to develop an understanding of chaos.

In recent years a great deal of progress has been made in the study of chaos in nonlinear mechanics. Broadly speaking, most of the dynamical systems studied to date can be categorized as either (1) forced response of stable (with respect to infinitesimal disturbances) dynamical systems or (2) self excited oscillations of autonomous systems. Representative of the first category is the study of the forced response of Duffing's equation with a negative linear stiffness and a positive cubic stiffness. This, for example, models the behavior of a buckled beam or plate under transverse dynamic excitation. Representative of the second category is the set of Lorenz equations or the panel flutter equations. As we shall see, however, the panel flutter equations and Duffing's equations are not unrelated.

A fundamental question is what mechanisms in these two categories of systems lead to chaotic oscillations. Are the mechanisms similar or are they distinct? In particular, the question of the relationship of the forced response of a buckled beam to the self-excited flutter oscillations of a buckled beam is examined. Another important question is that of modal convergence. Both Duffing's equations (or the panel flutter equations) and Lorenz' equations may be thought of as deriving from modal approximations to nonlinear partial differential equations: The question arises, how many modes are needed to converge?

7.1. On the understanding of chaos in Duffing's equation including a comparison with experiment*

Among the first mechanical systems where chaos was discovered was for the fluid flow over a buckled elastic plate. Using quasi-steady aerodynamics and Von Karman's nonlinear plate theory, a non-dimensional form for a two-mode model can be written as follows.

$$\ddot{A}_1 + \gamma \dot{A}_1 + A_1(1 + R_x) + A_1(A_1^2 + 4A_2^2) - \lambda A_2 = 0$$

$$(1.1a)$$

$$\ddot{A}_2 + \gamma \dot{A}_2 + 16A_2(1 + \frac{R_x}{4}) + 4A_2(A_1 + 4A_2^2) + \lambda A_1 = 0$$

---

*This section is a substantially modified version of E. H. Dowell and C. Pezeshki, "On the Understanding of Chaos in Duffings Equation Including a Comparison with Experiment," J. Applied Mechanics, Vol. 53, 1986, pp. 5-9, American Society of Mechanical Engineers.

Also see [1,2,3] for more detail. The form of the equations quoted here is slightly different from that of Ref. 1.

It has been established that for certain combinations of the parameters, R_x, λ, chaotic oscillations may occur in the solutions to these equations. A sufficiently large negative R_x (compressive in-plane load) leads to a buckled panel while a sufficiently large λ (flow velocity) leads to a limit cycle due to flutter. It is the interaction of the flutter limit cycle with the buckled panel which leads to chaos.

Effective means for observing the evolution of the chaotic oscillations and detecting their essential structural content have been developed. Nevertheless an improved understanding of why these chaotic oscillations occur is needed and the present discussion is directed toward that end.

Examining equations (1.1a) it is seen the chaotic oscillations are the result of a self-excited system where the modal amplitudes, A_1 and A_2 interact in a subtle and complicated way. It is known [1,2] that retention of additional equations and modal amplitudes, A_3, A_4, etc. lead to quantitative, but not qualitative, changes in the system's behavior. On the other hand, if one sets $A_2 \equiv 0$ and considers only A_1 it is known that no chaotic oscillations occur. Hence it would first appear that equations (1.1a) are not further reducible without loss of essential information. However, if one considers only the first of equation (1.1a) and retains A_2 as a known (prescribed) forcing function of time, then it becomes a Duffing equation of the type studied by Tseng and Dugundji [4], Holmes [5] and Moon [6]. Such a Duffing equation is known to exhibit chaos for certain parameters. Hence, it is not surprisinng that the panel flutter equations can exhibit chaos as well.

Conversely one could consider the second equation of (1.1a) as describing the behavior of A_2 when driven by a forcing function A_1. Of course it is the full interaction of the two equations that produces limit cycle oscillations and these in turn provide an oscillatory A_2 or A_1 to drive A_1 or A_2 into chaos via a Duffing equation mechanism.

The above discussion suggests the following single second order differential equation be studied

$$\ddot{A} + \gamma \dot{A} - \frac{A}{2} (1 - A^2) = F(\tau) \qquad (1.1b)$$

This is the particular form of Duffing's equation (with a negative linear stiffness) studied by Moon [6]. It is known this equation has solutions with chaotic oscillations under certain conditions. An improved understanding of why the chaotic oscillations occur can be obtained by first considering the initial value problem when $F \equiv 0$. These results are of substantial interest in their own right as well as leading to additional understanding of why chaotic oscillations occur. The present theoretical results are also compared to the physical experiments of Moon.

A physical model is helpful in the interpretation of Duffing's equation. Following previous authors [3-6], we interpret equation (1.1b) as describing a one mode oscillation of a buckled beam under the action of a prescribed lateral external force, $F(\tau)$. Other physical systems may also be described by this equation, but they will not be discussed here. As may be seen from equation (1.1b), when $F \equiv 0$ there are three static equilibrium solutions: $A = 0$, +1, and -1. It is easily shown that the first of these (an unbuckled beam) is dynamically unstable and the latter two are stable with respect to infinitesimal disturbances. It is of great help in understanding the onset of chaos to consider next the stability of the static equilibria, $A = +1$ or -1, with respect to finite disturbances. This is done in the following section and then chaos due to a harmonic force, $F(\tau)$, is considered directly in subsequent sections.

7.1.1. The initial value problem for the homogeneous Duffing's equation ($F \equiv 0$)

It is helpful to think first in physical terms. Consider the buckled beam at rest in one of its stable (with respect to infinitesimal disturbances) static equilibria, say $A = +1$. With prescribed initial conditions,

$$A(\tau \equiv 0) = A_0$$

$$\dot{A}(\tau \equiv 0) = \dot{A}_0$$

consider the transient solution and the final steady state solution as $\tau \to \infty$. Obviously $A(\tau \to \infty) \to +1$ or -1. The question is which of these two solutions is the correct one for given A_0, $\dot{A}_0$. As shall be seen, the answer is in a certain sense unknowable (or, to use a more technical term, uncertain). Once the reason for this is understood, the occurrence of chaos for certain $F \neq 0$ becomes more understandable, perhaps even expected.

It is possible to construct a diagram (which is called a shell plot because of its appearance) that summarizes compactly which of the two static equilibria solutions will be reached for a given set of initial conditions, A_0, $\dot{A}_0$. In the mathematical literature this is called an initial condition map.

To anticipate the form of the shell plot, consider a specific example of initial conditions and the subsequent solution trajectory. This is shown as a phase plane trajectory in Fig. 1 for $A_0 = 1$ and $\dot{A}_0 = 1$. Because $\dot{A}_0 > 0$, A increases for small time, but then decreases for larger time because of the nonlinear restoring stiffness. Indeed, A subsequently becomes negative (the beam moves from one buckled configuration, $A = +1$, to the other, $A = -1$, and beyond). The damping term, $\gamma\dot{A}$, leads to dissipation of energy; thus, the beam does not continue to oscillate between and about the two static buckled equilibria, but instead spirals into one of them as $\tau \to \infty$.

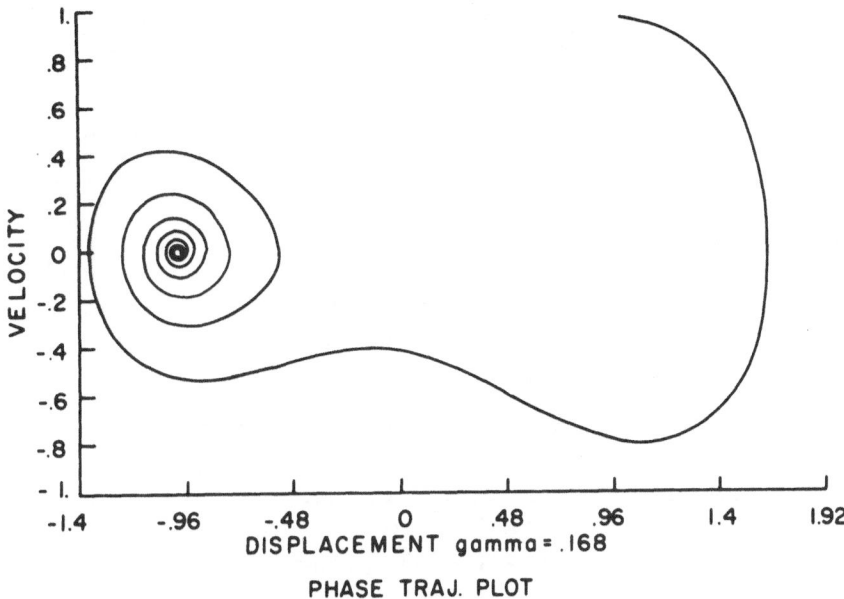

PHASE TRAJ. PLOT

Fig. 1. Phase plane trajectory.

From such phase plane trajectories, one can construct the shell plot, which shows the final state of
the system as $\tau \to \infty$, $A = +1$ or -1, for given initial
conditions, A_0 and $\dot{A}_0$. This is shown (partially) in
Fig. 2. Here $\dot{A}_0$ is plotted vs A_0 and various regions
are identified with integral values, 0, 1, 2, 3, 4,...
Note there are two disjoint regions associated with
each integer value. Consider first the integer zero
(0) regions. For definiteness consider the region
where $A_0 \geqslant 0$. If the system starts with A_0, $\dot{A}_0$ within
the zero region, the solution spirals into $A = +1$ as τ
$\to 0$ and crosses the $A = 0$ axis zero times. Consider
now the 1 region. A solution begun there moves clockwise and crosses the $A = 0$ axis one time and enters
the zero region for $A < 0$. Once there, it spirals into $A = -1$ as $\tau \to \infty$. To firmly establish the pattern,
finally consider the 2 region. For initial conditions
in the 2 region, the phase plane trajectory moves
clockwise, crosses the $A = 0$ axis the first time and
moves into the 1 region for $A < 0$. It then continues
to move clockwise and crosses the $A = 0$ axis a second
(and final time) and moves into the 0 region for $A >
0$. Once there it spirals into $A = +1$ as $\tau \to \infty$.

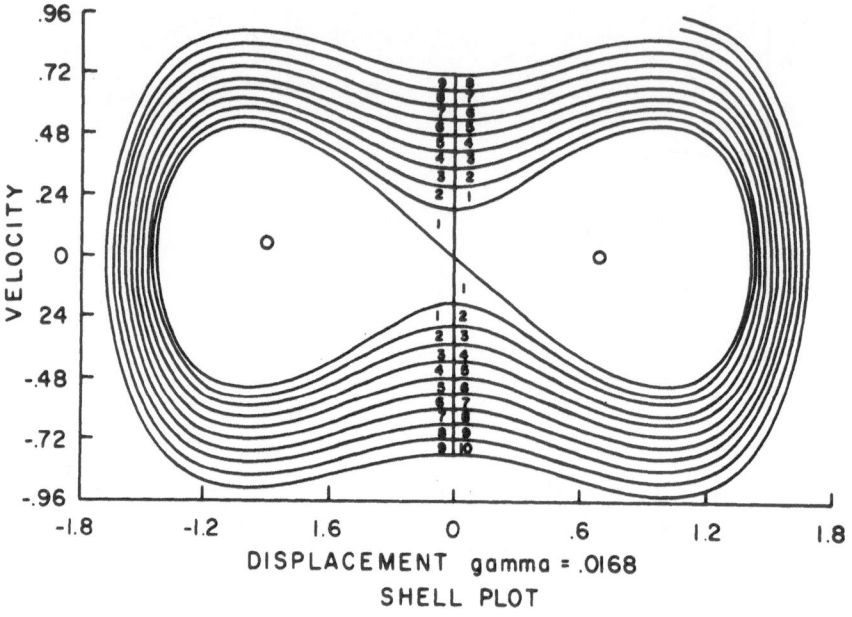

Fig. 2. Shell plot.

The pattern is now clear. For $A_0 > 0$, initial conditions in an even integer region reach a final state of $A = +1$. Those in an odd integer region reach a final state of $A = -1$. The integer number corresponds to the number of crossings of the $A = 0$ axis during the completion of the motion (phase plane trajectory). For $A_0 < 0$, a similar sequence of events occurs. For initial conditions that lie precisely on a shell boundary, the final configuration would be A, $\dot{A} \to 0$ as $\tau \to \infty$. In practice this will never occur, of course, because the shell boundary curves are of vanishing thickness.

It is interesting to note that a shell plot of any finite extent can be constructed from a single artfully chosen phase plane trajectory, once the zero region is known. The latter region is readily determined by direct calculation.

Now comes the central point. If there is <u>suf-ficient</u> uncertainty in the values of the initial con-ditions, A_0, $\dot{A}_0$, it is clear from an examination of the shell plot that the final system state, $A = +1$ or -1, is unpredictable, unknowable, or uncertain. This point is made all the more powerful by noting that as the damping becomes even smaller the width of each region of the shell plot (excluding the 0 region) becomes even smaller and vanishes as $\gamma \to 0$. Hence for any (finite) uncertainty in A_0 or $\dot{A}_0$ the final system state is unpredictable as $\gamma \to 0$.

Two additional points are worthy of note in con-cluding this discussion. First, the boundary contours of the shell plot are curves of essentially constant total (kinetic plus potential) energy. Second, as $\gamma \to 0$, the boundary curves for the two zero regions cor-respond to the separatrix of the undamped system.

Although not concerned with chaos, the reader will find the discussion of Ref. 7 on the initial value problem of interest.

7.1.2. The continuous oscillation problem for the inhomogeneous Duffing's equation ($F \neq 0$)

Here a simple harmonic external force is consid-ered.

$$F = F_0 \sin \omega\tau \qquad (1.2)$$

where F_0 is the force amplitude and ω is its frequency of excitation. This is not the only force-time his-tory which might be studied. It is, perhaps, the simplest periodic force.

As the reader may note, the initial value problem previously studied can be also thought of as an external force problem. For example, an initial velocity, $\dot{A}_0$, corresponds to an impulsive force,

$$F(\tau) = \dot{A}_0 \delta(\tau) \qquad (1.3)$$

This suggests that a study of continual impulses, periodically or randomly spaced in time, would be of interest. Nevertheless, only a simple harmonic force will be considered here.

The response of the system will be considered first for fixed frequency, $\omega = 1$, and increasing force amplitude, F_0. The frequency is normalized by the small amplitude natural frequency about the beam buckled equilibrium. For F_0 sufficiently small, it is expected that the response of the system will be a simple harmonic oscillation about one or the other of the two static equilibria, $A = +1$, or -1. For definiteness the initial conditions, $A_0 = 1$, $\dot{A}_0 = 0$ are chosen so that for small F_0 the harmonic response oscillation is about $A = +1$. It is anticipated that, as F_0 increases and the response phase plane trajectory approaches the zero region boundary of the shell plot, interesting response behavior will occur.

Note that for small F_0 the phase plane trajectory is an ellipse indicating a simple harmonic response oscillation. As F_0 increases, additional harmonics beyond the fundamental are detected and the phase plane trajectory is distorted from a simple ellipse. See Fig. 3(a) for the result for $F_0 = .177$. Also shown for reference is the boundary for the zero region from the previously discussed shell plot.

For $0 < F_0 < .177$ the response is termed 1 period motion. By that is meant, as the force oscillates through one period, the response also oscillates through one period. For $F_0 = .178$, however, as the force oscillates through one period the response oscillates through only one half a period. For the response to go through one period, the force must oscillate through two periods. Thus this is called 2 period motion. See Fig. 3(b). This change from 1 to 2 period response as $F_0 = .177 \rightarrow .178$ appears to be a bifurcation.

At a higher F_0, 4 period motion occurs. See Fig. 3(c) for example, and at yet higher F_0, 8 and 16 period motion occurs. Holmes [5] has suggested that 32, etc. period motions occur as F_0 increases further.

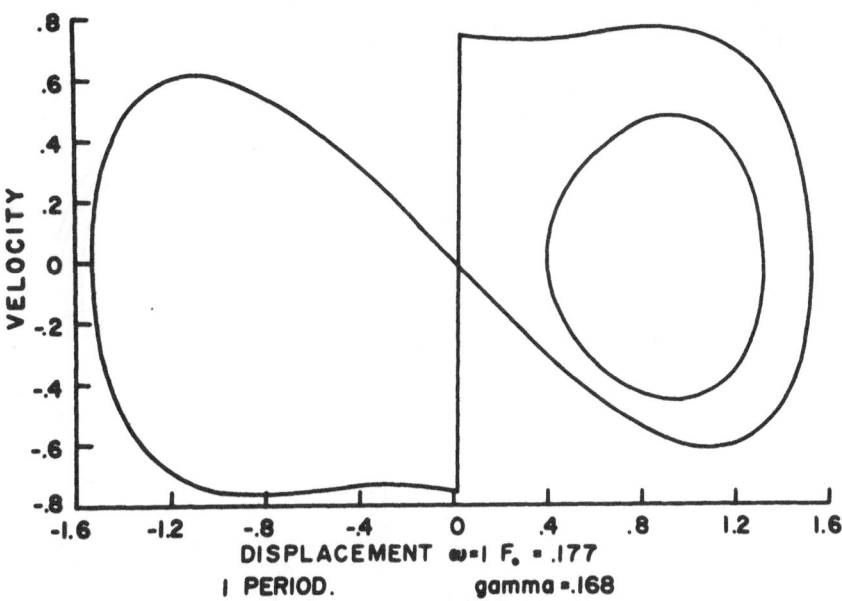

Fig. 3a. Phase plane trajectory (1 period motion).

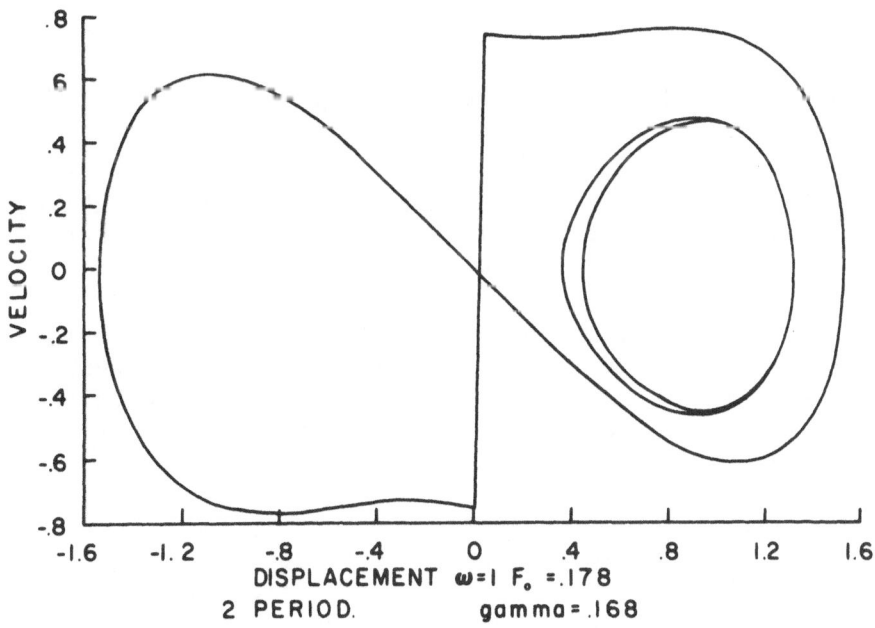

Fig. 3b. Phase plane trajectory (2 period motion).

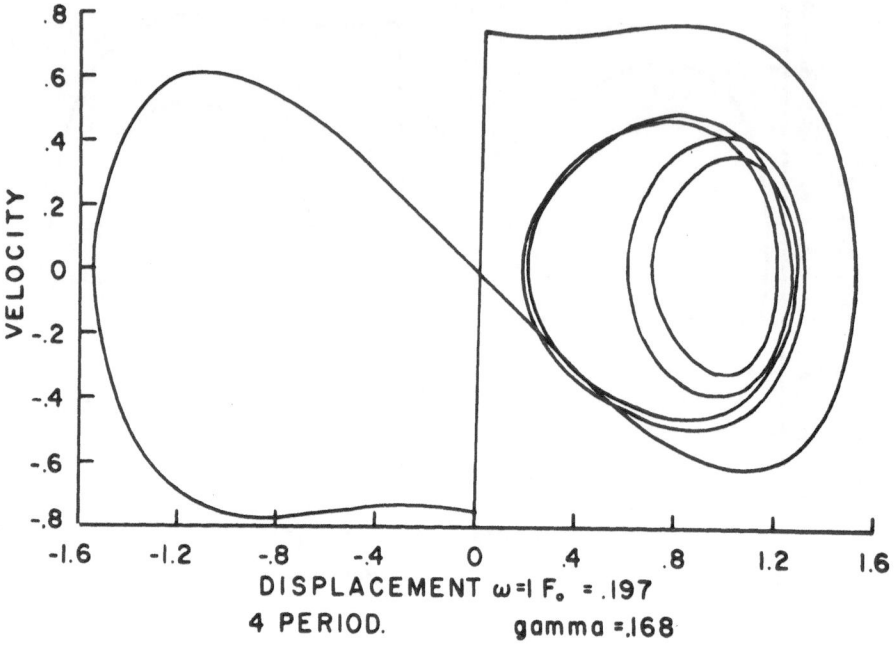

Fig. 3c. Phase plane trajectory (4 period motion).

This may well happen, but this behavior has not been observed by the present author. Possibly this is because the range of F_0 over which the higher period motions occur is very small. This period doubling behavior has been previously described and discussed by Feigenbaum [8] in a more general context.

For $F_0 \geqslant .205$ chaos is observed, i.e., no periodicity is apparent. See Fig. 3(d), which gives results for $F_0 = .21$. As Holmes has indicated, for yet higher F_0 the chaos no longer appears and periodicity returns.

It is clear that for F_0 just below the value where chaos first appears the periodic response phase plane trajectory approaches and slightly penetrates the boundary of the zero region shell plot. See Fig. 3(c). Moreover, it is clear that for this frequency, $\omega = 1$, chaos occurs when the motion is no longer about

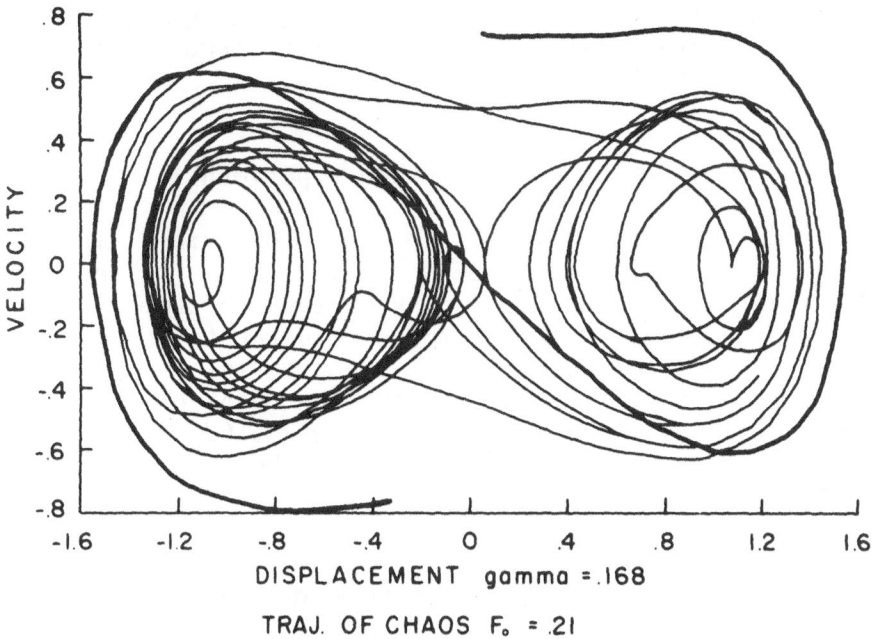

Fig. 3d. Phase plane trajectory (chaotic motion).

only one of the static equilibria points, say A = +1, but instead encircles both, A = +1 and -1. This is called snap-buckling. These observations suggest that the onset of chaos can be associated with periodic motions which penetrate the zero region boundary and thus lead subsequently to motion about both static equilibria points. Moon in an earlier paper [6] suggested a more restricted notion of this sort when he took as an empirical criterion for the onset of chaos that the periodic response maximum velocity (in his calculation he assumed one period motion) must exceed the maximum velocity of the system separatrix. Recall it has been shown here that the zero region boundry of the shell plot corresponds to the system separatrix as γ → 0.

For excitation frequencies well away from resonance, in particular for ω << 1, chaos was found to occur even without snap-buckling. The minimum F_0 for

the onset of chaos occurs for $\omega \cong .85$. This is close
to the characteristic frequency for free vibration
about both static equilibrium points.

7.1.3. Comparison of theory and experiment

Calculations similar to those described in the
previous section have been carried out for several
excitation frequencies, ω. From these a summary plot
may be constructed of the force required to cause snap
buckling, period doubling, or chaos versus the excitation frequency. Such a plot for the onset of chaos is
shown in Figs. 4(a) and 4(b) for $\gamma = .168$ and .0168,
respectively. The uncertainty in the data is less
than a diameter of a circle.

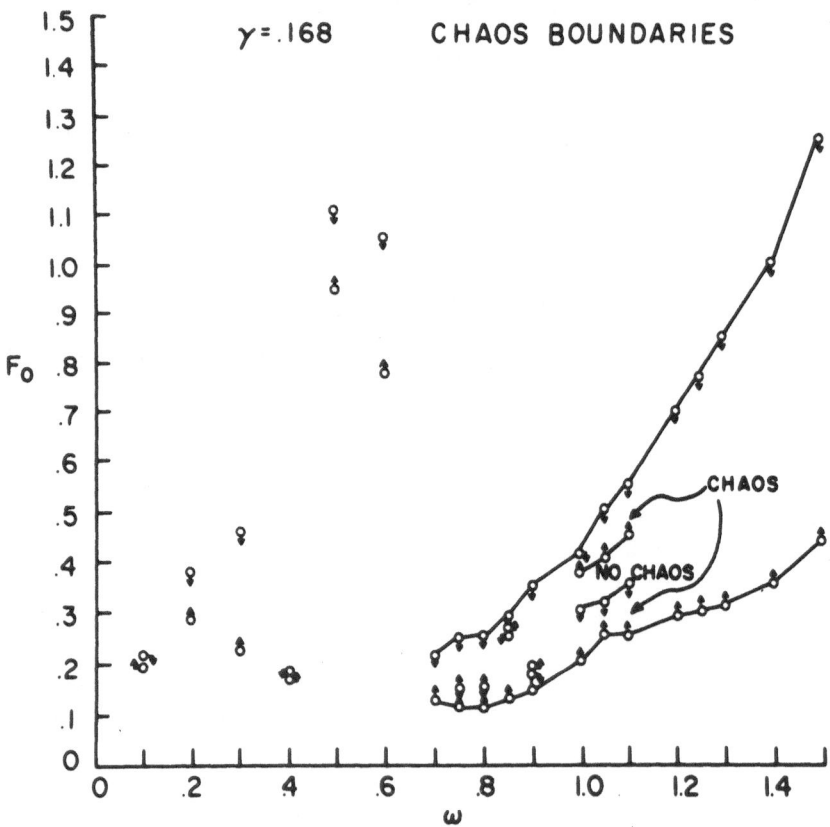

Fig. 4a. Force amplitude vs frequency of excitation.

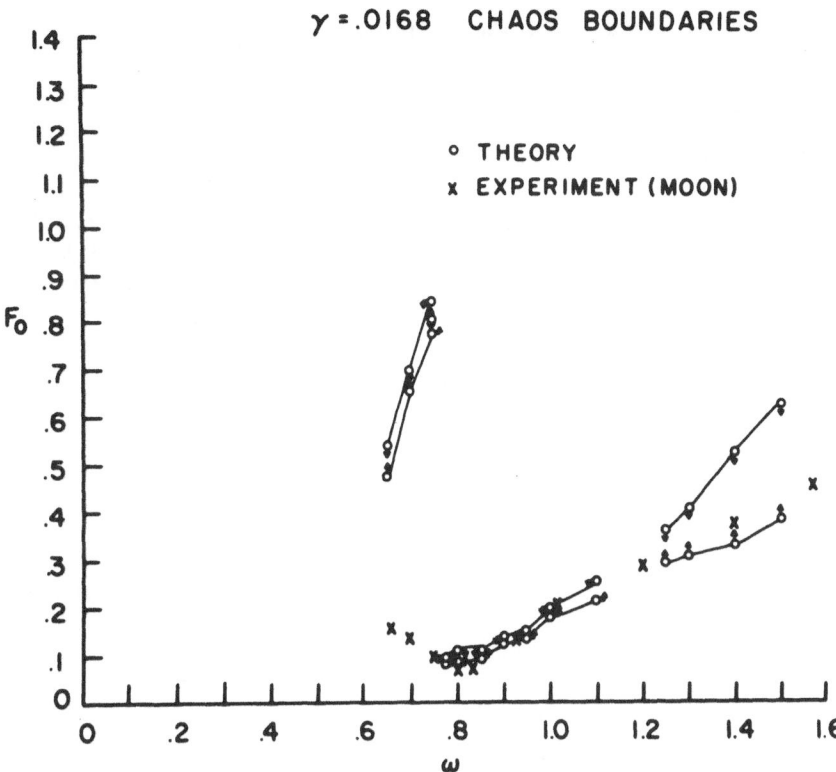

Fig. 4b. Force amplitude vs frequency of excitation.

Time integrations using the Runge-Kutta method were performed for frequency values ranging from .1 to 1.5 for varying force levels at damping levels of .168 and .0168. Principal lower and upper force chaos boundaries were found for a discrete series of frequencies by incrementing the first from zero until chaos was observed. See Fig. 4. Force increments of .01 and, where necessary, .001, were used.

All results shown are for simulations started from the initial condition values of one for the displacement and zero for the velocity. Other initial conditions were tried, but no substantial effect of initial conditions on chaos boundaries was detected. See the discussion in the next section however. Of course, the time history details do depend upon initial conditions, particularly in the chaotic regime.

A Poincaré map was used to gain further insight into the structure of the chaotic motion. Here a Poincare map is a set of points in the phase plane, A, Å, plotted at discrete intervals of time, once each period, $\omega\tau = 2n\pi$ for all n. For brevity, we have not shown the large number of phase-plane portraits and Poincaré maps that have been calculated. In Fig. 5(a) and 5(b), two representative Poincare maps are shown for $\omega = 1$ and the two damping values used in this study.

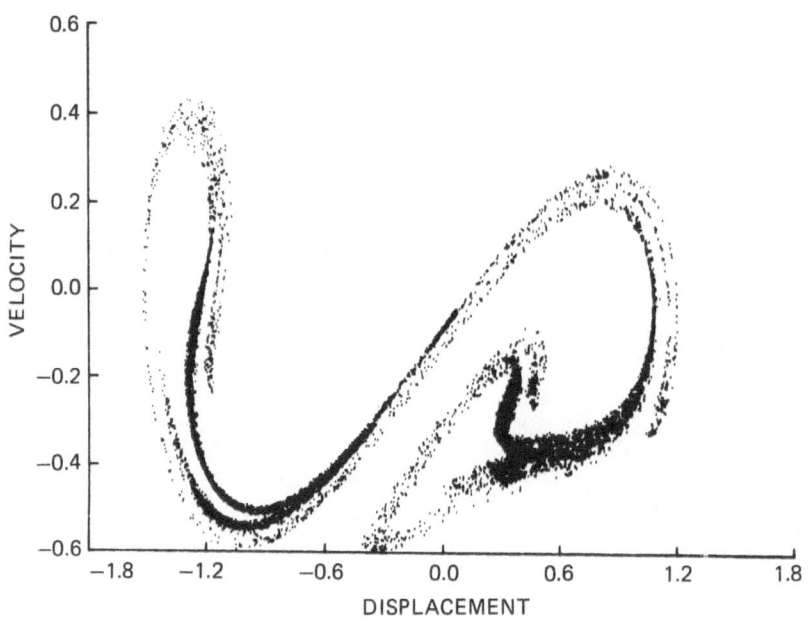

Fig. 5a. Poincare map; $\omega = 1$, $\gamma = .168$, $F_0 = .21$.

The types of chaos found in the simulations varied from frequency to frequency. However, the form of the Poincaré map for a given set of frequencies tended to be of the same type when the lower force steady state periodic phase plane portraits were shaped the same and possessed the same periodicity and when the corresponding upper force portraits were also

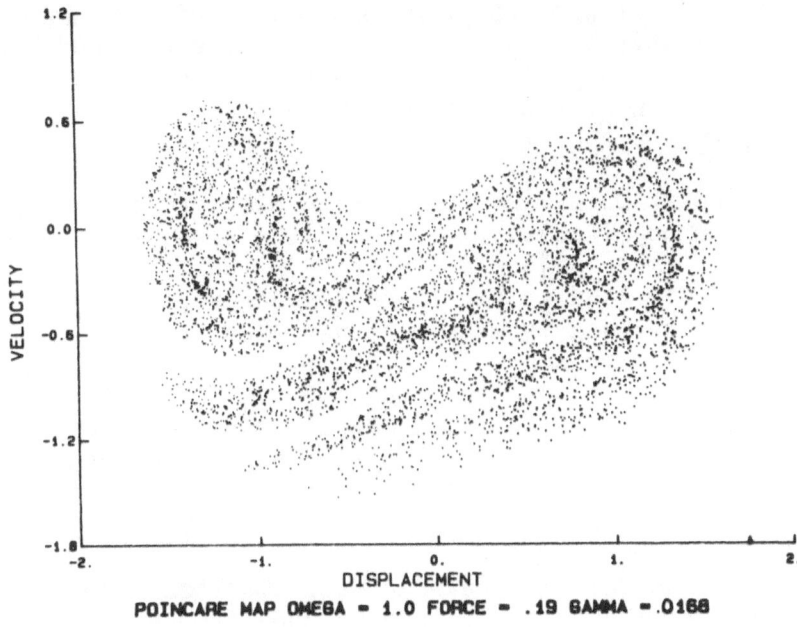

Fig. 5b. Poincare map; ω = 1, γ = .0168, F₀ = .09.

the same across the frequency band. The lower and
upper plane portraits were not identical.   Such identification or association of results at several frequencies has led us to connect some data points in
Fig. 4(a) and 4(b) by straight lines.   Such results
suggest that chaos lies in fragmented pockets in the
force-frequency plane.   These pockets can also have
smaller pockets of force-frequency combinations within
them that can lead to periodic phase plane orbits.

It is apparent from the situations simulated that
chaos can assume many forms. Some of the entries into
the chaotic region in the force-frequency plane for
certain frequency values were precipitated by beam
snap-through; others were not.   At certain frequencies, the system went into chaos even before the beam
snapped through. Period doubling was observed at some
frequencies, e.g., ω = 0.9 and 1.0, but not at others.
Chaos appeared at all frequencies simulated, although

this does not preclude the possibility of finding frequencies that are chaos-free. A simple boundary cannot be drawn in the force-frequency plane above which there is guaranteed chaos; in fact, the simulations point to the opposite.

The simulations run at high damping levels gave the same qualitative answer as the ones run at low damping. The major difference is that the width of the chaotic band in the force-frequency plane for the low damping case is much narrower than its counterpart for higher damping. The limit of zero damping may be pathological. As the damping is decreased, the Poincaré maps also lose their ordered structure. Another difference is that the higher damping case allows a much richer selection of equilibrium periodic phase plane orbits.

The correlation between data obtained from simulation and the data obtained by Moon from his physical experiment also appears to be generally good. See Fig. 4(b). The principal difference is that at $\omega \cong$ .65, the simulation predicts chaos at much higher force levels, $F_0 = .45-.55$, than those observed by Moon in his physical experiment, $F_0 \cong .17$. It is worthy of note that the simulation predicts that snap-through of the beam occurs at $F_0 \cong .12$ and it is possible that this snap-through was identified as chaos in the physical experiment. At higher frequencies, snap-through and chaos occur at force levels that are much closer together. Of course other factors may enter in including the effects of higher beam modes. Indeed, recent calculations including higher beam modes show even better agreement with the data from Moon's experiment [9].

7.1.4. On necessary and sufficient conditions for chaos to occur: a heuristic approach

A fundamental set of issues associated with chaos is the determination of necessary and sufficient conditions for chaos to occur. Indeed solid mathematical theorems would be most welcome here. However, they are not yet available and so a heuristic discussion may be helpful, not only in terms of interpreting and exploiting the available evidence, but also for its significance in providing guidance to more rigorous approaches. We shall use the Duffing oscillator as a concrete example, but the discussion is consistent with the available evidence for other dynamical systems, e.g., the Lorenz equations.

First we need to reconsider our discussion of the results for the Duffing oscillator and we begin with a discussion of the effect of initial conditions on the behavior of the underlined forced Duffings equation, $F \neq 0$. For a more detailed discussion, see [10].

Previously we had said, "... no substantial effect of initial conditions on chaos boundaries was detected." However, there is observed some effect of initial conditions and we shall explore that here. For concreteness consider the Duffing oscillator under sinusoidal excitation for various F_0 and a fixed ω, $\omega = 1$, and fixed γ, $\gamma = .168$. At each F_0 we shall consider a map of initial conditions, $\dot{A}_0$ vs A_0, and ask the question, within each map, what initial conditions lead to periodic motions and what initial conditions lead to chaotic motions? The results obtained from numerical simultations are as follows (see [10] for the detailed results).

- For sufficiently small F_0 ($F_0 < .119$), only periodic motions (there are two possibilities, one periodic motion about each buckled equilibrium position) are observed for all initial conditions.

- Above $F_0 \cong .119$, a few initial conditions lead to chaos, but the vast majority, say more than 98% lead to periodic solutions.

- As $F_0 \rightarrow .177$, the periodic motions began to period double and the percentage of initial conditions which lead to chaos begins to increase measurably.

● As $F_0 \rightarrow$ .205, the period of the period doubling goes to infinity and <u>all</u> initial conditions lead to chaos.

In Fig. 6 a plot of the percentage of initial conditions which leads to chaos vs F_0 is shown.

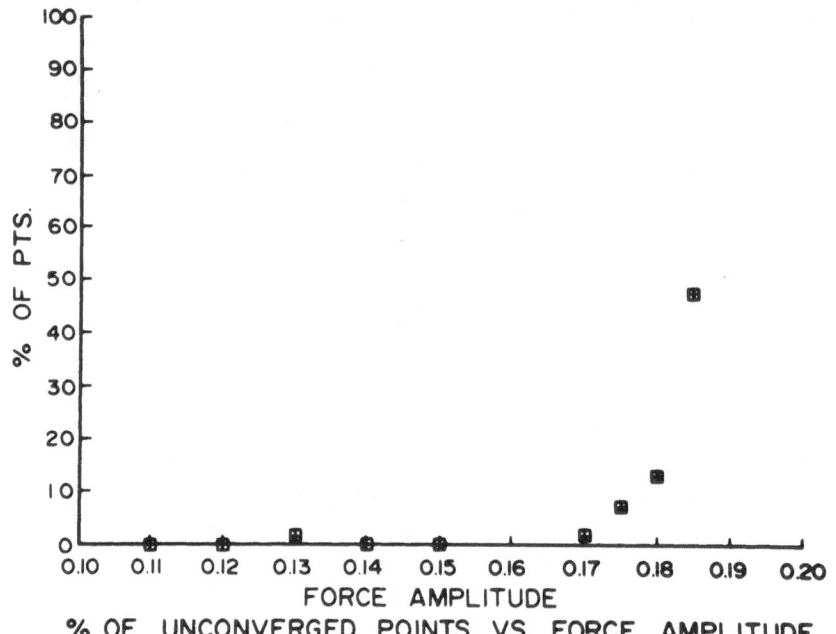

% OF UNCONVERGED POINTS VS. FORCE AMPLITUDE

Fig. 6. Percentage of initial conditions leading to unconverged trajectories vs force amplitude.

How do we interpret these results in terms of necessary and sufficient conditions for chaos? The answer to this question is presently a subject of lively discussion in the research community. Here we provide a set of answers consistent with presently known information.

Based on the results of numerical simulation, it does appear that there is a F_0 below which chaos will not occur for any initial condition and thus there is a <u>necessary</u> level of F_0 for chaos to occur. On the other hand, there is another larger F_0 above which all initial conditions lead to chaos and thus there is also a <u>sufficient</u> level of F_0 for chaos to occur. The later discussion will suggest that above F_0 = .119, <u>transient</u> chaos may occur from some initial conditions, but only for F_0 greater than .205 will steady state chaos occur for all initial conditions.

A criterion for a sufficient condition for chaos is somewhat easier to determine than a necessary condition and thus we shall begin there. What appears to happen as F_0 increases? At low F_0 (below the F_0 where chaos is assumed to occur) there are (many) initial conditions that lead to periodic motions. As F_0 increases these (stable*) periodic motions become unstable with respect to infinitesimal disturbances and are replaced by other stable periodic motions. Sometimes these new stable periodic motions are period doubled motions, but they also may be new families of periodic motions. With further increases in F_0, these new periodic motions themselves become unstable with respect to infinitesimal disturbances and the interval in F_0 between the onset of successive instabilities becomes progressively smaller. Again the period doubling (Feigenbaum) sequence is a well known example, but is not the only possible scenario. The possible sequence of periodic motions which become unstable and then lead to other periodic motions is very rich indeed. There is strong numerical evidence for this from the Duffing's equation (with either positive or negative linear stiffness terms) and from the Lorenz equations. Guckenheimer and Holmes [11] have emphasized this point on basic theoretical grounds.

Be that as it may, at some finite F_0 (usually not too far removed from the first instability of a periodic motion), <u>all</u> periodic motions appear to be unstable with respect to <u>infinitesimal</u> disturbances. Nevertheless it is possible to show (by Lyapunov's method and/or analytical integration of some appropriately simplified equations for large motions) that the motion of the system is bounded for large motions.

---

*There are, in general, unstable periodic motions as well.

If all periodic motions are unstable with respect to infinitesimal disturbances, but the motion can be shown to be bounded for sufficiently large motion, then chaos must occur. (A sufficient condition for chaos.)

Is this the only sufficient condition for chaos? Are there other equivalent conditions or ones that provide a sharper (or less sharp) bound on F_0? There are two we shall mention here. However, before leaving the above sufficient condition, it is worth mentioning a practical objection that has been raised. It is sometimes argued that it is very difficult to be assured that one has found all the periodic solutions and indeed there may even be an infinite number of them. Thus, it is further stated, to determine the stability of <u>all</u> periodic solutions may be a very large task indeed. There is some merit to this argument, of course, but there are several countervailing factors.

- Usually one is strongly motivated to determine all (stable) periodic solutins for reasons quite aside from any considerations of chaos.

- It appears the number of (unstable and stable) periodic motions is usually finite for a finite dimensional system. A good theorem would be helpful here! However, even if the number of periodic motions is infinite, one might hope that their behavior including their stability could be assessed by (analytical/ numerical) asymptotic means.

- The determination of the stability of periodic motions with respect to infinitesimal disturbances has a very well founded theoretical base (Floquet theory), and it is a very attractive alternative to exploit. Indeed from this point of view, a period doubling sequence is nothing more than the instability of a limit cycle due to parametric resonance. Also, by elementary means one can invoke this theory to display a fundamental connection between instability of periodic motions and (infinite) sensitivity of these motions with respect to system parameters including initial conditions. We briefly digress to show this.

7.1.4.1. Stability of periodic motion and sensitivity to system parameters

Consider again the following equation:

$$\ddot{A} + \gamma\dot{A} - \frac{1}{2} A(1 - A^2) = F(\tau) \qquad (1.4)$$

To consider the stability of (1.4) with respect to infinitesimal disturbances, let

$$A(\tau) = \bar{A}(\tau) + \hat{A}(\tau) \qquad (1.5)$$

where $\bar{A}(\tau)$ is a periodic solution of (1.4) and $\hat{A}(\tau)$ is an infinitesimal perturbation. Noting that $\bar{A}(\tau)$ satisfies (1.4) exactly, substituting (1.5) in (1.4) and linearizing in $\hat{A}(\tau)$ gives

$$\ddot{\hat{A}} + \gamma\dot{\hat{A}} + \frac{\partial f}{\partial A}(\bar{A})\hat{A} = 0 \qquad (1.6)$$

where $f(A) \equiv -\frac{1}{2} A(1 - A^2)$

This is a <u>linear</u> ordinary differential equation with a variable coefficient $\partial f(\bar{A})/\partial A$, which depends upon the periodic motion, $\bar{A}(\tau)$. Its stability is readily assessed by Floquet theory.

There is another line of argument that also leads to (1.6). Return to (1.4) and recall, for example, that we have initial conditions, say $A(\tau = 0) = A_0$ and $\dot{A}(\tau = 0) = 0$. Let us determine the change of A due to a change in the initial condition, A_0. For small changes,

$$\Delta A = \frac{\partial A}{\partial A_0}\bigg|_{A = \bar{A}} \Delta A_0 \qquad (1.7)$$

where $\bar{A}$ is some periodic motion associated with the original initial conditions. Hence we seek to determine the "sensitivity" to initial conditions which we define as $\frac{\partial A}{\partial A_0}\big|_{\bar{A}}$. Differentiating (1.4) and the initial conditions with respect to A_0, one obtains

$$\frac{\partial \ddot{A}}{\partial A_0} + \gamma \frac{\partial \dot{A}}{\partial A_0} + \frac{\partial f}{\partial A}\bigg|_{\bar{A}} \frac{\partial A}{\partial A_0} = 0 \qquad (1.8)$$

with initial conditions $\frac{\partial A}{\partial A_0} = 1$, $\frac{\partial \dot{A}}{\partial A_0} = 0$. We see that $\frac{\partial A}{\partial A_0}$ obeys the same equation as $\hat{A}$ by comparing (1.8) and (1.6). Hence instability of the periodic motion also implies extreme sensitivity to initial conditions (and also to other parameters, e.g. $\partial A/\partial \gamma$, etc., as a similar calculation will show). Note that since (1.6) and (1.8) are linear their stability does not depend on their initial conditions [unlike (1.4)].

The requisite numerical calculations for Duffing's equation using Floquet theory have not as yet been made. Such calculations would be most helpful.

7.1.4.2. Two other possible sufficient conditions for chaos

We briefly mention two other possible sufficient conditions for chaos. The first deals with Lyapunov exponents. See, for example, [12] and [13]. A Lyapunov exponent is essentially an exponential measure of the departure or convergence of two trajectories of an equation which start from slightly different initial conditions, averaged over all possible system trajectories. At least one positive Lyapunov exponent (the number of exponents is equal to the order of the system) is said to ensure chaos. A zero Lyapunov exponent is usually indicative of neutral stability of a periodic motion, although for a non-autonomous system which is converted to an autonomous form through introduction of an additional degree of freedom, one Lyapunov exponent will be trivially zero. For Duffing's equation (and some other equations) it has been shown [13] that a zero Lyapunov exponent occurs at the F_0 where a periodic motion becomes unstable and bifurcates to a new periodic motion (with a return to negative values of the Lyapunov exponent for further increases in F_0). However and moreover, when one of the Lyapunov exponents becomes positive, this occurs at the same F_0 where chaos exists for all initial conditions [13]. Lyapunov exponents may also be used to estimate the effective dimension of the system, but we shall not pursue that here, even though it is a very important topic for some systems, e.g. turbulent fluids [12,13].

Finally we consider one more possible sufficient condition for chaos. This is associated with the intersections of unstable and stable periodic trajectories in the phase plane (or more strictly speaking in the phase space or Poincare map). In simple terms the idea is as follows, thinking again of Duffing's equation for definiteness. In Fig. 7 we show for a certain F_0 the trajectories of the stable (solid lines) and unstable (dashed line) periodic motions. The latter is an unstable oscillation about the unstable straight beam configuration, $A = \dot{A} = 0$. As F_0 increases, the orbits grow, and for sufficiently large F_0 they intersect. If the intersection of the orbits occurs at the same instant of the force oscillation period (the corresponding A and $\dot{A}$ at that instant in time define a point from each orbit on a Poincaré map), then the subsequent motion of the system could continue on either the stable or unstable orbit and presumably this uncertainty will occur once (or more) each period of the force oscillation. The slightest deviation at the instant of intersection of the stable and unstable (Poincare map) orbits will lead to an essentially random choice between stable and unstable periodic motions and thus chaotic motion. In Fig. 8 we show a plot of A_{min} (see Fig. 7) vs F_0

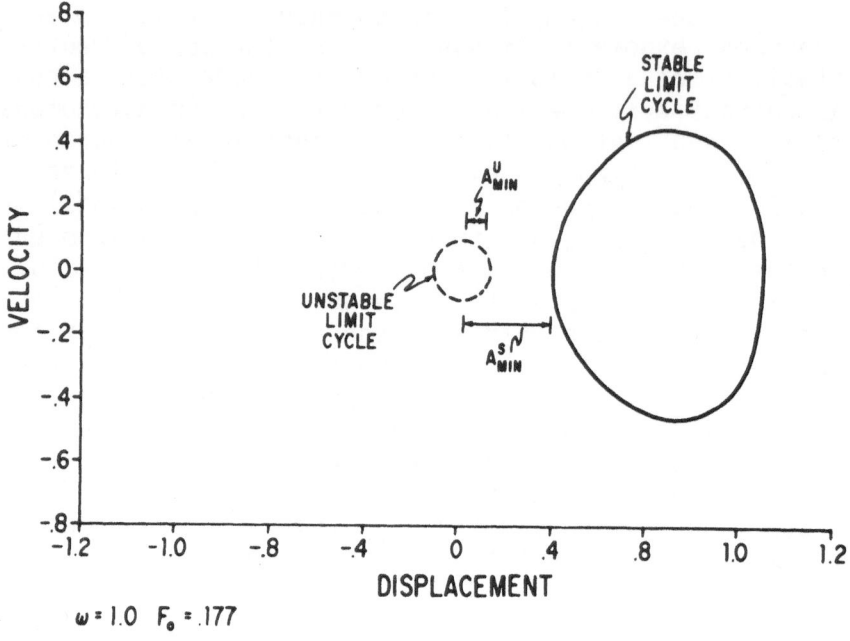

Fig. 7.   Stable   and   unstable   steady   state   limit
         cycles; F_0 = .177.

for the stable and unstable periodic motions.   At the
point of intersection of these two orbits in the phase
space, chaos is assured.   Because the phase space is
nearly two-dimensional, the intersection in the phase
plane, A, Ȧ, gives a   good representation of. the behavior in the three-dimensional space (A, Ȧ and τ).
This F_0 so determined correlates well with that determined from the previously discussed sufficient conditions.   An idea broadly similar to the above also has
been discussed by Ueda [14].

It is an interesting theoretical question as to
whether the phase space trajectories do, in fact,
intersect or simply become arbitrarily close.   This
suggests a close re-examination of the usual uniqueness theorems may be in order.

The F_0 corresponding to a sufficient condition
for chaos we shall call F_0^{SC}.

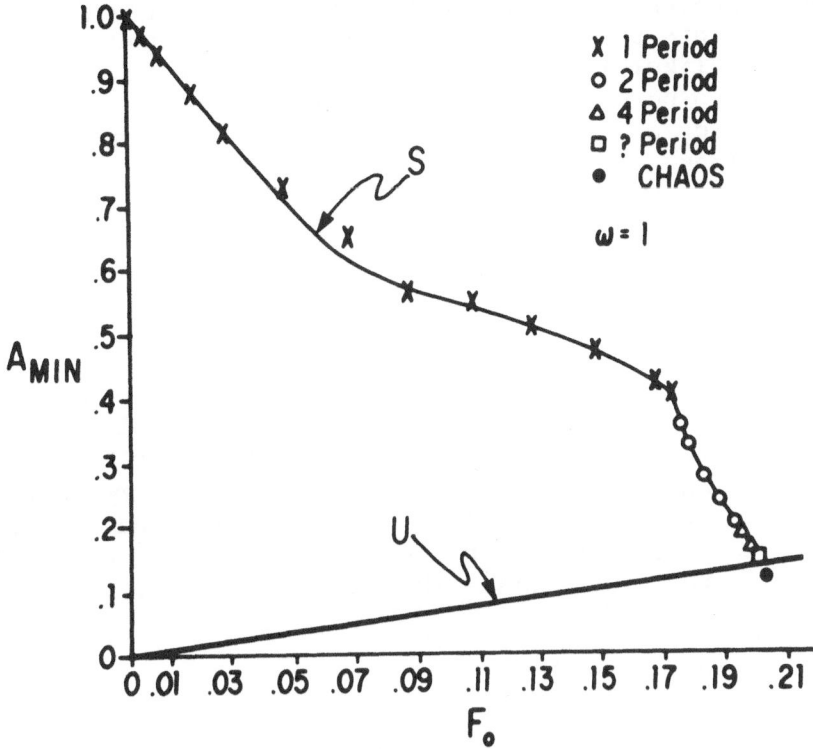

Fig. 8. Magnitudes of stable and unstable limit
cycles vs force amplitude.

7.1.4.3. Necessary conditions for chaos to occur:

A mathematically trivial, but physically important, necessary criterion for chaos to occur is that
the system be nonlinear. It is easy to show that
linear systems cannot exhibit chaos.

Given that the system must be nonlinear for chaos
to occur, is there a further necessary condition? The
answer is not rigorously known, but the work of
Holmes, Marsden, Melnikov, and others is suggestive.
See [11]. In a sense it is related to the last sufficient condition that we discussed.

Again consider the stable and unstable periodic
orbits, but now take a F_0 sufficiently small that

these periodic orbits do not intersect. Even if they do not intersect it may be possible (though in some sense unlikely) for the motion to move from near one orbit to near another. In general, this will only happen for a sufficiently large F_0 and certain initial conditions.

Following Holmes et al., consider two (actually three) families of initial conditions.

- Family 1

    This is the collection of initial conditions which will lead to a steady state oscillation in the unstable periodic orbit. For $F_0 \equiv 0$, these points are the boundary curves of the shell plot which separate the initial conditions which lead to one or the other of the two static (buckled) equilibrium as $\tau \to \infty$.

- Family 2

    This is the (much larger) collection of initial conditions (for $F_0 < F_0{}^{SC}$) which leads to stable periodic orbits as $\tau \to \infty$.

- Special Family 2

    But now consider a special subset of Family 2, namely those initial conditions that emerge from the unstable orbit and lead to a stable orbit as $\tau \to \infty$.

    It is conjectured by Holmes and others that,

    If Family 1 and Special Family 2 intersect in the phase plane (transversely), then chaos may ensue. If they do not intersect, chaos will not occur [a necessary condition for chaos].

One of the (perturbation) calculation procedures by which the intersection is shown to occur is due to

Melnikov, so this condition is usually called the Holmes-Melnikov criterion. It should be emphasized that this intersection in the phase plane is only a near intersection (entanglement) in the phase space.

Actually, one cannot quite prove this statement. Indeed, as discussed by Guckenheimer and Holmes [11], what can be proven is that very complicated orbits may exist when the intersection occurs. Intuitively, one can imagine a transient chaotic orbit that passes back and forth for a very long (but finite) time between and near families of initial conditions associated with the stable and unstable periodic orbits. However as $\tau \to \infty$, periodic motion appears to occur, unless and until one of the previously discussed sufficient conditions for steady state chaos is satisfied.

7.1.4.4. Concluding remarks

Although much remains to be proven in a strict mathematical sense, a reasonably clear picture is beginning to emerge from numerical evidence, intuition, and basic modern dynamical theory to better display necessary and sufficient conditions for chaos to occur.

7.1.5. What is strange about a strange attractor?

The term strange attractor has been introduced into the literature on chaos. But there is no universal agreement as to what is the best definition of the term, see Guckeheimer and Holmes [11]. The genesis of the phrase strange attractor is, at least in part, by analogy to ordinary attractors, e.g. fixed points and limit cycles. A fixed point that is stable with respect to infinitesimal disturbances is an attractor for a system with a stable static equilibrium configuration. A limit cycle that is stable with respect to infinitesimal disturbances is an attractor for a system with a stable periodic dynamic equilibrium configuration. By analogy, it has been hypothesized in the literature that a strange attractor exists for a system that undergoes chaotic motion. That is, the set of all points in phase space for the system undergoing chaotic motion is the attracting set of this strange attractor. However, the question of the stability of the strange attractor is a difficult one.

From our earlier discussion and a great deal of empirical (numerical) evidence, the trajectory of chaotic motion in phase space is known to be sensitive to small changes in parameters including initial conditions. Therefore, it seems extremely questionable as to whether one can say that a strange attractor is stable with respect to (even) infinitesimal disturbances. But if it is not, then of what value is it? Which then leads to the next question, can we understand chaos without the concept of a strange attractor? The answer is most assuredly yes.

Chaos can be understood in terms of ordinary attractors and their interaction. To recapitulate and elaborate on the discussion of the previous section

A sufficient condition for chaos is that all ordinary attractors be unstable with respect to infinitesimal disturbances, but that large motions remain bounded.

Thus to determine a sufficient condition for chaos, one can in principle examine each ordinary attractor and its stability separately. Moreover, recall that an alternative sufficient condition for chaos was obtained by considering the (near) intersection in the phase space or Poincaré map of a stable and an unstable ordinary attractor (limit cycle).

In addition, to determine a necessary condition (Holmes-Melnikov, for example) for transient chaos, it was also necessary to consider the interaction of a stable and an unstable ordinary attractor. When the initial condition map associated with each attractor intersects the Poincaré map of the other attractor in the phase plane, then transient chaos becomes possible (although, of course, not ensured). Hence, to consider a necessary, but not sufficient, condition for transient chaos more than one ordinary attractor must be considered.

It immediately follows that if all ordinary attractors are stable with respect to infinitesimal disturbances and do not intersect, then no steady state chaos will appear. It also follows that if the system has only one attractor (be it stable or unstable with respect to infinitesimal disturbances), then chaos will not occur. This last statement includes, of course, all linear systems.

Finally to return to the question that began this discussion of strange attractors, it can be said that what is strange about a strange attractor is that it does not exist (or to be somewhat more precise we need not define it to exist in order to understand chaos)!

7.1.6. Conclusions and future work

Among the conclusions reached based on the present work are the following:

1. The initial value problem is a key to the understanding of chaos.

2. Chaos is not difficult to find by numerical simulation; however, a Feigenbaum (period doubling) sequence may be difficult to find for some parameter conditions.

3. A comparison between theoretical results for Duffing's equation and (physical) experiments for a buckled beam shows generally good agreement.

4. A reasonably clear picture is beginning to emerge with respect to necessary and sufficient conditions for chaos to occur.

5. The concept of a strange attractor is not essential to the understanding of chaos.

6. Future theoretical studies should consider

- investigating the limit as damping approaches zero; setting the damping identically zero may lead to pathological results

- multimode convergence studies; based upon the results from panel flutter calculations [1,2], it is expected good convergence will occur and recent results support this expectation [9].

7. Future experimental work should attempt to study

- various damping levels

- determination of period doubling conditions

- identification of entire pockets of chaos

7.2. Self-excited, chaotic oscillations of an autonomous system: Lorenz equations*

7.2.1. Introduction

It might be thought that self-excited chaotic oscillations arise from a somewhat different mechanism than that associated with forced oscillations. In the case of the Lorenz equations [3], chaotic oscillations appear to be the result of all (static or dynamic) equilibrium solutions becoming unstable with respect to infinitesimal disturbances. However, recall the discussion of necessary and sufficient conditions for chaos to occur in Duffing's equation in Section 7.1; the differences between the Lorenz and Duffing's equations as far as chaotic behavior may be more apparent than real. It is desirable that for both the Duffings and Lorenz equations

*Section 7.2 is an extended version of E. H. Dowell and C. Pierre, "Chaotic Oscillations in Mechanical Systems," in Chaos in Nonlinear Dynamical Systems, edited by J. Chandra, © 1984, Society for Industrial and Applied Mechanics, Philadelphia, PA.

- a systematic study of the static and dynamic equilibria be made using recently developed techniques such as the incremental harmonic balance method [15,16]

- followed by a systematic study of the stability of such equilibria with respect to infinitesimal disturbances using conventional or Floquet stability theory

Such studies would further our understanding of the mechanism by which chaotic oscillations arise.

Here some of the important presently known characteristics of the solutions of the Lorenz equations are summarized. Also a new, recently developed [2] method is described for displaying and interpreting the results of computer simulations of these equations.

The Lorenz equations are obtained from a truncated modal expansion of the Navier-Stokes equations. They have the following form.

$$\dot{x} = \sigma(y - x) \qquad (2.1)$$

$$\dot{y} = -xz + rx - y \qquad (2.2)$$

$$\dot{z} = xy - bz \qquad (2.3)$$

It is clear from (2.1)-(2.3) that, if x,y,z is a solution, then so is -x,-y,z. Much of our present understanding of these equations and their solutions is summarized in the excellent book by Sparrow [17]. Originally it was hoped that these equations describe the Rayleigh-Benard flow of a fluid between heated plates. That no longer seems likely because of the poor convergence properties of the modal expansion. But the equations have retained interest for their possible prototypical value.

7.2.2. Static Equilibrium Solutions

It is easily shown that for $r < 1$, only a single static equilibrium solution exists, namely,

$$x_0 = y_0 = z_0 = 0 \qquad (2.4)$$

For $r > 1$, three static equilibrium solutions exist, one is that given by (2.4) and the two others are

$$x_0 = y_0 = \pm b^{1/2}(r - 1)^{1/2}$$

$$z_0 = r - 1 \qquad (2.5)$$

7.2.3. Stability of static equilibria with respect to infinitesimal disturbances

In the usual way, one assumes that:

$$x = x_0 + \hat{x}$$

$$y = y_0 + \hat{y} \qquad (2.6)$$

$$z = z_0 + \hat{z}$$

and substitutes (2.6) into (2.1)-(2.3). Linearizing the result in x, y, z and using (2.4) or (2.5), one can determine the stability or instability of the static equilibria, (2.4) or (2.5), by standard eigenvalue methods.

It is found that the equilibrium described by (2.4) is stable with respect to infinitesimal disturbances for $0 < r < 1$ and unstable for $r > 1$. The equilibria described by (2.5) do not exist for $0 < r < 1$, of course. For $r > 1$ they are stable until

some larger r = r* where r* depends on σ and b.   For
the commonly studied values of σ = 10 and b = 8/3, r*
= 24.74.

For r > r*, where all three static equilibria are
unstable with respect to infinitesimal disturbances,
chaos ensues.   One might reasonably ask, however,
whether there are any stable (large amplitude) dynamic
equilibria (limit cycles) for r > r*.   Apparently the
answer is no. The evidence for this last statement is
twofold.   First of all, numerical solutions obtained
by digital simulation show   chaotic (random-like)
motions and no evidence of periodic, limit cycle
motions.   Second,   Sparrow has used the so-called
"periodic shooting" technique to obtain solutions for
unstable periodic, limit cycle motions.   These appear
near r = 13.926 with infinite period in what Sparrow
calls a "homoclinic explosion." The period decreases,
as does the amplitude of this unstable limit cycle
motion, as r increases.   At r = 24.74, the amplitude
of this unstable limit cycle appears to go to zero.

Of course, these unstable limit cycle motions
would never be seen in a physical experiment or a
digital simulation (numerical experiment).   Nevertheless, the study of their birth and death does provide
additional understanding and insight into the transition to chaos as r increases.

7.2.4. Boundedness of large amplitude motion

When all (static or dynamic) equilibria become
unstable with respect to infinitesimal disturbances,
there is the question as to whether the motion will
remain bounded or not.   Lorenz and Sparrow have shown
that the motion will remain bounded by using Liapunov
function methods.   Here an alternative and simpler
approach is used.   (A similar calculation can be
carried out for Duffing's equation and the panel
flutter equations.)

For   sufficiently   large   motions,   equations
(2.1)-(2.3) simplify to

$$\dot{x} = \sigma(y - x) \qquad (2.1a)$$

$$\dot{y} = - xz \qquad (2.2a)$$

$$\dot{z} = xy \qquad (2.3c)$$

From (2.2a) and (2.3c), one concludes that

$$\dot{y}y + \dot{z}z = 0$$

$$\text{or } y^2 + z^2 = C, \text{ a constant} \qquad (2.7)$$

From (2.7), one sees that if y,z are bounded at $t = 0$, then y,z are bounded for all time. Moreover from (2.1a),

$$\text{if } x < y, \text{ then } \dot{x} > 0 \text{ and } x \text{ increases}$$

and   if $x > y$, then $\dot{x} < 0$ and x decreases

Thus x is bounded and x is of the same order of magnitude as y.

7.2.5. Some representative results

Several results will be shown for

- phase plane trajectories

- Poincaré maps

- the probability density of an event

In Fig. 9, a (two-dimensional cut of the) phase plane trajectory is shown for $r = 25$, $\sigma = 10$ and $b = 8/3$ in terms of z vs y. Each point is at a discrete instant of time. The phase plane trajectories spiral

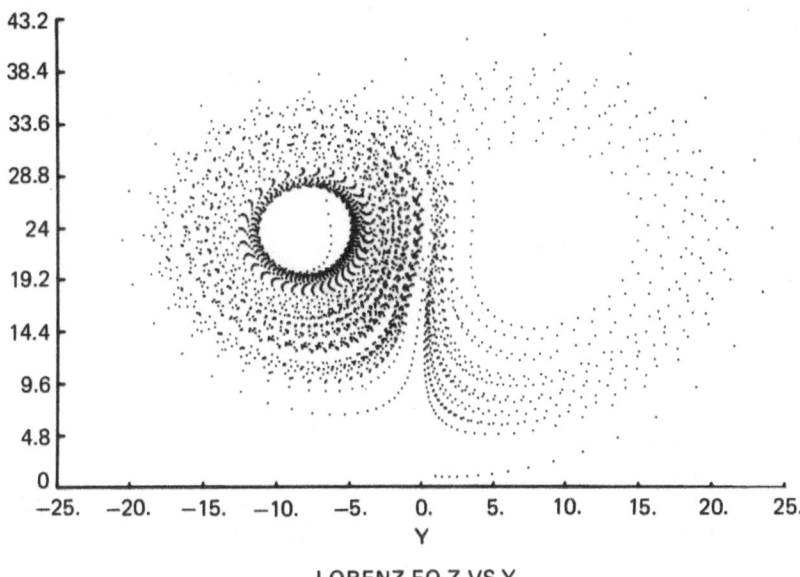

LORENZ EQ Z VS Y

Fig. 9. Phase plane trajectory; Z vs Y.

in a chaotic manner about the two non-trivial (and unstable) static equilibrium points given by (2.5).

In Fig. 10 a Poincaré map is shown in terms of x vs y. An _event_ is defined to occur when, say, $z = z_0$ ≡ $r - 1$. At those instants of time when an event occurs (which are separated in general by _random_ intervals of time), if x is plotted versus y, then a Poincaré map (or return map, i.e., z returns to its chosen value) is generated. As can be seen, the regions of possible values of x and y when an event occurs are relatively small. Such a map may be inter-preted as displaying a conditional probability. That is, given the occurrence of an event, $z = z_0$ ≡ $r - 1$, and a value of say x, it answers the question of what is the probable value of y.

Finally, in Fig. 11 the temporal statistics of an event are shown. Using the same parameters as before and the same definition of an event, the probability

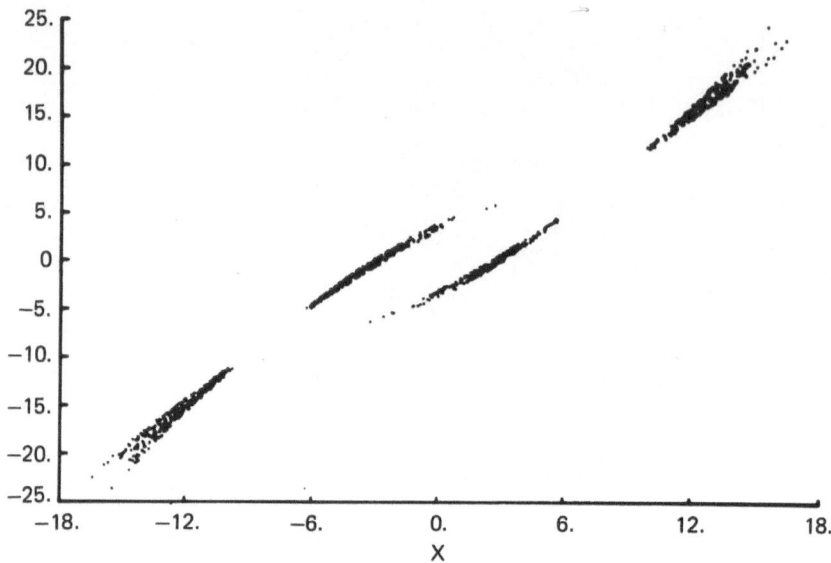

Fig. 10. Poincare map; Y vs X for Z = 24.

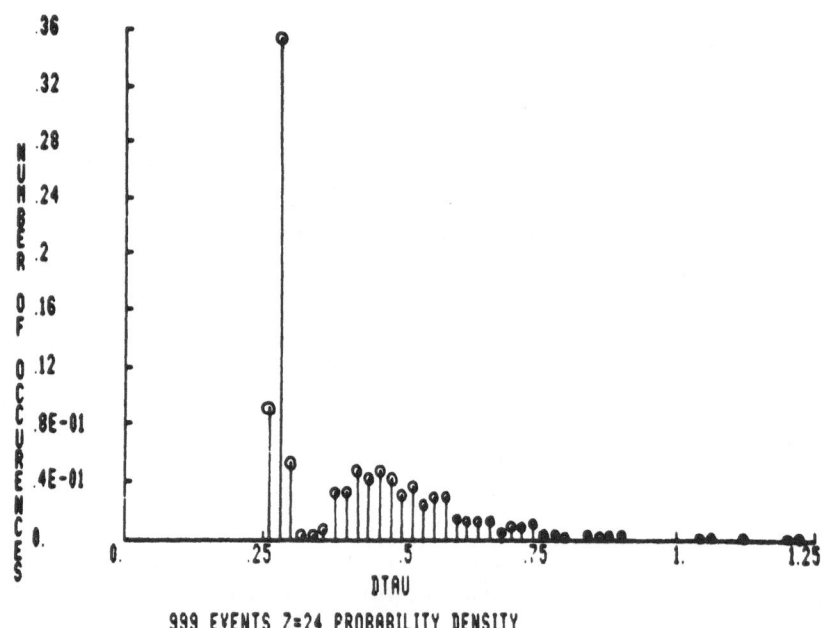

999 EVENTS Z=24 PROBABILITY DENSITY

Fig. 11. Probability of an event at time τ after a previous event.

density of the time interval between two successive
events is displayed. On the vertical axis the relative number of events that occur a certain time
interval apart is plotted and on the horizontal is
plotted the time interval itself. As can be seen
there is a well defined most probable time interval
between two successive events. This gives an important insight into the characteristic time scale of the
chaotic oscillation. For more details on the temporal
statistics of an event, see Ref. 2.

7.2.6. Concluding remarks

As Sparrow [17] has noted, "If finite-dimensional
equations are to be used to model infinite dimensional
systems, it seems important that similar behaviors
should be observed in low dimensional systems of dif-
fering dimension.: In the case of both the buckled
beam and the Lorenz equation models, there is a question of convergence or closure of the model equations.
That is, if additional terms are retained in the spatial modal (generalized Fourier) series that are used
to derive the model ordinary differential equations
from the original partial differential equations
(either beam theory or the Navier-Stokes equations),
does the series expansion converge and, if so, how
rapidly? Presently the answer to this question is not
known for the Navier-Stokes equations, but for the
beam models the situation is somewhat clearer [1,2,9].

It appears that the beam model converges rapidly
while there is considerable doubt about the Lorenz
model as Manley [18] and Miles [19], among others,
have discussed. Interestingly, the panel flutter
model is also known to have good convergence
properties. It appears that one way to achieve good
convergence is to have two controlling parameters, one
of which governs a static instability (divergence) and
other of which governs a Hopf bifurcation (flutter) or
a prescribed external forcing. This is the case for
panel flutter and may be the case for some fluid

dynamical phenomena as well. See Sparrow [17], pg. 188. In the generalized Lorenz (pure Benard) system, only a single controlling parameter exists and the convergence properties appear to be quite weak as suggested by the numerical work of Curry et al. [20]. Thus another parameter or physical effect, for example an electric field, might be added to moldify (improve) the convergence properties. This, in turn, might suggest an important physical effect, the control of chaos (turbulence). A similar thought has been expressed by Roux et al. [21].

References

Books

A. J. Guckenheimer and P. Holmes, Nonlinear Oscillations, Dynamical Systems and Bifurcations to Vector Fields, Springer-Verlag, 1983.

B. F. C. Moon, Experiments in Chaotic Oscillations, John Wiley and Sons, 1987.

C. J. M. T. Thompson and H. B. Stewart, Nonlinear Dynamics and Chaos, John Wiley and Sons, 1986.

Articles

1. E. H. Dowell, "Flutter of a Buckled Plate as an Example of Chaotic Motion of a Deterministic Autonomous System," J. Sound Vibration, Vol. 85, 1982, pp. 333-344.

2. E. H. Dowell, "Observation and Evolution of Chaos in an Autonomous System," J. Applied Mechanics, Vol. 51, 1984, pp. 664-673.

3. P. J. Holmes and F. C. Moon, "Strange Attractors and Chaos in Nonlinear Mechanics," J. Applied Mechanics, Vol. 108, 1983, pp. 1021-1032.

4. W. -Y. Tseng and J. Dugundji, "Nonlinear Vibrations of a Buckled Beam Under Harmonic Excitation," J. Applied Mechanics, Vol. 38, 1971, pp. 467-476.

5. P. J. Holmes, "A Nonlinear Oscillator with a Strange Attractor," Phil. Trans. of Royal Society, London, Vol. 292, 1979, pp. 419-448.

6. F. C. Moon, "Experiments on Chaotic Motions of a Forced Nonlinear Oscillator: Strange Attractors," J. Applied Mechanics, Vol. 47, 1980, pp. 638-644.

7. E. L. Reiss and B. J. Matkowsky, "Nonlinear Dynamic Buckling of a Compressed Elastic Column," Quart. Appl. Math., Vol. 29, 1971, pp. 245-260.

8. M. J. Feigenbaum, "Quantitative Universality for a Class of Nonlinear Transformations," J. Stat. Physics, Vol. 19, 1978, pp. 25-52.

9. D. M. Tang and E. H. Dowell, "On the Threshold Force for Chaotic Motions," J. Applied Mechanics, to be published.

10. C. Pezeshki and L. H. Dowell, "An Examination of Initial Condition Maps for the Sinusoidally Excited Buckled Beam Modeled by the Duffings Equation," J. Sound Vibration, Vol. 117, 1987, pp. 219-232.

11. John Guckenheimer and Philip Holmes, Nonlinear Oscillations, Dynamical Systems and Bifurcations of Vector Fields, Springer Verlag, New York, 1983.

12. A. Wolf, J. B. Swift, H. L. Swinney, and J. A. Vastano, "Determining Lyapunov Exponents from a Time Series," Physica 16D, 1985, pp. 285-317.

13. C. Pezeshki and E. H. Dowell, "Generation and Analysis of Liapunov Exponents for the Duffings Equation," to be published.

14. Y. Ueda, "Explosion of Strange Attractors Exhibited by Duffings Equation," in <u>Nonlinear Dynamics</u>, edited by Robert H. G. Helleman, New York Academy of Sciences, New York, 1980, pp. 422-434.

15. S. L. Lau, Y. K. Cheung and S. Y. Wu, "A Variable Parameter Incrementation Method for Dynamic Instability of Linear and Nonlinear Elastic Systems," J. Applied Mechanics, Vol. 49, 1982, pp. 849-853.

16. C. Pierre and E. H. Dowell, "A Study of Dynamic Instability of Plates by an Extended Incremental Harmonic Balance Method," J. Applied Mechanics, Vol. 52, 1985, pp. 693-697.

17. C. Sparrow, <u>The Lorenz Equations: Bifurcations, Chaos, and Strange Attractors</u>, Applied Mathematical Sciences, Volume 41, Springer-Verlag, New York, 1982.

18. O. Manley, "Determining Modes and Fractal Attractors of Turbulent Flows," in <u>Chaos in Nonlinear Dynamical Systems</u>, Edited by J. Chandra, SIAM, 1984.

19. J. Miles, "Strange Attractors in Fluid Dynamics," in <u>Advanced in Applied Mechanics</u>, edited by J. W. Hutchinson and T. Y. Wu, Academic Press, 1984.

20. J. H. Curry, J. R. Herring, J. Loncaric, S. A. Orszag, "Order and Disorder in Two-and Three-Dimensional Benard Convection," J. Fluid Mechanics, Vol. 147, 1984, pp. 1-38.

21. J. C. Roux, P. Richetti, A. Arnedo, and F. Argoul, "Chaos in a Chemical System: Toward a Global Understanding," J. de Mecanique Theorique et Applique, Vol. 3 Numero Special, 1984, pp. 77-100.

CHAPTER VIII

THE EFFECTS OF COMPLIANT WALLS ON

TRANSITION AND TURBULENCE*

Summary

Normally aeroelastic phenomena are considered deleterious or even destructive. However Ashley (Chapter 0, Ref. 7) has emphasized the possible beneficial effects of aeroelasticity. Perhaps the most exciting possibility for a favorable effect of aeroelasticity is that of reduction of drag on a body through the aeroelastic (or hydroelastic) interaction of a compliant wall and the surrounding fluid stream. Not surprisingly, perhaps, since the full understanding of such a possible effect requires nothing less than an understanding of fluid turbulence and its interaction with a compliant wall, progress has occurred in fits and starts and the subject is far from completely understood. In this chapter our present knowledge of this fascinating topic is considered.

A review is undertaken to identify the major findings of the research community and to suggest future directions. An annotated, selective bibliography is also included. A fundamental theoretical approach to the problem is proposed. Two possible thrusts are identified. One is the study of the limit cycle oscillations of the Navier-Stokes equations for a boundary layer flow over a compliant wall or Poiseuille flow in a channel with compliant walls. The other is a complementary study of a simpler, sequen-

---

*This chapter is based on a review article prepared at the request of Dr. Michael Reischman of the Office of Naval Research. It first appeared in Shear Flow-Structure Interaction Phenomena, edited by A. Akay and M. Reischman, NCA-Vol. 1, ASME, November, 1985. The several other articles in these proceedings are highly recommended to the reader for their complementary coverage and views.

tial theoretical model which partially decouples the fluid and wall motions. The first study is more rigorously based but will be initially limited to Reynolds numbers in the transition region. The second, more approximate approach will allow consideration of fully developed turbulent flow. It is a principal premise of this chapter that our understanding of turbulence transition and structure will be substantially advanced when a soundly based theoretical model is available to guide, interpret, and suggest experiments. These experiments should, in turn, validate or lead to improvements in the theoretical models.

Introduction

A number of years ago Kramer suggested that a compliant wall might lead to drag reduction over that achieved with a rigid wall. Years later his intriguing experiments [1-3] with compliant coatings on bodies captured the imaginations of many engineers and scientists. As is often the case with a complex set of physical phenomena, the subsequent investigations raised as many questions as were answered. However, an understanding of at least the questions to ask has emerged, and we now seem to have reached the point where a fundamental and systematic construction of self-consistent theoretical models should pay real intellectual dividends and provide the guidance so badly needed for the definitive experiments that have yet to be performed.

Comprehensive work on this topic has been conducted by the NASA Langley Research Center where primary emphasis was given to conceptual theoretical modeling and key experiments. Moreover the recent program of research supported by the Office of Naval Research has further advanced our understanding of the subject.

A selected and annotated bibliography is provided at the end of this chapter.

8.1. A brief, historical overview

8.1.1. Drag reduction mechanisms:

There are three mechanisms by which it has been suggested that a compliant wall may lead to drag reduction. (1) First of all the critical Reynolds number at which the laminar boundary layer becomes unstable may increase. (2) Second, even if this does not happen, the amplification rates of the unstable laminar flow may be reduced and hence the completion of transition from laminar to fully turbulent flow delayed. (3) Third, even if neither of the first two occurs, then perhaps the turbulence intensity level and, hence, Reynolds stress and, hence, drag may be decreased.

From a physical point of view (1), (2), and (3) are in increasing order of required sophistication of modeling. From a theoretical point of view they also represent a hierarchy of difficulty. (1) may be addressed using classical linear hydrodynamic stability theory with the boundary condition modified to take into account the compliant wall. (2) requires at least the calculation of amplification rates of the unstable laminar flow rather than simply the stability boundary, and possibly the calculation of the actual (nonlinear, limit cycle) motions of an unstable laminar flow. (3) clearly requires calculation of the nonlinear, time dependent, turbulent flow.

The preponderance of the evidence available today suggests that

(1) does not happen to any significant degree for physically realizable compliant walls [4,5]

(2) may well happen but the best available theoretical model [6] is semiempirical and the experimental data [7], although promising, are inconclusive

(3) may or may not happen although there is a promising, though incomplete, theoretical model [8] and some tentative experimental data [7].

8.1.2. Experiment

The general state of the available experimental
data for drag reduction by compliant walls deserves
some further discussion. It is safe to say that most
research workers find it unsatisfactory. Some data
show drag increases, some drag reduction for compliant
walls vs. rigid walls. Some data show turbulence
intensity level increases, others decreases. Some
show transition delay, others do not. It is of significance, perhaps, that no data show substantial
increases in critical Reynolds number, i.e., the
laminar flow still becomes unstable at very nearly the
same Reynolds number for a compliant wall or a rigid
wall. See References 9-11 for a detailed discussion
of the literature.

Perhaps the most consistent experimental finding,
which is disconcerting but not surprising, is that at
sufficiently high flow velocities many compliant walls
have suffered an aeroelastic instability [7,11-15].
Indeed, it is no exaggeration to say that more compliant walls have shown an aeroelastic instability in
experiments than have shown a drag reduction. Of
course, the standard hydrodynamic stability theory
when modified to include the effect of a compliant
wall does predict an aeroelastic instability [4,5].
In the fluid mechanics literature this is called a
type (c) instability after the nomenclature of Benjamin [4]. In the aeroelastic literature this would be
called plate or membrane flutter [16].

It is generally agreed that the difficulties with
the experimental data for drag reduction by compliant
walls are greatly aggravated, if not caused, by the
large number of parameters which may affect the out-
come and the lack of a theoretical model with quantitative predictive power to guide the experiments. As
discussed below, however, there are sound theoretical
models for predicting aeroelastic instabilities and
pressure distributions over wavy walls which have been
confirmed by experiment. Also see Appendix 8A.

8.1.3. Theory

Let us turn therefore to a discussion of the available theories with primary attention devoted to mechanisms (2) and (3). There is an available, apparently satisfactory theory for (1) [4,5] and it shows that mechanism is not effective for drag reduction.

8.1.3.1. Transition delay

Gyorgyfalvy [6] has calculated amplification rates for the unstable laminar boundary layer and shown that for certain compliant wall-fluid parameter combinations they indeed may be substantially decreased. However there are many parameter combinations which will lead to increased amplification which has an obvious impact on the probability of an experiment unguided by theory being successful. Relating amplification rates to transition per se is, at present, empirical.

One of the interesting consequences of the results of Reference 6 is the clear demonstration of the importance of aeroelastic instabilities in limiting the amount of transition delay one can achieve. In Fig. 1, which is constructed from theoretical data given in Reference 6, see Figure 10 of Reference 6, the position at which transition occurs, X_T, is plotted versus flow velocity. As flow velocity increases the compliant wall actually delays transition, i.e., X_T increases. However, at a sufficiently high flow velocity, the compliant wall is aeroelastically unstable.

8.1.3.2. Decreased turbulence intensity

There are two, somewhat different types of relevant theoretical models available. There are those that attempt nothing less than computation of turbulence from the nonlinear time dependent Navier-Stokes equations [17-20]. However, to date, they have not included the effect of a compliant wall in any realistic way. Clearly it would be desirable to do so even though such methods will be limited to relative small Reynolds numbers in the short term by computer speed and numerical algorithm constraints.

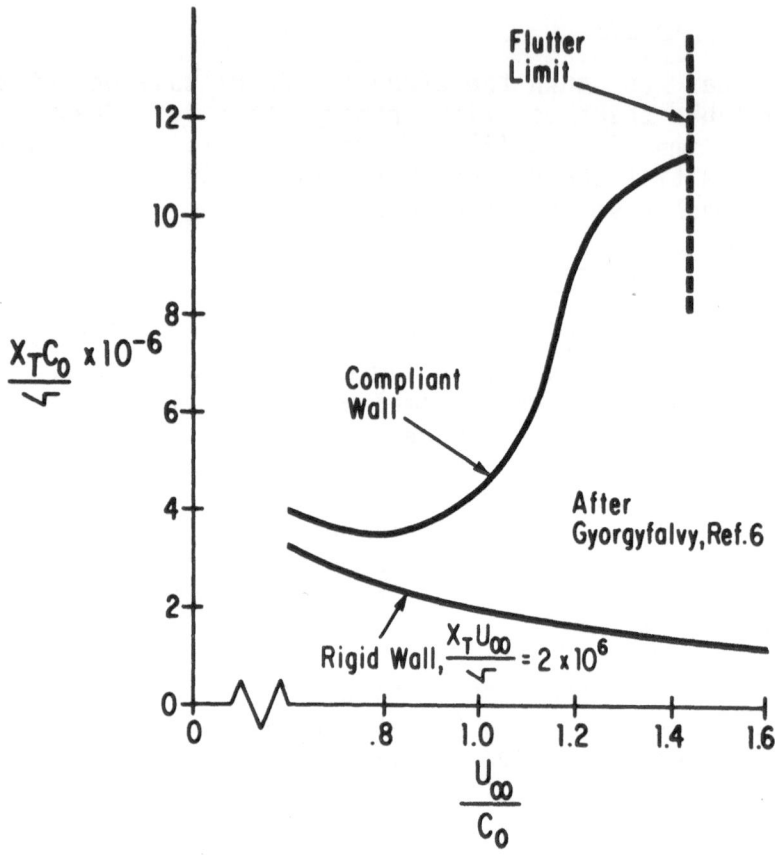

Fig. 1. Transition location vs. flow velocity: x_T - transition location; ν - kinematic viscosity; C_0 - natural wavespeed of membrane; U_∞ - mean flow velocity.

A somewhat different approach such as that of Tsahalis [8] is more modest in its goal, but allows a larger range of Reynolds numbers to be considered. This method as exemplified by Tsahalis proceeds as follows:

(1) A measured turbulent pressure power spectra is taken as given.

(2) The wall motion is calculated in response to the given turbulent pressure fluctuations.

The fluid forces on the wall due to its own motion are neglected, even though it is known that as one nears a condition of aero-elastic instability these forces are important [16]. Moreover, the physical model of the wall assumes that it is point reacting, i.e., a force at one point on the wall only produces at deflection at that point and nowhere else. For many walls of scientific and technological interest that will be too sweeping an approximation.

(3) From the known wall motion, the viscous velocities are calculated from (a linearized set of) the Navier-Stokes equations. Whether linearization is permissible here is an open question.

(4) From the viscous velocities, the Reynolds stress is calculated which, in turn, allows the drag to be estimated.

This approach is promising, although as indicated above, there remain some important open questions. Of course, the fundamental question with this model, which the other approach attempts to answer, is whether it is permissible to determine the wall and then the fluid motion sequentially or whether one must determine them simultaneously (a fully coupled analysis)? Recent work suggests that at least a partial decoupling is possible. To place this approach in context a conceptual summary of various theoretical models is given in the next section of the paper.

8.2. The total model and various component models

8.2.1. The total model

For definiteness, consider a compliant wall, initially quiescent, over which there is a turbulent fluid flow. The time dependent fluid pressure (which will be called Fluid Pressure I) due to the instability of the laminar flow and the consequent turbu-

334

lent fluid motion excites the compliant wall into motion. As a result of the wall motion a further fluid pressure is generated, which shall be called Fluid Pressure II. The compliant wall motion is further modified by Fluid Pressure II and thus a feedback between compliant wall motion and fluid pressure exists. See Fig. 2.

DYNAMICS OF FLUID - STRUCTURE INTERACTION

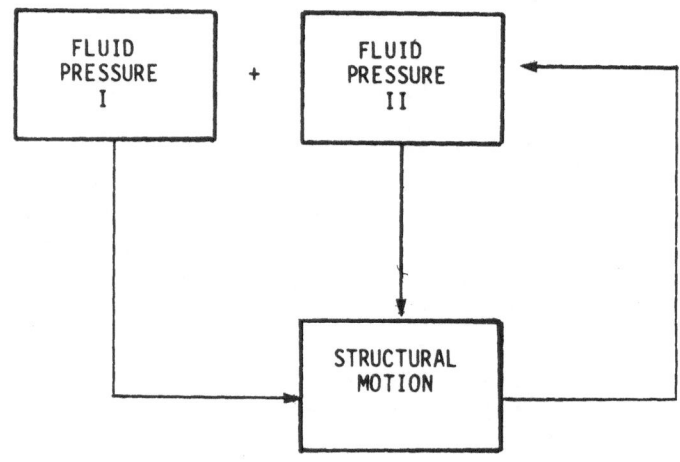

FLUID PRESSURE I

The time dependent fluid pressure due to the instability of the laminar flow and the consequent turbulent (chaotic, deterministic) fluid motion

FLUID PRESSURE II

The time dependent fluid pressure due to the structural motion

Fig. 2. The total problem.

We have spoken here of fluid pressure, but fluid shear stresses are also present. Indeed, these are crucial to the overall assessment of drag. However, the shear stresses will not usually play an important role in determining the compliant wall motion per se. This is a key point with regard to fluid pressure modeling. The wall motion may significantly influence the shear stresses, however, and indeed it is this effect that offers the prospect of a reduction in drag.

Several simpler component physical models have been studied in the past in other contexts. They shall be briefly reviewed here for their relevance to the Total Model. In particular, the degree to which the component theoretical models have been verified by experiment is noted.

8.2.1.1. (Classical, aeroelastic) flutter model

Here the turbulent flow time dependent pressure (Fluid Pressure I) is ignored and the stability of the fluid-structure feedback system is addressed. See Fig. 3 and Refs. 16, 21, and 22. Perhaps most note-worthy of the results of such studies is the establishment that for the determination of aerodynamic pressure due to structural wall motion and flutter the feedback Fluid Pressure II can often be determined by a potential flow model [16,21] (or at most a shear flow model [22]). More will be said of this later. Suffice it to say here that a substantial body of theoretical-experimental correlation exists which supports the above conclusion [16,21,22]. Also see Parts I and II of Appendix 8A.

8.2.1.2. (Structural response to fluid) noise model

Here Fluid Pressure II is neglected (the change in the fluid pressure due to compliant wall motion) and the wall motion is determined due to a prescribed fluid pressure, Fluid Pressure I. This is a standard problem in structural dynamic analysis. See Fig. 4 and Refs. 10, 23 and 24. The theory has been verified by experiment in many contexts.

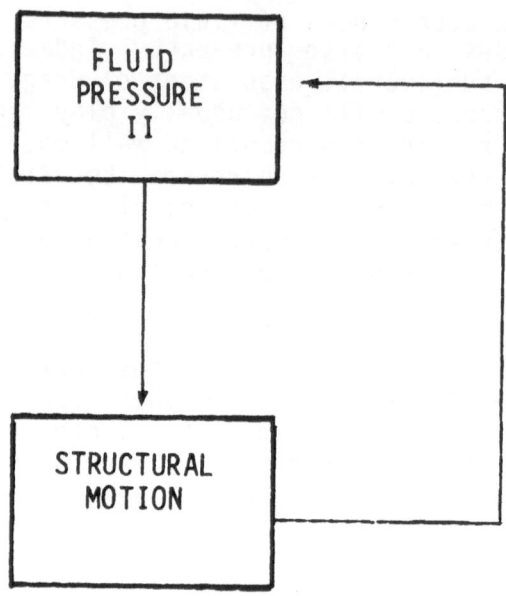

Fig. 3. (Classical, aeroelastic) flutter problem.

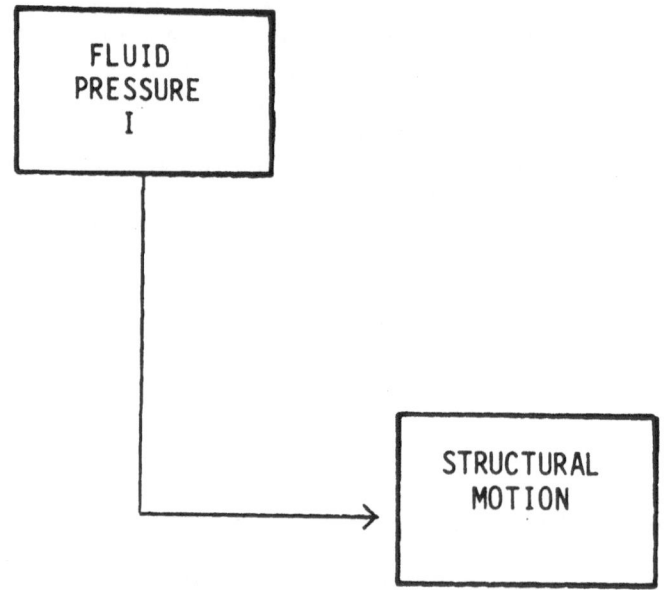

Fig. 4. (Structural response to fluid) noise problem.

8.2.1.3. Fluid pressure (II) induced by (prescribed, structural) wall motion model

This model has received attention from various authors. See Fig. 5 and Refs. 8, 25 (especially the article by Balasubramian and Orszag), and 26. The key idea that emerges is how and at what level of fluid modeling viscous effects must be included in order to estimate skin friction drag. The results of Ref. 26 are particularly noteworthy. Also see Part III of Appendix 8A. Theory and experiment are in good agreement on fluid pressure, but the results appear less satisfactory for shear stress, at least for the linear models.

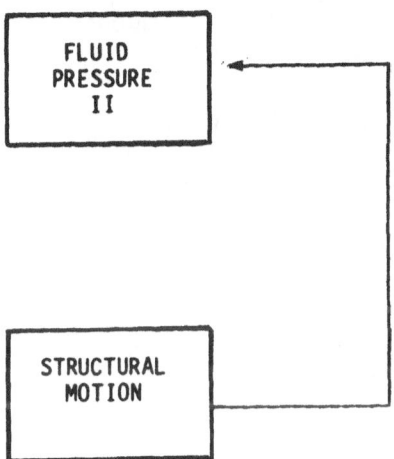

Fig. 5. Fluid pressure (II) induced by (prescribed, structural) wall motion.

8.3. Proposed methods for addressing the total model

There are two possible thrusts. One would extend the fundamental approach of Dowell [17]; George and Hellums [18], Zahn, Toomre, Spiegel and Gough [19];

and Murdock [20] (after Orszag), which determines nonlinear oscillations of the Navier-Stokes equations for
a rigid wall, to include the effects of a compliant
wall. This has recently been suggested independently
by Buckingham et al. [27]. Although this approach
will be limited formally by computational constraints
to relatively low Reynolds numbers in the short term,
it should produce results of fundamental interest.
Moreover, if Tani's [28] contention that after laminar
flow breakdown into turbulence the flow is only weakly
dependent on Reynolds number should prove correct,
then this approach may achieve results of technological interest sooner than would otherwise be expected.
Boundary layer and Poiseuille flow, for example, could
be considered.

The second thrust would seek to improve the more
modest, but still challenging, model of Tsahalis [8].
In the present context, this means constructing the
Total Model from suitable component models. Primary
emphasis would be given to

(1) including the effect of fluid force feedback* on the wall motion (initially using
the shear flow model previously developed
and exploited for related problems [22])

(2) improving the structural wall portion of the
model with particular emphasis on the
effects of distributed mass, damping and
stiffness

and (3) exploring the effects of fluid nonlinearities in calculation of the Reynolds
stresses.

Specifically, referring to the Total Model as
described by Fig. 2,

---

*Without the fluid feedback forces, the model
cannot predict the important aeroelastic instabilities
which may occur.

- Fluid Pressure I would be taken from empirical data for turbulent data over <u>rigid</u> walls as described by Ash [10] and Buckingham [27].

- Fluid Pressure II would be modeled by the shear flow method of Dowell, et al. [22]. In this fluid model the flow is separated into a mean flow over a rigid surface which includes viscosity and a perturbation flow due to the moving compliant wall in which the direct effects of viscosity are neglected. The perturbation flow field depends on the mean flow, however, and thus viscous effects are partially accounted for indirectly. This flow model is known to predict the pressure field over moving walls accurately for sufficiently small wall motion. See Refs. 22 and also 25 and 26.

- The structural wall portion of the model would be treated by any standard method which properly accounts for mass, stiffness and damping, e.g., modal analysis or finite element methods [23,24,27].

- Possible solution techniques for the equations of this version of the Total Model are described in Refs. 16, 23 and 24.

It is important to note that the above Total Model will predict the compliant wall motion accurately, but will not per se be sufficient to predict the skin friction drag. To complete the model, once the wall motion is known the method of Sengupta and Lekoudis [26], for example, may be used separately to compute skin friction (and pressure) drag. The pressure drag can be computed by either the method of Ref. 22 or 26.

It is interesting to note, as a glimpse of the future results to be obtained by such an approach, that Sengupta and Lekoudis [26] have concluded that

"For the types of wall motion considered [traveling simple harmonic waves], the [theoretical]

technique predicted no drag reduction, except for the case of wave speeds approaching the free-stream speed."

Some years ago Dugundji, Dowell, and Perkin [21] and Dowell [29] showed under what conditions the wavespeed of a traveling wave due to the flutter of a panel would approach the freestream speed.

Finally, we note that it may be desirable to avoid the complications of a fluttering panel or even one whose motion is significantly modified by fluid pressures. Instead one could design a wall which is hydroelastically rigid, but elastically compliant, in order to achieve wall shapes and motions that will lead to drag reduction. This somewhat distinct and simpler topic is discussed in Appendix 8B.

The two Total Model thrusts discussed here are complementary and results from each should serve to reinforce and strengthen the other. The first approach has almost certainly all the ingredients necessary to compute quantitatively accurate results. The second approach is in the spirit of building a theory which is simpler for both computation and interpretation, yet still sufficient to describe the physical phenomena of interest. It is a principal premise of this discussion that our understanding of turbulence transition and structure will be substantially advanced when a soundly based theoretical model is available to guide, interpret and suggest experiments. Those experiments, in turn, should validate or suggest needed improvements in the theory.

8.4. Concluding remarks

Several interesting questions have been asked by reviewers of the material of this chapter. Such questions may have also occurred to the reader and we close this chapter by addressing several of these.

Question: Is it useful to think about a driven wall motion to reduce drag (ala Kendall), since the power required to drive the wall might well be used more efficiently in a more conventional propulsive device?

Answer: There are two parts to the answer.

- It is of fundamental interest to see if propulsion by a moving wall is possible.

- If one had a passive energy storage device on the vehicle, this device might be used to move the wall over short periods of time to provide bursts of high speed performance. During longer periods of low speed performance the storage device could be recharged. Whether this is an attractive alternative to more conventional propulsion devices is an open question.

Question: Is it thermodynamically possible that a compliant wall which undergoes a (traveling wave) flutter motion (at wave speeds approaching the flow free stream velocity) could provide drag reduction?

The answer to this is not simple. But it does seem that the same question could be raised about any compliant wall which leads to reduced drag. Net thrust would seem to be precluded, but not drag reduction.

Question: Are not flutter motion amplitudes much larger than those desired to obtain drag reduction?

Answer: By adjusting the compliant wall parameters, one can obtain a wide range of amplitudes (and wavelengths) of flutter motion. Typically for conventional aircraft compliant wall (skin panel) construction, the flutter motion amplitudes are on the order of the panel thickness and the

flutter wavelength on the order of the
smaller of the panel length or width. In
addition, one can modulate the flutter
amplitude by changing the tension (or compression) in the compliant wall and modify
the flutter wavelength by mounting the
compliant wall panel on an elastic foundation. Increasing the tension will reduce
the flutter amplitude and conversely.
Increasing the elastic foundation stiffness
will decrease the wavelength and conversely.

Question: What   is flutter and how does it relate to
the Type A, B, C stability classification
of Benjamin?

Answer: The   basic definition of flutter used in
the field of aeroelasticity is, "a self-
excited oscillation (dynamic stability) due
to the mutual interaction between a structure and a fluid".   As such, this definition of flutter includes the Type A, B, C
instabilities of Benjamin plus many other
"types" of flutter which have been identified in the literature.

In aeroelasticity various types of flutter are
distinguished for at least two reasons:

• The type of flutter may have implications for
the nature of the mathematical model needed to
describe the flutter, e.g., "transonic buzz"
requires the shock wave motion to be included
in the mathematical model; "stall flutter"
requires flow separation to be modeled; plate
and shell flutter imply certain structural
models beyond simple beam theory are required;
single-degree-of-freedom flutter implies only
a single structural mode is necessary in the
model (and also usually that the flutter is
the result of loss of damping due to the presence of the flow), etc. Of course, these
names are not mutually exclusive.   For
example, one can have "single-degree-of-

freedom, stall flutter". In Benjamin's case he considers traveling wave motion over an infinitely long compliant wall and low flow speeds. Those assumptions already limit the kind of dynamic instabilities that can be encountered to what he calls Types A, B, C.

• A second reason for identifying types of flutter is because some are more sensitive to certain parameters, e.g. damping, than others. Presumably this was part of the motivation for Banjamin's classification.

References

1. Kramer, M. O., "Boundary Layer Stabilization by Distributed Damping," J. Aeronautical Soc., Vol. 24, June 1957, pp. 459-560.

2. Kramer, M. O., "Boundary Layer Stabilization by Distributed Damping," J. Am. Soc. Naval Engineers, Vol. 72, Feb. 1960, pp. 25-33, also J. Aerospace Sciences, Vol. 27, 1960, p. 69.

3. Kramer, M. O., "The Dolphins' Secret," J. Am. Soc. Naval Engineers, Vol. 73, Feb. 1961, pp. 103-107.

4. Banjamin, T. B., "Effects of a Flexible Boundary on Hydrodynamic Stability," J. Fluid Mech., Vol. 9, 1960, pp. 513-532.

5. Landahl, M. T., "On the Stability of a Laminar Incompressible Boundary Layer Over a Flexible Surface," J. Fluid Mech., Vol. 13, 1962, pp. 609-632.

6. Gyorgyfalvy, D., "Possibilities of Drag Reduction by Use of Flexible Skin," J. of Aircraft, Vol. 4, June 1967, pp. 186-192.

7. Kawamata, S., Kato, T., Matsumura, Y., and Sato, T., "Experimental Research on the Possibility of Reducing the Drag Acting on a Flexible Plate," Technical Information Service, A74-27950, 1974.

8. Tsahalis, D. T., "On the Theory of Skin Friction Reduction by Compliant Walls," AIAA Paper 77-686, Presented at the 10th Fluid and Plasmadynamics Conference, Albuquerque, N.M., June 1977.

9. Fischer, M. C. and Ash, R. L., "A General Review of Concepts for Reducing Skin Friction, Including Recommendations for Future Studies," NASA TM X-2894, March 1974.

10. Ash, R. L., Bushnell, D. M., Weinstein, L. M., and Balasubramanian, R., "Compliant Wall Surface Motion and Its Effect on the Structure of a Turbulent Boundary Layer," Fourth Biennial Symposium on Turbulence in Liquids, U. of Missouri, Rolla, September 1975.

11. Gad-el-Hak, M., Blackwelder, R. F., and Riley, J. J., "On the Interaction of Compliant Coatings with Boundary Layer Flows," J. Fluid Mechanics, Vol. 140, 1984, pp. 257-280.

12. Hansen, R. J. and Hunston, D. L., "An Experimental Study of Turbulent Flows Over Compliant Surfaces," J. Sound Vibration, Vol. 34, 1974, pp. 297-308.

13. McAlisiter, K. W. and Wynn, T. M., "Experimental Evaluation of Compliant Surfaces at Low Speeds," NASA TM X-3119, October 1974.

14. Lissaman, P. B. S. and Harris, G. L., "Turbulent Skin Friction on Compliant Surfaces," AIAA J., Vol. 17, August 1969, pp. 1625-1627.

15. Chu, H. H. and Blick, E. F., "Compliant Surface Drag Reduction as a Function of Speed," J. Spacecraft Rockets, Vol. 6, June 1969, pp. 763-764.

16. Dowell, E. H., "Aeroelasticity of Plates and Shells," Noordhoff International Publishing, 1975.

17. Dowell, E. H., "Nonlinear Theory of Unstable Plane Poiseuille Flow," J. Fluid Mechanics, Vol. 38, September 1969, pp. 401-414.

18. George, W. D., Hellums, J. D. and Martin, B., "Finite-Amplitude Neutral Disturbances in Plane Poiseuille Flow," J. Fluid Mechanics, Vol. 63, 1974, pp. 765-771.

19. Zahn, J.-P., Toomre, J., Spiegel, E. A. and Gough, D. O., "Nonlinear Cellular Motions in Poiseuille Channel Flow," J. Fluid Mechanics, Vol. 64, 1974, pp. 319-345.

20. Murdock, J. W., "A Numerical Study of Nonlinear Effects on Boundary-Layer Stability," AIAA Journal, Vol. 15, August 1977, pp. 1167-1173.

21. Dugundji, J., Dowell, E. and Perkin, B., "Subsonic Flutter of Panels on Continuous Elastic Foundations," AIAA Journal, Vol. 1, May 1963, pp. 1146-1154.

22. Williams, M. H., Chi, M. R., Ventres, C. S., and Dowell, E. H., "Aerodynamic Effects of Inviscid Parallel Shear Flows," AIAA Journal, Vol. 15, August 1977, pp. 1159-1166.

23. Lin, Y. K., "Probabilistic Theory of Structural Dynamics," McGraw-Hill, New York, N.Y., 1967.

24. Dowell, E. H., Curtiss, H. C., Jr., Scanlan, R. H., and Sisto, F., "A Modern Course in Aeroelasticity," Sijthoff and Noordhoff, The Netherlands, 1978.

25. Hough, G. R., Editor, "Viscous Flow Drag Reduction," Vol. 72, Progress in Astronautics and Aeronautics, Martin Summerfield, Series Editor, 1980.

26. Sengupta, T. K., and Lekoudis, S. G., "Calcula-
    tion of Two-Dimensional Turbulent Boundary Layers
    Over Rigid and Moving Surfaces," AIAA Journal,
    Vol. 23, April 1985, pp. 530-536.

27. Buckingham, A. C., Hall, M. S., and Chun, R. C.,
    "Numerical Simulations of Compliant Material
    Response to Turbulent Flow," AIAA Paper 84-0537,
    AIAA 22nd Aerospace Sciences Meeting, January
    1984.

28. Tani, I., "Boundary Layer Transition," Annual Re-
    view of Fluid Mechanics, Vol. 1, 1969, pp. 169-
    196.

29. Dowell, E. H., "Flutter of Infinitely Long Plates
    and Shells. Part I: Plate," AIAA Journal, Vol.
    4, August 1966, pp. 1370-1377.

Appendix 8A

Correlation between theory and experiment for aerodynamic and flutter characteristics of wavy walls

When theory and experiment are in substantial
agreement, one may have confidence in the understanding of the physical phenomena. Here brief summaries
are given of some of the theoretical-experimental correlation studies which provide the basis for the
state-of-the-art assessments made in the text. The
discussion is an edited version of the original references. The reader is referred to the original references for further details.

(I) Subsonic Flutter of Panels on Continuous Elastic
    Foundations (Ref. 21)

The subsonic aeroelastic stability of a two-
dimensional panel resting on a continuous elastic
foundation was investigated. Tests were conducted
experimentally on a 104-x, 24-x 0.018-in. rectangular

aluminum panel in a low-speed wind tunnel. Definite flutter of a traveling-wave-type was observed. Films and oscillograph records were taken. Theoretically, a finite-panel, two-mode, standing-wave analysis was shown to give essentially the same behavior as the infinite-panel, traveling-wave analysis of Miles for this panel. Although a mild, divergence-type instability exists for these panels, the principal instability was shown to be of a traveling-wave, flutter-type. Comparison of experiment and theory showed good agreement in flutter speed and wavelength but poor agreement in wave speed and frequency at flutter. This latter discrepancy was attributed to limitations in the test set-up as well as to the general difficulty of predicting the wave speed and frequency as accurately as the flutter speed. See Table I for a quantitative comparison between theory and experiment.

Table I. Summary of Experimental and Theoretical Results

	Experiment	Theory (infinite)	Theory (finite)
Flutter speed, mph	95	105.5	103.1
Wavelength, in.	27	25	24.5
Wave speed, fps	35	11.4	6.9
Frequency, cps	14	5.5	3.4
Divergence speed, mph	...	101.5	101.5
Wavelength, in.	...	25	26

(II) Aerodynamic Effects of Inviscid Parallel Shear Flows (Ref. 22)

We consider the interaction of a boundary layer on a flat plate with a local deformation of the plate surface. If the Reynolds number based on the characteristic wavelength of the deformation is large, but the growth of the (undisturbed) boundary layer over a wavelength is small, then the boundary-layer flow can be treated, locally, as an inviscid parallel shear flow.

For simplicity we shall suppose that the local velocity profile is given by

$$\bar{u}(z)/U_\infty = (z/\delta)^{1/N} \qquad z < \delta$$

which models a turbulent boundary layer of thickness δ under adiabatic conditions. It will be noted that this profile is invalid for $z \ll \delta$ since no attempt has been made to describe the laminar sublayer. For reasons discussed below this neglect is crucial to the success of the theory.

We first consider the interaction of the flow with an infinite sinusoidal wall of wavelength $2\pi/\alpha$ in the flow direction. In the interest of clarity we consider only the two-dimensional, steady case, although three-dimensional, unsteady problems are not substantially more difficult to solve.

For any steady flow with $\bar{u}(0) = 0$, the governing equation (G.E.) of the wavy wall problem has a singularity at $z = 0$. This singularity is a consequence of the fact that the linearization invoked to obtain the G.E. cannot be valid near any point where $\bar{u} = 0$. Nevertheless, it can be shown that the G.E. does have a solution satisfying the far-field conditions and the surface boundary condition, provided that $\bar{u}(z)$ vanishes sufficiently slowly that $1/\bar{u}^2$ is integrable at $z = 0$. For the present flow, this constraint limits consideration to values of $N > 2$. Since experimental observations of turbulent boundary layers indicate that $N > 7$, this limitation has little importance to the present considerations. However, it should be borne in mind that the simple linear inviscid theory given here cannot work for a laminar profile or for a turbulent profile with an accurate description of the laminar sublayer. For such problems, either nonlinear interactions or direct viscous forces (or both) must be included in the formulation. See Section (III) below. One of the consequences of the present work is

that such refinements are not necessary for the turbulent case in order to predict surface pressure.

For <u>subsonic</u> external Mach numbers, the surface
pressure, induced by the flow on a steady two-
dimensional wavy wall, varies with wavelength and Mach
number as illustrated in Fig. 1 (from Ref. 1). [Fig-
ure and reference numbers refer to those at the end of
this section.] The boundary layer always tends to
reduce the pressure amplitude below its potential flow
value (but with no shift in phase, theoretically).
The decrease becomes more pronounced as $M_\infty \to 1$. At M_∞
= 1 the pressure remains finite for all nonzero δ.

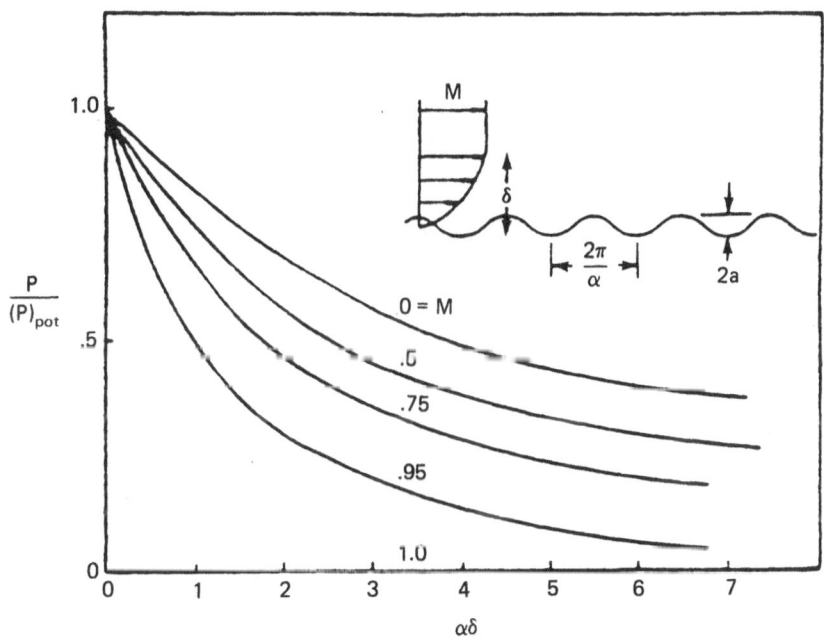

Fig. 1.   Subsonic wavy wall pressure.

The above results are compared in Fig. 2 with the
experimental results of Muhlstein and Beranek [2] at
M_∞ = 0.8, 0.9, and 0.95. The comparisons are made by
equating the theoretical and measured displacement
thicknesses. The experimental data include only the
fundamental spatial harmonic at the smallest amplitude

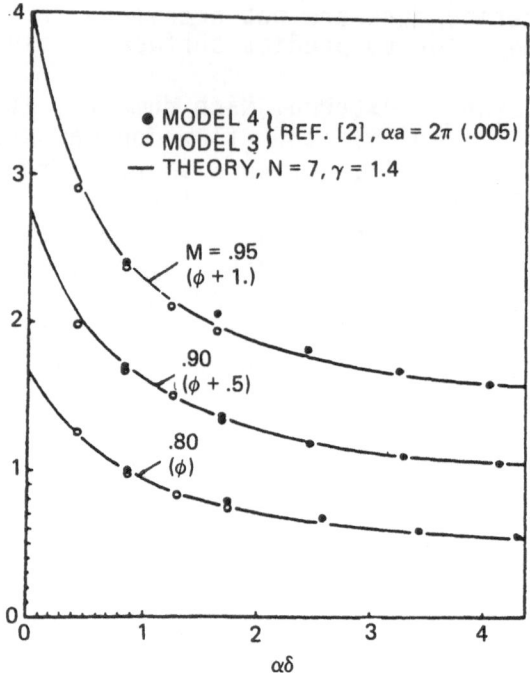

Fig. 2. Comparison of theory and experiment: subsonic
wavy wall pressure $\phi = |C_p|/2\alpha a$.

to wavelength ratio (0.005) considered in Ref. 2. It
is of some interest that according to potential flow
theory the critical Mach number of this configuration
is 0.92, although no distinct transonic phenomena are
observed even at $M_\infty = 0.95$. The agreement between
theory and experiment shown in Fig. 2 is remarkable
considering the relatively crude boundary-layer profile used in the calculation. This fact suggests that
a precise description of the undisturbed boundary-
layer profile and, in particular an accurate modeling
of the laminar sublayer, is not necessary for the
quantitative prediction of surface pressure at wavelengths which are long compared to the laminar sublayer thickness.

In a supersonic external stream the boundary layer not only decreases the magnitude of the induced pressure, but also increases the phase lag toward 180°. These effects have been studied by Yates [3] using a formulation similar to that considered here but with an approximation valid for small $\alpha\delta$. The exact solution for the flow (19) at $M_\infty = 1.1$ is shown in Figs. 3 and 4. The measured results from Ref. 2 are, as in the subsonic case, in excellent agreement with the theory over the reported range of $\alpha\delta$.

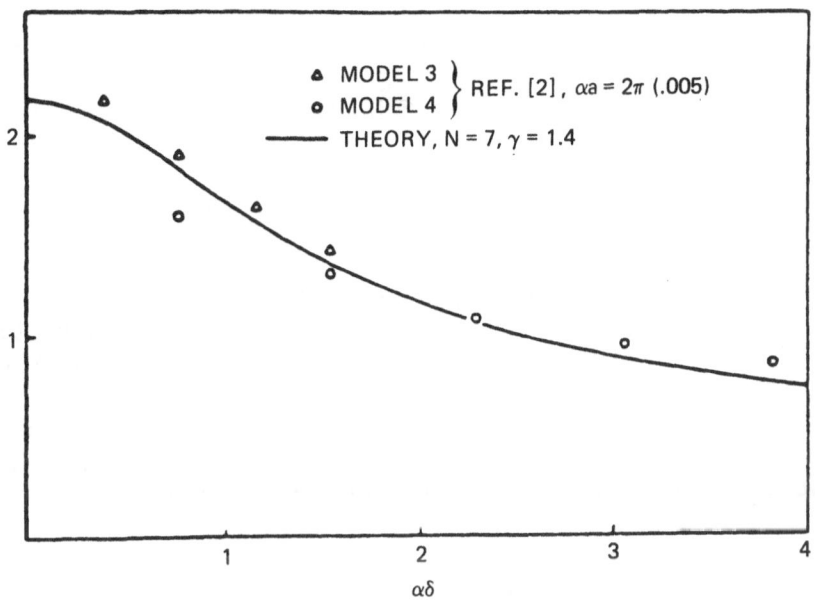

Fig. 3. Wavy wall pressure, M = 1.1, magnitude.

As shown by Yates [3] for the supersonic case, and more generally in Ref. 1, the behavior of the solution for small $\alpha\delta$ is controlled by various integral properties of the boundary-layer profile. This fact is largely responsible for the close agreement shown in Figs. 2-4. However, the behavior of the solution for large $\alpha\delta$ is controlled by the structure of $\bar{u}(z)$ very near the surface. Thus, the theory can be expected to be increasingly inaccurate as $\alpha\delta$ increases to large values, although both theory and experiment (if such existed) would show the pressure amplitude dropping to zero in this limit.

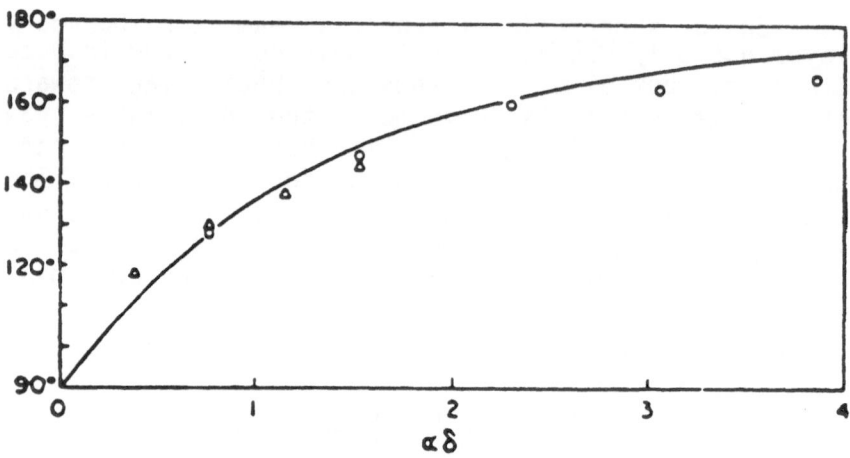

Fig. 4. Wavy wall pressure, M = 1.1, phase.

Three-dimensional and/or unsteady solutions to the wavy wall problem can either be computed directly [4], or, in some cases, approximated from the two-dimensional steady solution. For instance, if the Strouhal number based on boundary-layer thickness $\omega\delta/U_\infty$ is small, then we can expect significant unsteady effect to arise only when $\alpha\delta < \omega\delta/U_\infty < 1$, i.e., for wavelengths that are long compared to δ. Conversely, at these wavelengths the boundary layer will have very little effect on the pressure. Thus a uniformly valid (for all α) approximation to the pressure at small $\omega\delta/U_\infty$ can be constructed from the steady shear flow and the unsteady potential flow solutions by using the method of matched asymptotic expansions. Similarly, the principal effect of three dimensionality (in the steady case) is to reduce the effective freestream Mach number [1].

The pressures induced by an arbitrary wall deformation can be constructed from the wavy wall solution by simple superposition. This has been done for flexible rectangular panels in low supersonic streams and used in conjunction with an appropriate structural model to predict aeroelastic stability boundaries. The resulting flutter boundaries are shown in Figs. 5 and 6 (from Ref. 4), which illustrate the effect of

353

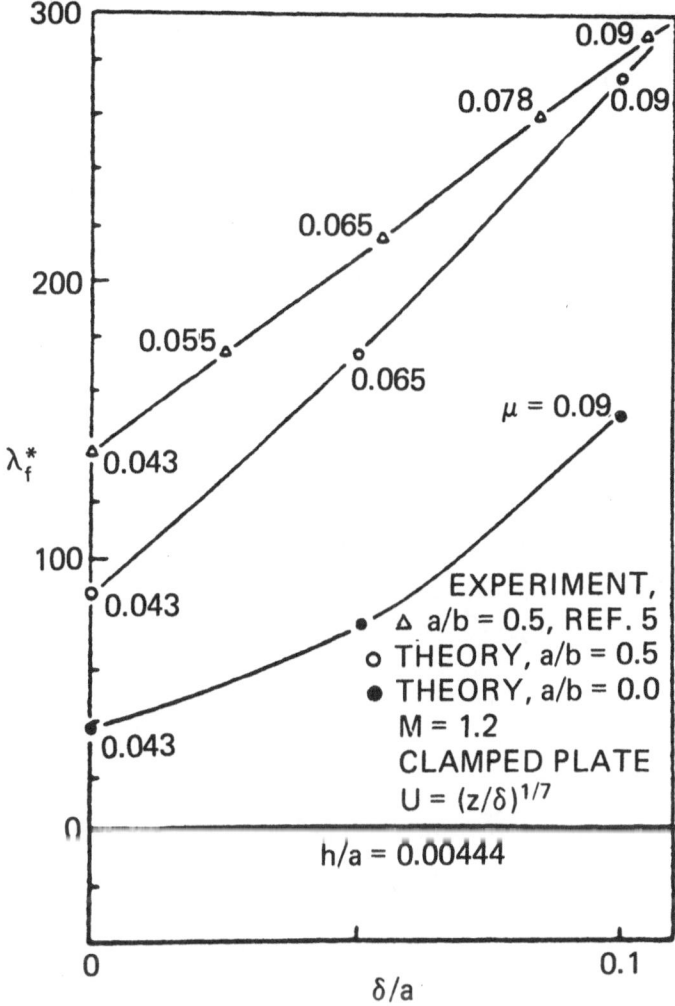

Fig. 5. Panel flutter dynamic pressure as function of
shear layer thickness: M = 1.2.

boundary-layer thickness on flutter speed at M_∞ = 1.2
and of Mach number on flutter speed for fixed
boundary-layer thickness. These results are compared
to the stability boundaries determined experimentally
in Ref. 5. The agreement is again quite good, although not as precise as in the wavy wall problem,
presumably because of both numerical and experimental
inaccuracies in this more complex physical situation.

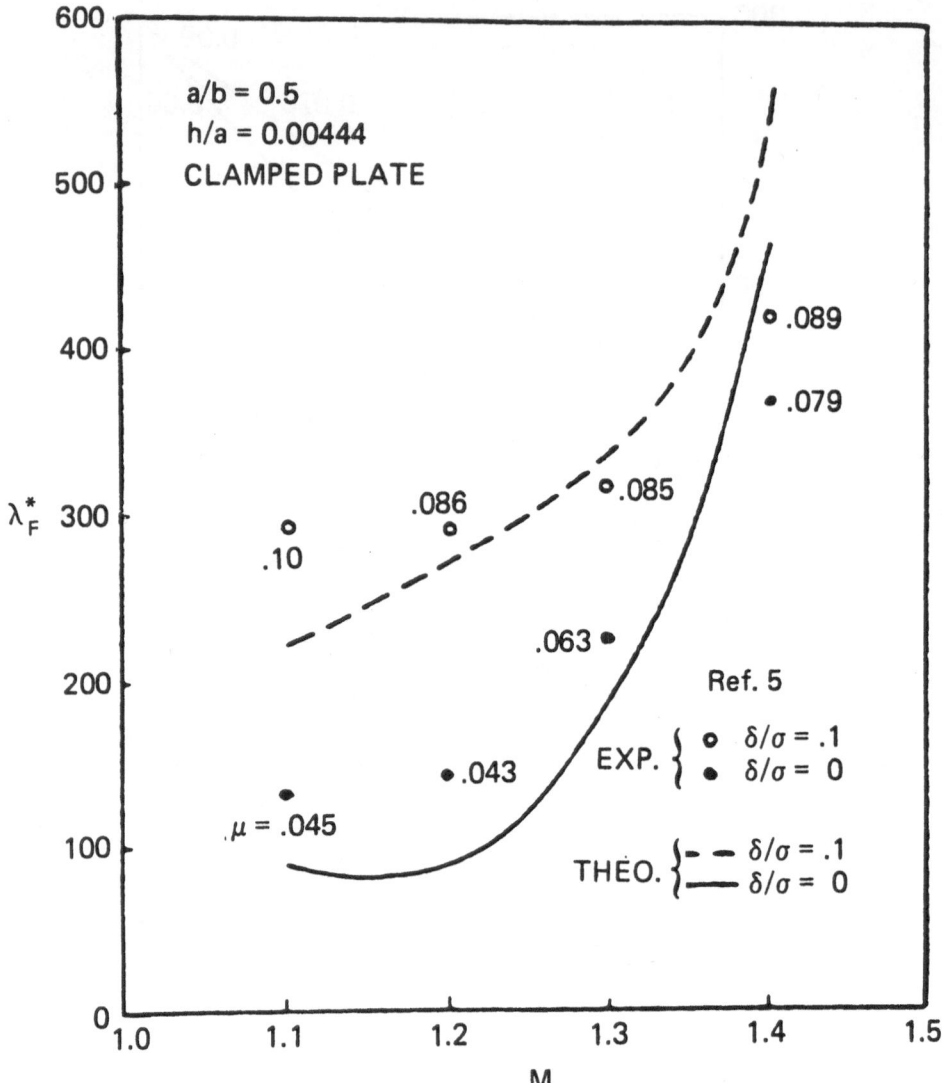

Fig. 6. Panel flutter dynamic pressure as a function
function of Mach number, M.

References

1. Williams, M. H., "General Theory of Thin Wings in Inviscid Parallel Shear Flows," Princeton University AMS Rept. 1269, March 1976.

2. Muhlstein, L. and Beranek, R. G., "Experimental Investigation of the Influence of the Turbulent Boundary Layer on the Pressure Distribution over a Rigid Two-Dimensional Wavy Wall," NASA TN-D-6477, Aug. 1971.

3. Yates, J. E., "Linearized Integral Theory of Three-Dimensional Unsteady Flow in a Shear Layer," AIAA Journal, Vol. 12, May 1974, pp. 596-601.

4. Dowell, E. H., "Generalized Aerodynamic Forces on a Flexible Plate Undergoing Transient Motion in a Shear Flow with an Application to Panel Flutter," AIAA Journal, Vol. 9, May 1971, pp. 834-841.

5. Muhlstein, L., Gaspers, P. A., and Riddle, D. W., "An Experimental Study of the Influence of the Turbulent Boundary Layer on Panel Flutter," NASA TN-D-4486, March 1968.

(III) Calculation of Two-Dimensional Turbulent Boundary Layers over Rigid and Moving Wavy Surfaces (Ref. 26)

A technique is presented for computing turbulent boundary layers over rigid and moving sinusoidal wavy surfaces. The technique consists of an iterative coupling between solutions of the boundary layer, and solutions of an Orr-Sommerfield-type system. The predicted pressure distributions and total drag agree well with recent measurements for rigid wavy surfaces. Comparisons are also made with measurements of airflow over deep water waves.

A. Flows Over Rigid Wavy Surfaces

A comparison of results from the present theory with the measurements of Ref. 1 is shown in Figs. 1 and 2. [Figure and reference numbers refer to those at the end of this section.] The agreement is reasonable only for the pressure. However, these data are for amplitude to wavelength ratios of 0.028 and the effect of linearization [of the boundary condition on the moving wall] is evident in the inaccurate prediction of the shear.

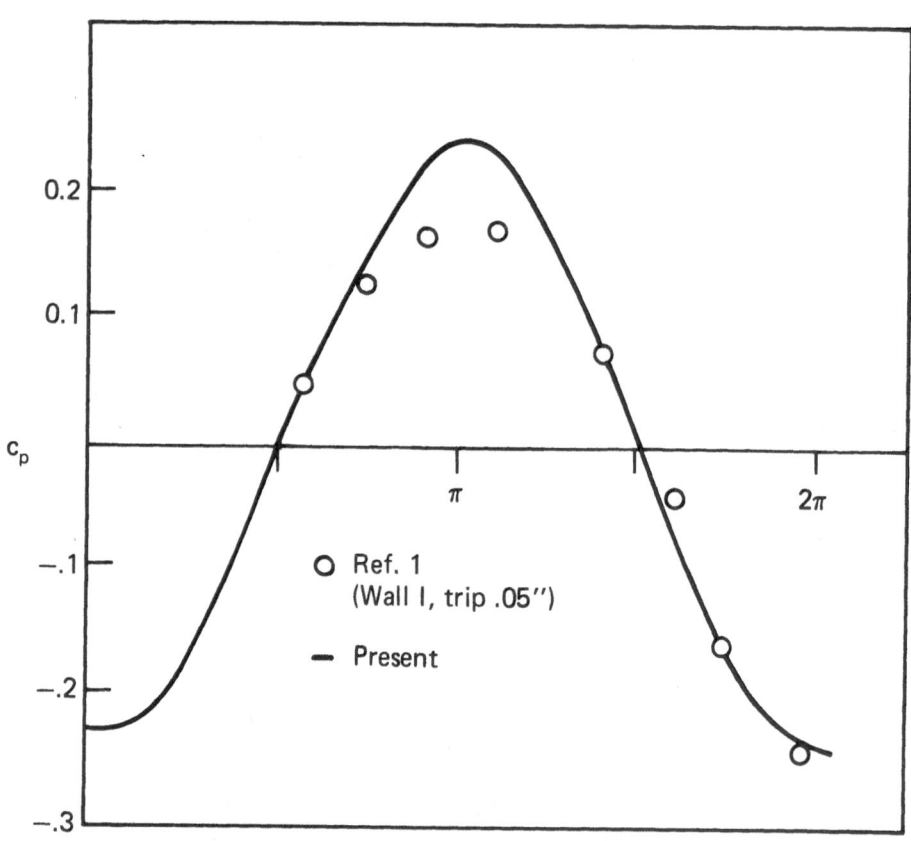

Fig. 1. Calculated and measured C_p for Sigal's experiment[1].

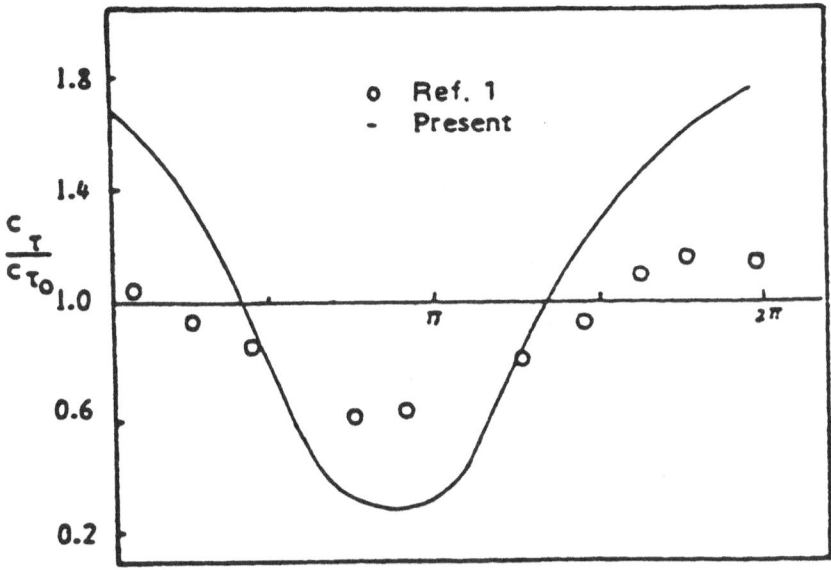

Fig. 2. Calculated and measured skin friction for Sigal's experiment.[1]

Several geometries were examined at NASA Langley [2]. Figure 3 shows predicted and measured pressure coefficients. The measured data were reproduced from Fig. 7 of Ref. 2. The agreement is good. Figure 4 shows predicted and measured drag coefficients. The measured data were reproduced from Fig. 5 of Ref. 2. The level of agreement is of the same quality as that obtained from Navier-Stokes solvers [2]. This lends credibility to the assumptions used in the developed formulation. Moreover, it makes the predictions for the case of moving walls, where experiments are scarce, more credible. The developed procedure does reproduce the observed reduction in the mean skin friction, something impossible to obtain with a [purely] linear theory.

B. Flows Over Moving Wavy Surfaces

The authors also investigated flows over moving surfaces. The variation of the amplitude and phase of

358

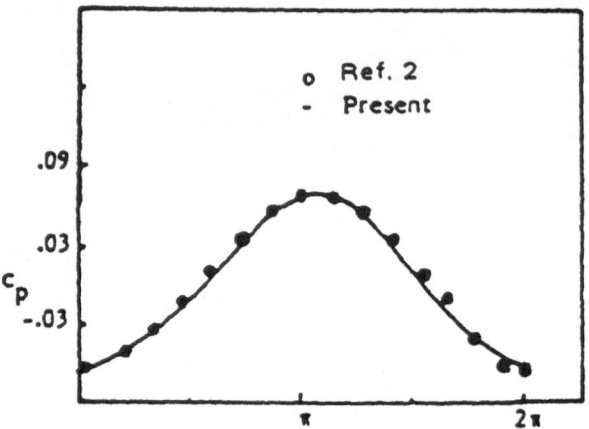

Fig. 3. Calculated and measured pressure coefficient
for NASA Langley's experiment.

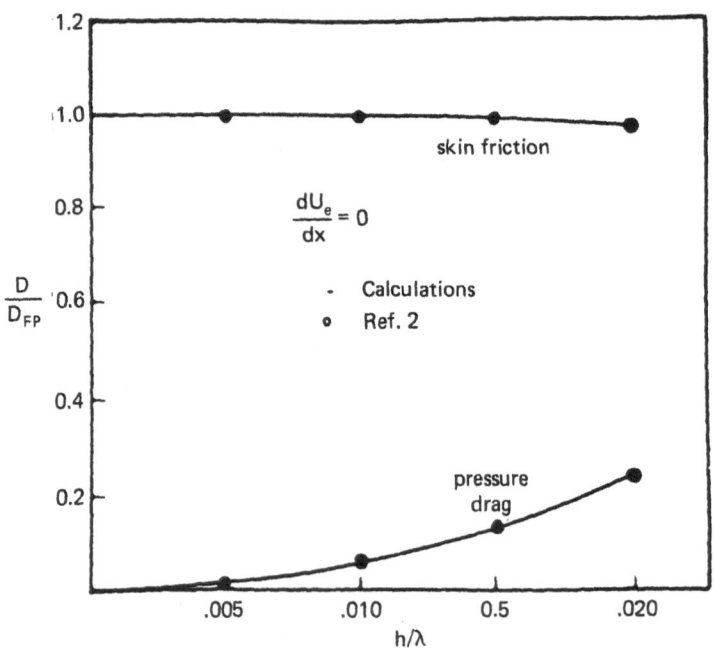

Fig. 4. Calculated and measured drag components.

the pressure are plotted in Fig. 5. Although the amplitude of the oscillating pressure decreases with phase speed, the variation of the phase is drag producing. However, as the phase speeds approach the freestream speeds, the phase of the pressure produces thrust. The variation of the amplitude and phase of the oscillating shear is shown in Fig. 6. These results are for a surface motion simulating deep water wave and with the viscous boundary condition. The results shown in Figs. 5 and 6 are in qualitative agreement with Kendall's measured data [3]. However, because these data are for an amplitude to wavelength ratio of 0.03, the present formulation indicated separation and direct quantitative comparison is not possible.

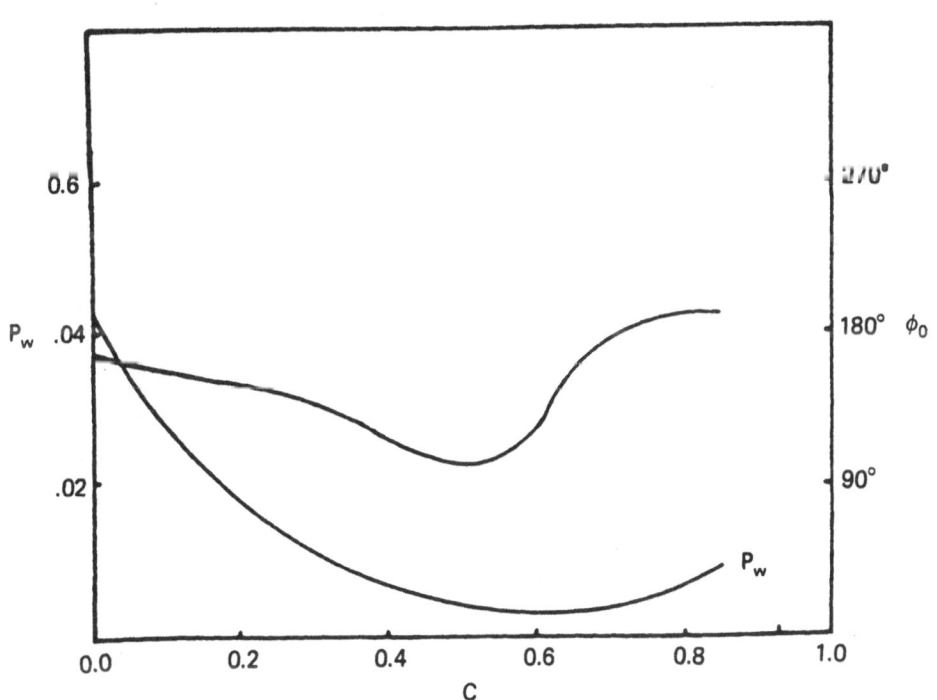

Fig. 5. Variation of the amplitude and phase of the pressure with phase speed.

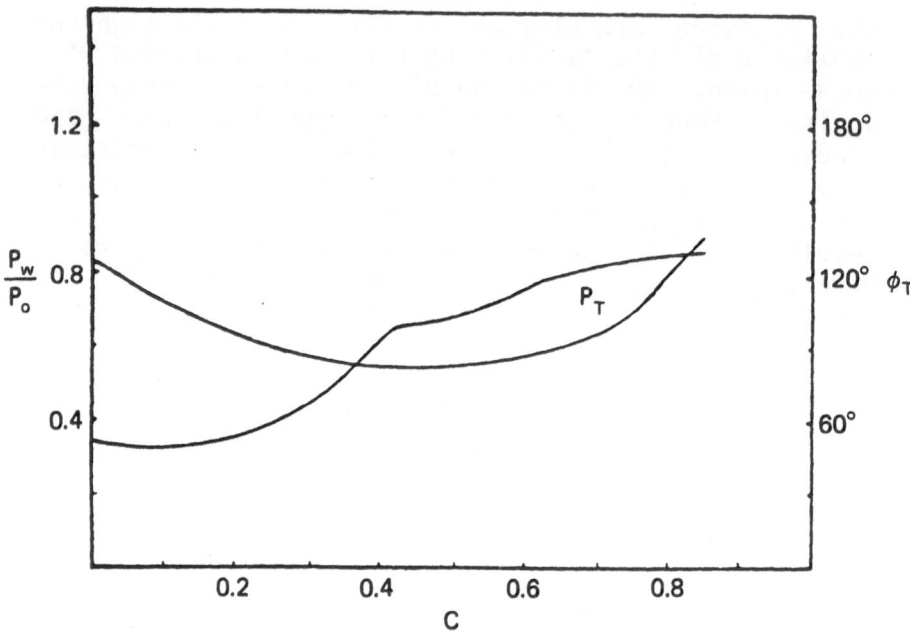

Fig. 6. Variation of the amplitude and phase of oscillatory shear with phase speed.

The drag values normalized with those of the equivalent flat plate are plotted in Fig. 7. The plate geometry and flow conditions are identical with those of Langley's experimental setup [2] and the amplitude to wavelength ratio of the waves used is 0.02. The skin friction reduction diminishes at the higher wave speeds but the phase of the pressure indicates thrust. Because the turbulence model was developed for zero phase speeds, the high-speed results need experimental verification. This prompted the authors to examine some new experiments [4,5].

The experiments reported in Refs. 4 and 5 deal with air turbulent boundary layers over deep water waves. Detailed measurements of the perturbation components of the velocity are given. The agreement

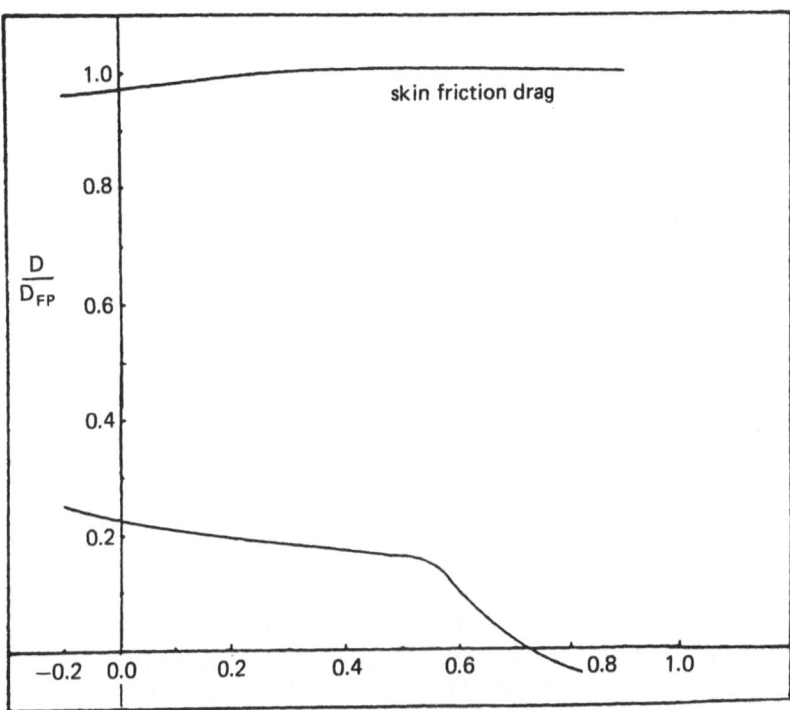

Fig. 7. Variation of the skin friction and pressure
drag components with phase speed.

between the predictions and measurements is reasonable. [The results are not shown here. See the
original reference.] In general, the comparison
between calculations and experiments becomes worse at
the higher phase speeds. The good agreement between
the calculations and measured data for phase speeds up
to about 0.7 times the freestream speed gives confidence in the drag predictions for this range of
speeds.

References

1. Sigal, A., "An Experimental Investigation of the Turbulent Boundary Layer Over a Wavy Wall," Ph.D. Thesis, California Institute of Technology, Pasadena, Calif., 1971.

2. Lin, J. C., Walsh, M. J., Watson, R. D., and Balasubramanian, R., "Turbulent Drag Characteristics of Small Amplitude Rigid Surface Waves," AIAA Paper 83-0228, 1983.

3. Kendall, J., "The Turbulent Boundary Layer Over a Wall with Progressive Surface Waves," Journal of Fluid Mechanics, Vol. 41, Pt. 2, 1970, pp. 259-281.

4. Hsu, C. T., Hsu, E. Y., and Street, R. L., "On the Structure of Turbulent Flow Over a Progressive Water Wave: Theory and Experiment in a Transformed Wave-Following Coordinate System," Journal of Fluid Mechanics, Vol. 105, 1981, pp. 87-117.

5. Hsu, C. T. and Hsu, E. Y., "On the Structure of Turbulent Flow Over Progressive Water Wave: Theory and Experiment in a Transformed Wave-Following Coordinate System, Part 2," Journal of Fluid Mechanics, Vol. 131, 1983, pp. 123-153.

Appendix 8B

Elastic compliant walls for drag reduction using statically and dynamically unstable structures

Several wavy (non-planar) wall configurations may be useful for drag reduction studies. These include

- a Rigid Wall

- an Elastic Compliant Wall

- a Hydroelastic Compliant Wall

These are shown schematically in Fig. 1. [Figure and reference numbers refer to those at the end of this Appendix.]

- **RIGID WALL**

- **ELASTIC COMPLIANT WALL**

P → ⟶ — — — — — — ← ← P

Euler Buckling

- **HYDROELASTIC COMPLIANT WALL**

U ⟹

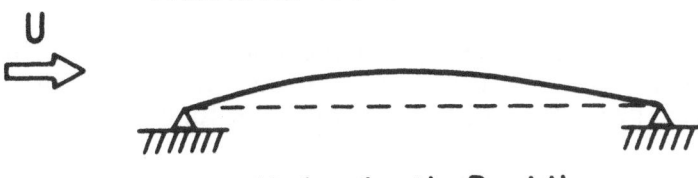

**Hydroelastic Buckling
(Divergence)**

Fig. 1. Wavy walls for drag reduction.

• A rigid wavy wall has a non-planar shape but does not deform under the action of mechanical or hydrodynamic forces.

• An elastic compliant wall does deform under mechanical forces, e.g., in-plane compression, but does not deform significantly due to hydrodynamic loads.

• A hydroelastic compliant wall does deform significantly due to hydrodynamic loads.

The motivation for studying an elastic compliant wall is twofold. It provides a possible means for designing (adaptive) wavy walls of various wavelengths and amplitudes. It is also a natural, simpler precursor to the study of hydroelastic compliant walls.

A more detailed discussion of the work summarized in this Appendix is given in Ref. 1 and 2.

Brief Descriptive Summary of Significant Technical Results

Theoretical and numerical studies have been undertaken of statically and dynamically buckled plates. Significant results obtained include the following:

- A determination has been made of the conditions under which spanwise bending of a compliant plate can be minimized with respect to chord-wise bending.

- An algorithm has been developed for designing compliant statically buckled plates with a desired wave amplitude and wavelength.

- A derivation has been made of the governing equations for a dynamically buckled plate which will allow the study of compliant moving walls.

Numerical results have been obtained for the dynamic case and a systematic parameter study undertaken.

Technological Significance of the Research

One may now design a compliant wall with a desired static geometric (sinusoidal) shape for use in drag reduction studied. The dynamic, moving wall case has proven to be much more challenging and, while analyzable, a simple design procedure is not yet available.

Analysis and Design

Here, as a simple example, a brief outline of the analysis and design of a statically buckled plate (with no spanwise bending) is presented. The equation of equilibrium for an initially flat plate under a compressive end-load, P, is

$$D \frac{\partial^4 w}{\partial x^4} + Kw + (P - N_{x_f}) \frac{\partial^2 w}{\partial x^2} = 0 \tag{1}$$

where w is the plate deflection, D its bending stiffness, K its foundation spring stiffness and the induced in-plane tension is

$$N_{x_f} = \frac{Eh}{2a} \int_0^a (\frac{\partial w}{\partial x})^2 \, dx \tag{2}$$

A series solution is assumed of the form

$$w = \sum_n f_n \sin \frac{n\pi x}{a} \tag{3}$$

Substituting (3) into (2) and (1), f_n may be determined as follows

$$(\frac{f_n}{h})^2 = \frac{2}{3(1 - v^2)} [P/P_{min} - 1] \tag{4}$$

where P_{min} is the minimum buckling load as given by

$$P_{min} = 2(KD)^{1/2} \qquad (5)$$

and the corresponding buckling mode number $n = n_{min}$ is given by

$$n_{min}^4 = \frac{Ka^4}{D\pi^4} \text{ or } \frac{n_{min}\pi}{a} = (\frac{K}{D})^{1/4} \qquad (6)$$

From these equations one may adjust D, h, K and P to obtain a buckled shape of desired ampliitude, f_n, and wavelength

$$\lambda = 2a/n$$

In Table I a summary of data for wavelength, amplitude, Reynolds number, and fluid from various sources is given. These data are for fluid dynamic experimental studies over rigid wavy walls.

In Table II a variety of configuration possibilities is given for an elastic compliant wall as determined by the present analysis. Note the range of possible wavelengths. Note also that the amplitude can be adjusted to various multiples of plate thickness, h, by varying the in-plane compressive load, P. See equation (4). Clearly an elastic compliant wall provides a possible means of achieving wavy wall configurations of appropriate wavelengths and amplitudes.

In practice it may be desirable to build the elastic plate so as to be initially buckled into the desired shape, with some modest variation in compressive load, P, available to adjust the amplitude of the buckled shape.

References

1. E. J. Cui and E. H. Dowell, "Postbuckling Behavior of Rectangular Orthotropic Plates with Two Free Side Edges," International Journal of Mechanical Sciences, Vol. 25, No. 6, pp. 429-446, 1983.

2. C. Pierre, and E. H. Dowell, "A Study of Dynamic Instability of Plates by an Extended Incremental Harmonic Balance Method," Journal of Applied Mechanics, Vol. 52, pp. 693-697, 1985.

TABLE I

Investigator	Wavelength λ	Amplitude/ Wavelength $2f/\lambda$	Reynolds Number $\lambda u/\nu$	Fluid	Source*
Stanton (1932)	2.6 cm	0.4	23400	Air	I-1
Motzfield (1937)	30 cm	0.05 0.10	330000 330000	Air Air	I-2
Zagustin (1966)	91.44 cm 60.96 cm	0.042 0.021	147000- 317000	Water Water	I-3
Cook (1970)	5.08 cm	0.05		Water	I-4
Kendall (1970)	10.16 cm	0.062	19000- 64000	Air	I-5
Sigal (1971)	15.24 cm 30.48 cm	0.052 0.056	154000- 306000	Air	I-6
Hsu & Kennedy (1971)	25.39 cm 50.81 cm	0.044 0.022	238000- 476000	Air	I-7
Beebe (1972)	10.67 cm	0.17 0.40	21400- 85600	Air	I-8
Zilker (1976)	5.08 cm	0.0312 0.05	11000- 64000	Water	I-9
Zilker (1977)	5.08 cm	0.012	11000- 64000	Water	I-10
Thorsness (1978)	5.08 cm	0.012	11000- 64000	Water	I-11

*(I-1)   Proc. Roy. Soc. A 137 p. 283
 (I-2)   Z. Angew. Math. Mech. 17 p. 193.
 (I-3)   Dept. Civil Eng. Stanford Univ. Tech. Rep. No. 60.
 (I-4)   Ph.D. Thesis, Univ. of Illinois.
 (I-5)   J.F.M. 41. p. 259.
 (I-6)   Ph.D. Thesis, Cal. Inst. Tech.
 (I-7)   J.F.M. 47, p. 481.
 (I-8)   Ph.D. Thesis, Colorado State Univ.
 (I-9)   Ph.D. Thesis, Univ. of Illinois.
 (I-10) J.F.M. 82, p. 29.
 (I-11) Chem. Eng. Sci., 30, p. 579.

See Zilker and Hanratty (I-5) for a fuller discussion and

TABLE II

$$a = 300 \text{ mm}, \quad E = 6.895 \times 10^{10} \text{ n/m}^2$$

$$\sigma_{yield} = 275.8 \times 10^6 \text{ n/m}^2$$

h mm	K $\times 10^{-6}$ n/m^3	n_{min}	P_{min} m/m	σ_{min} $\times 10^{-3}$ n/m	$\lambda = \dfrac{2a}{n_{min}}$ mm	$\dfrac{h}{\lambda}$
.16	.2	5.03	143.8	899	119	.0013
.32	.2	3	406.8	1271	200	.0016
.64	.2	1.78	1151	1798	227	.0019
.64	.4	2.12	1627	2545	283	.0023
.64	.8	2.52	2301	3596	238	.0026
1.	.6	1.68	3892	3893	357	.0028
1.	1.2	1.99	5505	5505	300	.0033
1.	2.4	2.37	7786	7786	253	.0040

Note: In theory, only integer values of n are permissible. Also each n value may occur with
other values of n not present. In practice,
all values of n will be present to some degree,
but those near n_{min} may be expected to domi-
nate.

A SELECTED AND ANNOTATED CHRONOLOGICAL BIBLIOGRAPHY

1. J. W. Miles, "On the Aerodynamic Stability of Thin Panels," J. Aerospace Sci., Vol. 23, No. 8, August 1956, pp. 771-791.

   An analytical investigation of the instability of an infinitely long, one-dimensional structural panel in a two-dimensional potential flow. While this represented an over-idealized model, incapable of physical realization, it led directly to the work described below in Ref. 5.

2. J. W. Miles, "On Panel Flutter in the Presence of a Boundary Layer," J. Aerospace Sci., Vol. 26, No. 2, February 1959, pp. 81-93.

   A shear flow model (viscous effects are included in the mean flow, but neglected in the flow perturbations due to panel motion) is used in place of the potential flow model of Ref. 1. This work was later extended by Dowell to plates of finite length, see Ref. 22 and 26.

3. T. B. Benjamin, "Shearing Flow Over a Wavy Boundary," J. Fluid Mechanics, Vol. 6, Part 2, August 1959, pp. 161-205.

   An early, seminal viscous fluid theoretical model. Classical linear hydrodynamic stability theory with a wavy wall.

4. M. T. Landahl, "On the Stability of a Laminar Incompressible Boundary Layer over a Flexible Surface," J. Fluid Mechanics, Vol. 13, Part 4, August 1962, pp. 609-632.

   Follows Benjamin. Shows no substantial theoretical effect of a compliant wall on the onset of laminar flow instability, i.e., the critical Reynolds number determined by small perturbation (linear) hydrodynamic stability theory.

371

5. J. Dugundji, E. Dowell, and B. Perkin, "Subsonic
    Flutter of Panels on Continuous Elastic Founda-
    tions," AIAA Journal, Vol. 1, No. 5, May 1963,
    pp. 1146-1154.

    The analytical work of Miles, Ref. 1, was ex-
    tended to include the effect of an elastic foun-
    dation beneath the panel. This has the important
    physical consequence that the wavelength of the
    critical flutter motion is finite (rather than
    infinite as in the case of Ref. 1). The analyti-
    cal results were verified by experiment.

6. E. H. Dowell, "Flutter of Infinitely Long Plates
    and Shells. Part I: Plate," AIAA Journal, Vol.
    4, No. 8, August 1966, pp. 1370-1377.

    The extension of Ref. 1 (and 5) to plates of
    finite width (two-dimensional plates in a three-
    dimensional flow). The consequence of finite
    width is also to produce a finite wavelength for
    the critical flutter motion. (Although of less
    immediate interest here, it is now known that for
    supersonic flow, the infinitely long panel model
    fails because of structural boundary layer
    effects at the leading and trailing edges of a
    panel in supersonic flow. See Ref. 22.)

7. Dezso Gyorgfalvy, "Possibilities of Drag Reduc-
    tion by the Use of Flexible Skin," J. Aircraft,
    Vol. 4, No. 3, May-June 1967, pp. 186-192.

    Follows Benjamin and Landahl. Classical hydro-
    dynamic stability theory is used not only to
    determine the critical Reynolds number at which
    the laminar boundary layer becomes unstable, but
    also the rate of amplification of the unstable
    disturbances above this Reynolds number. The
    amplification rate is, in turn, related empiri-
    cally after Smith to the Reynolds number at which
    transition to turbulence occurs. The transition
    Reynolds number determined by this approach can
    be substantially increased by a flexible wall. A
    highly suggestive, although nonrigorous, result.

8. M. T. Landahl, "A Wave-Guide Model for Turbulent Shear Flow," J. Fluid Mechanics, Vol. 29, Part 3, September 1967, pp. 441-460.

    An interesting approach via forced response of Orr-Sommerfeld equation (classical hydrodynamic stability theory) for turbulent mean velocity profiles; it would be interesting to try the same approach using the shear flow equation in view of large Reynolds numbers and stable eigenvalues, see Williams, Chi, Ventres and Dowell, Ref. 26 in this bibliography.

9. E. R. Blick and R. R. Walters, "Turbulent Boundary-Layer Characteristics of Compliant Surfaces," J. Aircraft, Vol. 5, No. 1, Jan.-Feb. 1968, pp. 11-16.

    The basic experimental paper by Blick et al.; discusses both drag and turbulence intensity measurements. Has a good set of references including Laufer, Landahl, Benjamin.

10. R. L. Smith and E. F. Blick, "Skin Friction of Compliant Surfaces with Foamed Material Substrate," J. Hydronautics, Vol. 3, No. 2, April 1969, pp. 100-102.

    A fairly complete literature survey including negative results (i.e., no drag reduction) by Laufer, Benjamin, etc.

11. H. H. Chu and E. F. Blick, "Compliant Surface Drag Reduction as a Function of Speed," J. Spacecraft and Rockets, Vol. 6, No. 6, June 1969, pp. 763-764.

    A continuation of a series of publications; shows significant drag reduction. Apparently observed an aeroelastic instability.

12. P. B. S. Lissaman and G. L. Harris, "Turbulent Skin Friction on Compliant Surfaces," AIAA J., Vol. 7, No. 8, August 1969, pp. 1625-1627.

An experiment. Only a modest reduction in drag, ~10%, was observed. Authors were pessimistic about prospects for drag reduction and flatly disagreed with Blick and Walters, see Reference 9. Apparently they observed some form of aero-elastic instability of the compliant surface, either divergence or flutter.

13. E. H. Dowell, "Nonlinear Theory of Unstable Plane Poiseuille Flow," J. Fluid Mechanics, Vol. 38, September 1969, pp. 401-414.

The nonlinear, time-dependent Navier-Stokes equations are transformed using Galerkin's method in the spatial coordinates to obtain a set of nonlinear ordinary differential equations in time. The latter are solved numerically by time marching to obtain limit cycle oscillations and also to determine sensitivity to amplitude of initial disturbances.

14. I. Tani, "Boundary-Layer Transition," Annual Review of Fluid Mechanics, Vol. 1, 1969, pp. 169-196.

Argues that the breakdown into turbulence (after an extended transition region) is only weakly dependent on Reynolds number.

15. J. T. Stuart, "Nonlinear Stability Theory," Annual Review of Fluid Mechanics, Vol. 3, 1971, pp. 347-370.

A very useful review of the now standard fluid dynamic/applied mathematical approach; narrow gap cylinder problem is considered in some detail.

16. M. E. McCormick, B. Johnson and W. Miles, "Effects of a Flexible Surface on Surface Shear-Stress Fluctuations beneath a Turbulent Boundary Layer," J. Hydronautics, Vol. 6, No. 1, Jan. 1972, pp. 62-64.

An experimental study. The shear stress fluctuations were increased by flexible surface. Also see references to earlier work.

17. G. R. Inger and E. P. Williams, "Subsonic and Supersonic Boundary-Layer Flow Past a Wavy Wall," AIAA J., Vol. 10, No. 3, May 1972, pp. 636-642.

    A theoretical-experimental aerodynamic study for a rigid wavy wall. Also see Lekoudis, Nayfeh and Saric, Ref. 23, and also Williams, Chi, Ventres and Dowell, Ref. 26, below.

18. W. D. George, J. D. Hellums, and B. Martin, "Finite-Amplitude Neutral Disturbances in Plane Poiseuille Flow," J. Fluid Mechanics, Vol. 63, Part 4, 1974, pp. 765-771.

    References 18 and 19 agree reasonably well on the lower bound of Reynolds number for which laminar flow may become unstable as a result of finite amplitude-initial disturbances. A finite difference numerical solution technique is employed.

19. J.-P. Zahn, J. Toomre, E. A. Spiegel, D. O. Gough, "Nonlinear Cellular Motions in Poiseuille Channel Flow," J. Fluid Mechanics, Vol. 64, Part 2, 1974, pp. 319-345.

    The latest and best of the papers on the Dowell-Hellums approach to nonlinear stability theory and fluid limit cycle oscillations.

20. M. C. Fischer and R. L. Ash, "A General Review of Concepts for Reducing Skin Friction, Including Recommendations for Future Studies," NASA TM X-2894, March 1974.

    Places the compliant wall in context with alternative concepts for drag reduction. Has an extensive list of references.

21. R. L. Ash, D. M. Bushnell, L. M. Weinstein and R. Balasubramanian, "Compliant Wall Surface Motion and Its Effect on the Structure of a Turbulent Boundary Layer," Fourth Biennial Symposium on Turbulence in Liquids, September 1975, U. of Missouri, Rolla.

A description of the NASA Langley Research Center program which has been among the most substantial and comprehensive of recent years. A heuristic qualitative theory due to Bushnell is used to define experiments and theoretical models which hopefully will provide the quantitative predictive power needed for further progress.

22. E. H. Dowell, "Aeroelasticity of Plates and Shells," Sitjhoff-Noordhoff, The Netherlands, 1975.

   The standard reference on flutter and divergence of elastic plates and shells.

23. S. G. Lekoudis, A. H. Nayfeh, W. S. Saric, "Compressible Boundary Layers Over Wavy Walls," Physics of Fluids, Vol. 19, April 1976, pp. 514-519.

   A numerical solution to the Navier-Stokes equations, linearized in the wavy wall amplitude. Includes viscosity directly in the perturbation equations and compares to the experiments of Inger and Williams, Ref. 17.

24. D. T. Tsahalis, "On the Theory of Skin Friction Reduction by Compliant Walls," AIAA Paper 77-686, Presented at the AIAA 10th Fluid and Plasmadynamics Conference, Albuquerque, N.M., June 1977.

   The approach is as follows:

   (1) A measured turbulent pressure power spectra is given.

   (2) The wall motion is calculated (the fluid forces on the wall due to its own motion are omitted; also a point-reacting wall model is used).

   (3) From the now known wall motion, the viscous velocities are calculated from (a linearized set of) the Navier-Stokes equations.

(4) From the viscous velocities, the Reynolds stress is calculated which, in turn, allows the drag to be estimated.

As discussed in the text, this approach has promise, but there are some gaps in the physical model described above.

25. J. W. Murdock, "A Numerical Study of Nonlinear Effects on Boundary-Layer Stability," AIAA Journal, Vol. 15, No. 8, August 1977, pp. 1167-1173.

The flat, rigid wall case using the spectral algorithm of Orszag, which is a more mathematically sophisticated version of the Dowell-Hellums-Toomre approach.

26. M. H. Williams, M. R. Chi, C. S. Ventres, and E. H. Dowell, "Aerodynamic Effects of Inviscid Parallel Shear Flows," AIAA Journal, Vol. 15, No. 8, August 1977, pp. 1159-1166.

A summary of this team's results on the use of an inviscid shear layer model to determine the effect of a boundary layer on aerodynamic pressures over stationary and oscillating wavy walls and airfoils. In this model viscous effects are included in the mean flow, but not in the perturbation flow due to wall motion. Agreement with experiment for aerodynamic pressures and flutter boundaries is good.

27. Many Authors, "Special Course on Concepts for Drag Reduction," AGARD Report No. 654, 1977.

The articles by Zimmerman and Dinkelacker, describing the German experience, are interesting but inconclusive. The tutorial article by Bushnell, Hefner and Ash gives a distillation of the Langley experience and their assessment of the work of others.

28. R. Balasubramanian, "Analytical and Numerical Investigation of Structural Response of Compliant Wall Materials," NASA CR 2999, May 1978.

     One of the Langley "progress reports." Also see Ref. 30 below.

29. H. L. Swinney and J. P. Gollub, "The Transition to Turbulence," Physics Today, Vol. 31, No. 8, August 1978, pp. 41-49.

     A readable review of the flat, rigid wall case.

30. G. R. Hough, Editor, "Viscous Flow Drag Reduction," Vol. 72, Progress in Astronautics and Aeronautics, Martin Summerfield, Series Editor, 1980.

     A collection of papers from the Langley and ONR programs. See especially Chapter III, Nonplanar Geometries, and Chapter VII, Compliant Surfaces. It is generally concluded that it is difficult to compute accurately drag in turbulent flows over nonplanar geometries and also difficult to detect any drag reduction with the compliant walls examined experimentally. Calculation of pressure distributions over wavy walls seems to agree reasonably well with experiment.

31. C. T. Hsu, E. Y. Hsu, and R. L. Street, "On the Structure of Turbulent Flow Over a Progressive Wave: Theory and Experiment in a Transformed Wave-Following Coordinate System," J. Fluid Mechanics, Vol. 105, April 1981, pp. 87-117. Also see Part 2, Vol. 131, June 1983, pp. 123-153.

     These experimental results were later used by Sengupta and Lekoudis to compare with their theoretical results.

32. M. M. Reischman, Editor, "Drag Reduction Symposium-Abstracts," held at the National Academy of Sciences, Sponsored by ONR, NASA, AFOSR, NASC, September 13-17, 1982.

A brief overview and summary of work active as of the date of publication. The formidable nature of the issues is apparent.

33. E. A. Caponi, B. Fornberg, D. D. Knight, J. W. McLean, P. G. Saffman, H. C. Yuan, "Calculation of Laminar Viscous Flow Over a Moving Wavy Surface," J. Fluid Mechanics, Vol. 124, November 1982, pp. 347-362.

The steady laminar, incompressible flow over a periodic wavy surface with a prescribed surface-velocity distribution is found from a [numerical] solution of the two-dimensional Navier-Stokes equations. Results showed excellent agreement for small-amplitude wavy surfaces with the analytical small perturbation solutions of Benjamin, Ref. 3.

34. L. P. Kozlov, V. I. Korobov, and V. V. Babenko, "The Effect of an Elastic Wall on a Boundary Layer," Reports of Ukranian Academy of Sciences, Series A: Physical, Mathematical and Technical Sciences, Jan. 1983, pp. 45-47.

A report on Soviet experimental work which is said to show drag reduction.

35. J. C. Lin, M. J. Walsh and R. D. Watson, "Turbulent Drag Characteristic of Small Amplitude Rigid Surface Waves," AIAA Paper 83-0228, January 1983.

A careful experimental-theoretical study of the rigid wall case for both sinusoidal and nonsinusoidal wall shapes. No net drag reduction found experimentally. Skin friction drag reduction is found experimentally. Skin friction drag reduction is overwhelmed by pressure drag increase. Theory suggests some wall shapes might give a net drag reduction.

36. J. W. McLean, "Computation of Turbulent Flow Over a Moving Wavy Boundary," Physics of Fluids, Vol. 26, August 1983, pp. 2065-2073.

Numerical solutions to the Reynolds-averaged Navier-Stokes equations for rigid and moving boundaries. Results agree well with experiments for small amplitude waves. This approach complements that of Sengupta and Lekoudis, Ref. 40, and, of course, preceded their work.

37. J. H. Duncan and C. C. Hsu, "The Response of a Two-Layer Viscoelastic Coating to Pressure Disturbances from a Turbulent Boundary Layer," AIAA Paper 84-0535, presented at the AIAA 22nd Aerospace Sciences Meeting, January 1984, Reno, Nevada.

An extension of Ref. 1 and 5 to include (1) a more elaborate compliant wall model and (2) a prescribed idealized external fluid forcing excitation.

38. A. C. Buckingham, M. S. Hall and R. C. Chun, "Numerical Simulations of Compliant Material Response to Turbulent Flow," AIAA Paper 84-0537, presented at the AIAA 22nd Aerospace Sciences Meeting, January 1984, Reno, Nevada. Also see an abbreviated version of this paper in AIAA Journal, Vol. 23, No. 7, July 1985, pp. 1096-1052.

An ambitious program is described to model elaborate compliant wall structures by finite element methods and to model the fluid flow by a series of increasingly sophisticated procedures, i.e.

    (I)   The fluid excitation is taken as prescribed from experiments over rigid walls. All of the numerical results described in this paper are obtained using this procedure.

    (II) In the second level of fluid modeling, the change in fluid pressures due to compliant wall motion is to be accounted for with a potential flow theory. This is as in Ref. 1, 5, 6, and 37.

(III) In the final level of fluid modeling it is
proposed to compute the change in fluid
pressures (and shear stresses) due to compliant wall motion by using the Navier-
Stokes equations and the spectral algorithm of Orszag, et al.

If the goals of this brave program are realized,
a major advancement in theoretical modeling
should occur.

Some results from (III) may have been presented
without publication in M. S. Hall and A. C. Buckingham, "Calculations of the Interaction Between
a Compliant Material and an Unsteady Flow," presented at the XVIth International Congress of
Theoretical and Applied Mechanics, Lyngby, Den-
mark, August 1984.

39. M. Gad-el-Hak, R. F. Blackwelder, and J. J.
Riley, "On the Interaction of Compliant Coatings
with Boundary Layer Flows," J. Fluid Mechanics,
Vol. 140, 1984, pp. 257-280.

An experimental study using a compliant material
of viscoelastic plastisol gel produced by heating
a mixture of polyvinyl chloride resin, a plasti-
cizer and a stabilizer, and allowing them time to
gel. No drag reduction or significant modifica-
tion of the turbulent flow observed. A hydroelastic instability was found. Contains a nice
summary of earlier literature.

40. T. K. Sengupta and S. G. Lekoudis, "Calculation
of Two-Dimensional Turbulent Boundary Layers Over
Rigid and Moving Wavy Surfaces," AIAA Journal,
Vol. 23, No. 4, April 1985, pp. 530-536.

In this important paper, an effective means is
devised for computing the pressure and skin friction on a moving surface of prescribed form. (A
mean boundary layer and perturbation viscous Orr-
Sommerfeld system is used.) The method agrees
reasonably well with experimental data over rigid
walls. Not surprisingly, pressure is predicted

better than skin friction.   Among the   important
conclusions reached are

> "For accurate estimations of the total drag,
> the most important quantities are the ampli-
> tude and phase of the pressure and the mean
> shear. ...
>
> For the types of wall motion considered, the
> [theoretical] technique predicted no drag
> reduction, except for the case of wave
> speeds approaching the free stream speed.
> In all other cases the pressure drag over-
> powers the mean skin friction reduction gen-
> erated by the waves."

This work may be thought of as a generalization
of the work on moving surfaces of Ref. 2 and 26
to include viscous effects in the perturbation
equations which allows the computation of skin
friction changes due to wall motion.   Also see
Inger, Ref. 17, and Lekoudis, Ref. 23, for the
rigid wavy wall problem with viscous effects
included, and the earlier publication of McLean,
Ref. 36.

41. M. Gad-el-Hak, "Boundary Layer Interactions with
     Compliant Coatings," Applied Mechanics Reviews,
     Vol. 39, No. 4, April 1986, pp. 511-523.

     A recent, readable review which emphasizes the
     work of the last five years.   The perspective is
     that of an experimentalist with an abiding inter-
     est in turbulence.

APPENDIX A

OBSERVATION AND EVOLUTION OF CHAOS
FOR AN AUTONOMOUS SYSTEM*

Summary

Time histories, phase plane portraits, power spectra, and Poincaré maps are used as descriptors to observe the evolution of chaos in an autonomous system. Although the motions of such a system can be quite complex, these descriptors prove helpful in detecting the essential structure of the motion. Here the principal interest is in phase plane portraits and Poincaré maps, their methods of construction, and physical interpretation. The system chosen for study has been previously discussed in Chapter V, i.e., the flutter of a buckled elastic plate in a flowing fluid.

Introduction

Several authors including Holmes [1], Holmes and Marsden [2], Thompson [3], and Dowell [4] have discussed the complicated motion that may occur due to the flutter of buckled plates. Indeed chaotic motions are now known to occur for a variety of relatively simple deterministic systems [3-5]. In the present paper the flutter of a buckled plate is again considered, and various means for observing the evolution of chaos are employed.

In particular, the Poincaré maps that are presented are the first obtained for the particular example studied and among the few extant for any autonomous system. A simple physical motivation is provided for the manner in which the Poincaré maps are constructed. Also an interpretation of their physical significance is given.

---

*This appendix is based on the paper, E. H. Dowell, "Observation and Evolution of Chaos in an Autonomous System," J. Applied Mechanics, Vol. 51, 1984, pp. 664-673.

In the following, a general discussion of observation and evolution of chaos is presented first and
then the example of flutter of a buckled plate is considered in some detail.

Observation of chaos

To observe chaos, and to better understand it,
several descriptors are available. Among these are:

Time histories;

Phase plane portraits;

Power spectral densities (or their Fourier
transforms, correlation functions);

Poincaré maps

The first three are standard and require no
further elaboration here. The last is not and a few
words of explanation are in order. Also see Reference
[5].

A (two-dimensional) Poincaré map is a plot in the
(two-dimensional) phase plane at chosen discrete
instants of time. For example, one might plot a displacement coordinate versus its velocity. However one
might also plot two coordinates versus each other for
systems of third or higher order. The crucial question, however, is the choice of the discrete instants
of time. A normal phase plane plot, of course, is for
all (continuous) instants of time.

For a nonautonomous system an appropriate choice
is rather obvious. The system is forced by an input
of prescribed frequency and, hence, period. Thus to
construct a Poincaré map one normally chooses to plot
data in the phase plane at instants of time separated
by equal time intervals corresponding to a period of
the input. For an autonomous system, however, which
is our interest here, the choice is less obvious.
Nevertheless Poincare maps may also be constructed for
such systems and the method for doing so may be con-

sidered a generalization of that used for nonautonomous systems.

The key concept is that of a clock, since the
Poincaré care map based on the occurrence of an event.
In general, these events will <u>not</u> occur at <u>equal</u> time
intervals. Indeed they may occur at <u>chaotic</u> time
intervals. For example, and to be concrete, consider
an elastic plate whose deflection, W, depends on one
(or more) spatial variable(s) and time. Choose a
point on the plate and observe the plate deflection
and its velocity there. Define the event as being the
passage of the velocity of the plate at that point
through zero (or through any constant value, although
some values of velocity will not occur and hence are
inappropriate choices). Now, at those time instants
(which may occur periodically or chaotically) when the
plate velocity is zero, plot the deflection of the
selected point on the plate versus the deflection at
some other selected point on the plate. The result is
a Poincaré map. There are (infinitely) many
variations on this same idea, of course. For example,
plotting the deflection versus velocity of some
selected point on the plate at those time instants
when the velocity of some <u>other</u> selected point on the
plate is zero would give a more conventional
(chaotically stroboscopic) phase plane plot, which is
another Poincaré map. Indeed, for technical reasons,
in some applications, yet other choices of Poincaré
map coordinates may be desirable. This will be
discussed further. Also, of course, higher (than two)
dimensional Poincaré maps may be constructed. Indeed
in the mathematical literature, a Poincaré map of
dimension less than N-1, where N is the order of the
system, is referred to as a reduced Poincaré map [6].
The unifying feature of all such maps, however, is the
concept of a (stroboscopic) clock based on the
occurrence of an event.

Using the four descriptors listed in the fore-
going, including the Poincaré map, one may observe and
better understand the evolution of chaos.

Evolution of Chaos

There are several distinct phenomena that occur during the evolution of chaos. While the compilation discussed in this section is not thought to be exhaustive for all examples that may be of interest, it does appear reasonably complete for the example discussed in detail in the following section. Moreover, and perhaps more importantly, because of their intrinsic nature, these phenomena are likely to appear in many other examples as well.

It is convenient to divide the evolution of chaos into three phases; the beginning, the onset, and the maturing process.

The Beginning. It is presumed that a limit cycle of a dominant single frequency has been established. (Often the motion will be nearly simple harmonic in time.) This may be a result of a Hopf bifurcation, for example [5]. As some parameter in the governing equations of motion is changed, several characteristic changes in the motion occur. These include:

large changes in frequency or period of motion near some parameter values for small changes in the parameter; the period generally becomes larger as chaos is approached.

changes in the number of maxima and minima in the time history of the motion during one period of motion; the number may increase as chaos is approached.

In the beginning (by definition), the motion remains periodic.

The Onset. The onset of chaos may be identified in several ways:

The failure of the time history to repeat itself over any finite time interval. Operationally, this may be tedious to determine.

The diffusion of the phase plane plot [ideally diffusing from a curve of infinitesimal thickness that represents periodic motion to a set of points that nearly (completely) fills a finite region of the phase space when the motion is observed over a very large (infinite) time]. This is usually readily observed.

The broadening of the power spectral peaks (ideally from delta functions, which are characteristic of periodic motion or even aperiodic motion with discrete frequency content to a continuous distribution characteristic of chaotic motion). This is usually readily observed.

The Maturing Process. It is in the maturing process that Poincaré maps become of special interest. Well beyond the onset of chaos, diffusion in the phase plane has become so pronounced that now a phase plane portrait is only indicative of the minima and maxima of the chaotic motion. However, Poincaré maps may be used to detect order in the chaos, although eventually such maps may undergo diffusion as well.

One interpretation of a Poincaré map is as a form of conditional probability. That is, it answers the question, given some event and the value of one system coordinate, what is the likely value of another system coordinate? Not surprisingly, at least in retrospect, the answer to this question possesses considerably more structure than the answer one would obtain if the coordinates were observed at any choice of equal time intervals rather than at those instants determined by the occurrence of an event. Choosing instants of equal time intervals would simply produce conventional phase plane portraits, unless such time intervals should happen to correspond to an event, e.g., the (periodic) input force reaching a maximum for a nonautonomous system. Note that using the concept of a Poincaré map based on the occurrence of an event, in the nonautonomous case it is the force achieving a certain value (occurrence of an event), rather than the passage of a certain time interval, which triggers the clock of the Poincaré map.

387

An Example

The physical system consists of an elastic plate embedded in an otherwise rigid surface and the adjacent fluid that flows over the top of the plate. A (compressive) in-plane load is applied to the plate.

The behavior of this system is governed by a (set of) partial differential equation(s), which here is considered in its simplest form. More elaborate mathematical-physical models are considered in reference [7]. For simplicity, in the present paper the plate geometry is taken to be one-dimensional and the plate deflection, w, is therefore a function of a single spatial position variable, x, and time, t. The nonlinear partial differential equation is reduced to a set of nonlinear ordinary differential equations (NODE) in time by expanding w in a set of chosen spatial functions, e.g.,

$$w(x,t) = \sum_{n=1}^{n} a_n(\tau)\sin \frac{n\pi x}{a} \qquad (1)$$

where a is the plate length. The a_n satisfy a set of N NODE. These equations are solved by a numerical time marching algorithm. For further details see Chapter V and [4]. τ is a non-dimensional time. The equations that govern the a_n are given in Chapter V.

There are two important (control) parameters in the equations. λ is a nondimensional flow velocity and R_x a nondimensional mechanical in-plane load. Negative values of R_x denote an applied compressive load. It is known that certain values of λ, R_x will lead to chaotic motion. It is our purpose here to observe the evolution of chaos in some detail. Before reading further, a reader interested in more detail about the mathematical-physical model, the solution techniques, and the global behavior of the system in λ-R_x space should consult Chapter V or [4].

We will consider, in turn, the four descriptors; time histories, phase plane diagrams, power spectra, and Poincare maps in order to view the beginning, onset, and maturing of chaos.

Time Histories

For a fixed value of λ, R_x/π^2 was varied over a range of values from -3 to -8. Only two terms were retained in equation (1), i.e., N = 2. The time histories were computed through the transient to determine one period of motion, except of course where periodic motion does not exist. For brevity, the time histories per se are not shown.

However, it is of interest to summarize the information contained in the time histories by presenting the period of the motion versus R_x. This is done in Fig. 1. The vertical arrows indicate the period is very large, certainly in excess of $\tau = 200$, and probably infinite. This motion generally appears chaotic. For $-R_x/\pi^2$ less than 2.3 no (steady state) motion exists because the system is dynamically stable. Between 2.3 and 3, a predominantly single frequency limit cycle occurs which is a result of a Hopf bifurcation [1-4]. Near 3, a sudden increase in the period of the motion occurs and the motion now contains two dominant frequencies. The two frequencies are in the ratio 1:3 as one might expect with a cubic nonlinearity. For a very small interval near $-R_x/\pi^2 = 3$, the frequencies are not strictly in the ratio 1:3 and the motion is aperiodic. From 3 to 5.25 the motion retains two dominant frequencies. At $-R_x/\pi^2 = 5.5$, the motion has another dramatic increase in its period and additional frequency components become significant. These appear to be, in addition to the first and third harmonics, fifth and seventh harmonics as well. At 5.625 chaos appears; at 5.75 it disappears; it reappears at 5.875 and disappears at 7.75. Beyond 7.75 there is no chaos until at least 12 where computations were halted.

In these calculations $\lambda = 150$ and the mass ratio parameter, μ/M, was selected as 0.1.

389

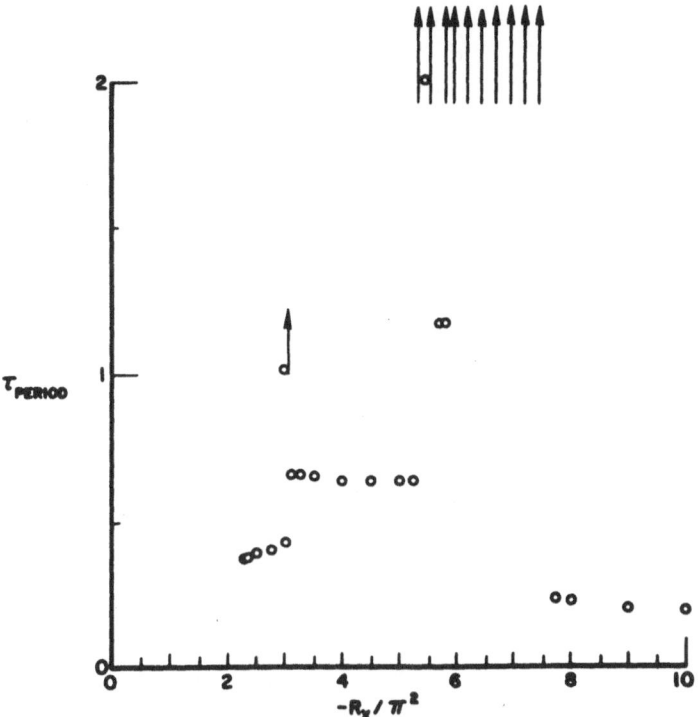

Fig. 1. Period of motion vs. in-plane load.

Phase Plane Portraits

These are shown in Fig. 2 for selected R_x. As expected, the portraits become increasingly complex as $-R_x/\pi^2 = 2.5 - 5$ with many loops appearing immediately prior to the onset of chaos. Of course, chaos appears at 5.625, disappears at 5.75, reappears at 5.875, and disappears again at 7.75. Although the portraits are quite diffuse when chaos occurs, one may sometimes detect the "shadows" of periodic motion. For example, compare the results for 6 and 5.75.

Typically, the (chaos and/or) periodic motion reaches a (statistical) steady state at $\tau = 10$ or so. However when chaos appeared at 6 and then subsequently disappeared at 7.75, the emergence of the periodic motion sometimes took a remarkably long time; a "chaotic" transient of the order of $\tau \simeq 100$ might occur before the periodic motion established itself. For example, the transient for $-R_x/\pi^2 = 10$ persisted until approximately $\tau = 105$. (The phase plane

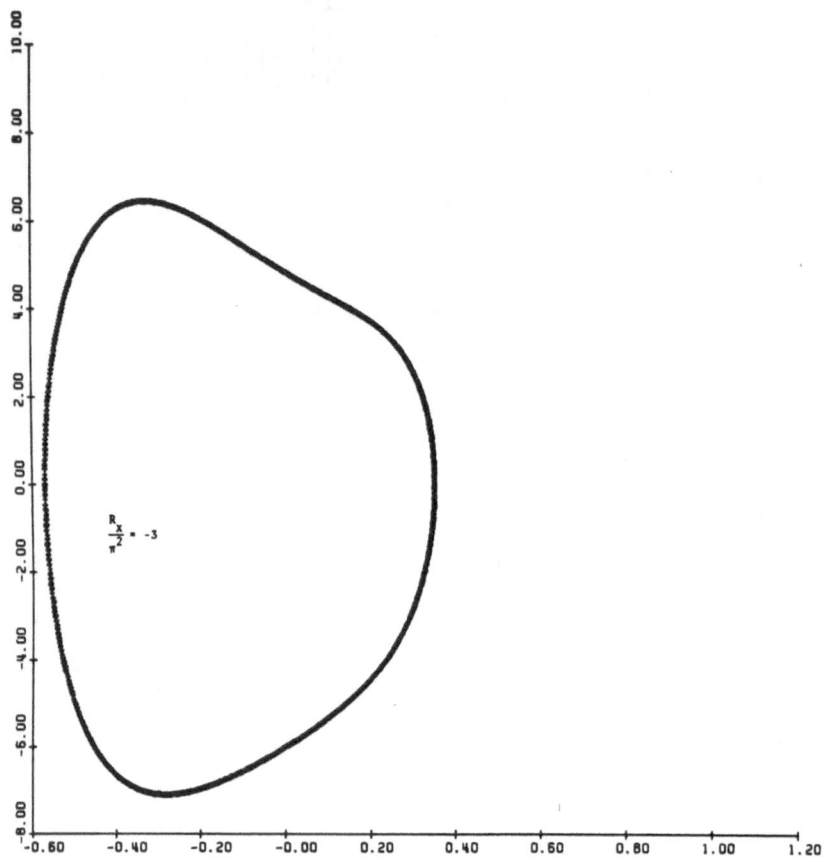

Fig. 2a. Phase plane diagram, velocity vs. displacement.

portraits represent points plotted for τ = 100 → 140
for time intervals of 0.01.)   One is tempted to asso-
ciate this behavior with the concept of "intermittency" where both periodic and chaotic motions exist
over extended periods of time.

The maturing of the chaos may be observed by
noting the diffusion of the phase plane portraits.
See particularly the results   for $-R_x/\pi^2$ =   5.625 and
6.0.   However additional structure of the motion may
be observed by considering the corresponding power
spectra and Poincaré maps.

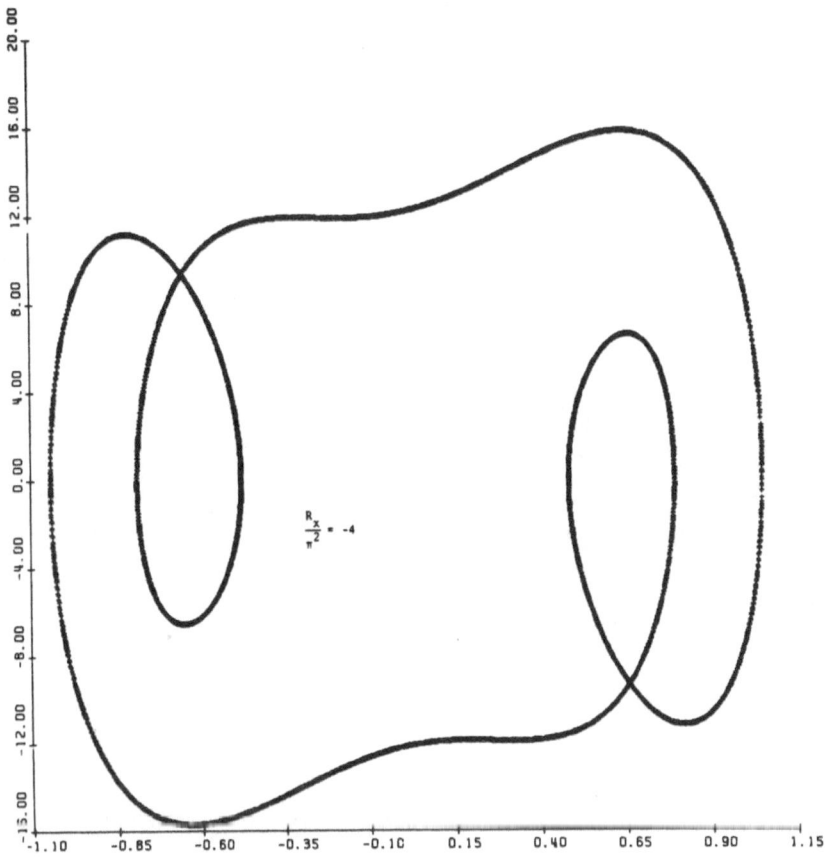

Fig. 2b.   Phase plane diagram, velocity   vs. displace-
           ment.

    The   phase plane portraits presented here   are in
terms of velocity versus displacement for   a representative point on the plate, i.e., x = 3/4a.

Power Spectra

    Power   spectra have   been presented in   Chapter V
and [4] and the reader is referred to that discussion.
As   noted previously, periodic motion will   be characterized by   sharp peaks   and chaotic   motion   by broad
band behavior. Consider now Poincaré maps.

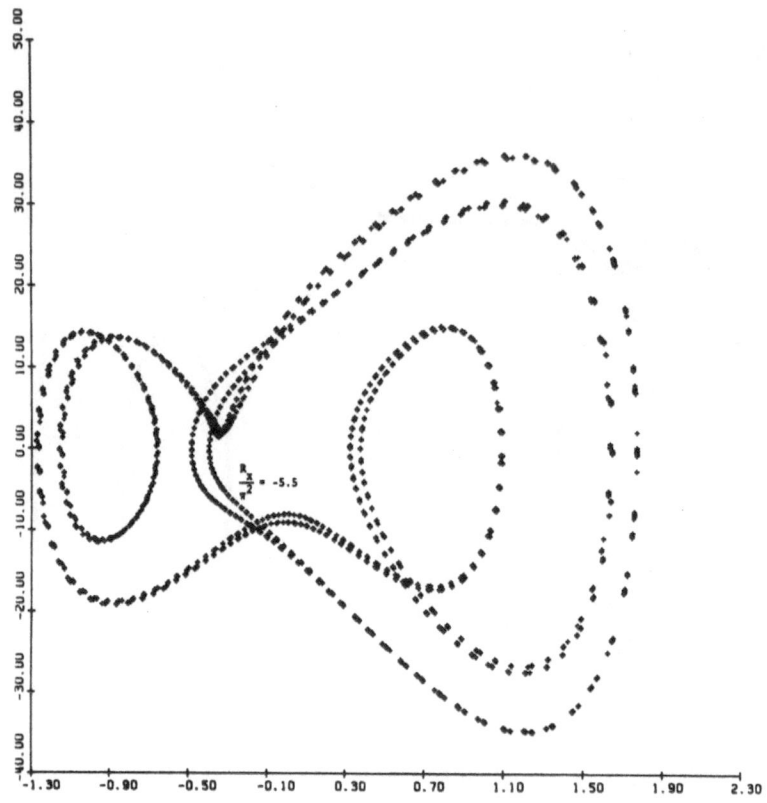

Fig. 2c. Phase plane diagram, velocity vs. displacement.

Poincaré Maps

Various Poincaré maps may be constructed based
on:

The definition of the event that determines those
instants of time when the coordinates are
selected for plotting;

The choice of coordinates themselves.

Here the event selected is the vanishing of the velocity of a selected point on the plate, namely x = 3/4a.
The coordinates selected for the map are the first two

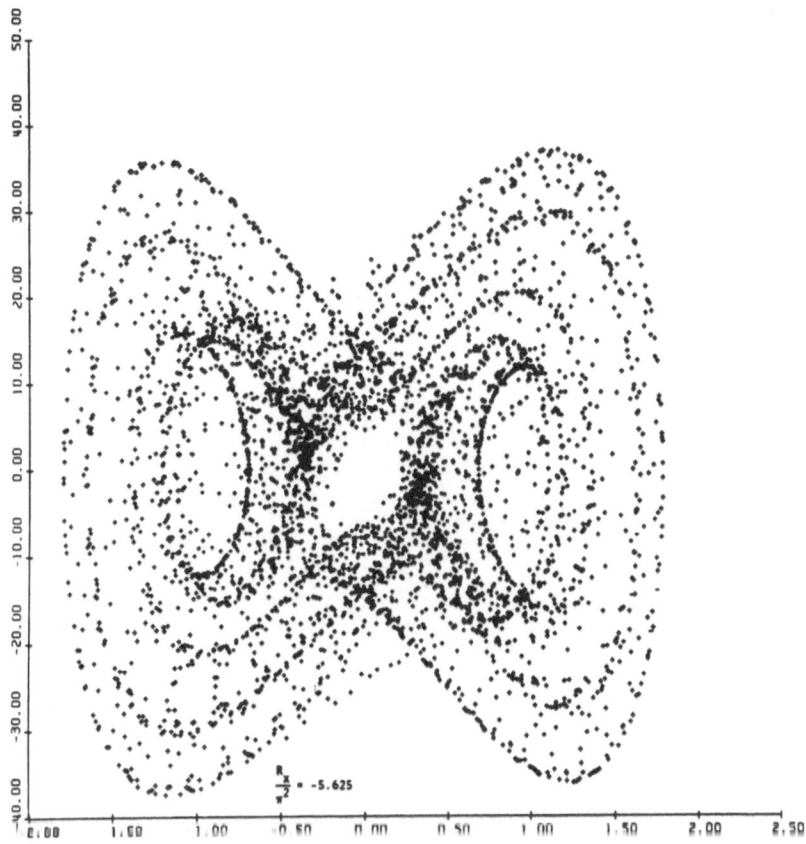

Fig. 2d.   Phase plane diagram, velocity vs. displace-
            ment.

Fourier coefficients of equation (1) (nondimensional-
ized as discussed in Chapter V).   The Poincaré maps
are presented in Fig. 3 for selected R_x.

     For periodic motion, a Poincaré map will consist
of a finite number of (local) extrema over one period
of motion. Thus for $-R_x/\pi^2 = 2.5 \rightarrow 5.5$, only a finite
number of discrete points appear in the Poincaré map.

     However, when chaos appears at, for example,
$-R_x/\pi^2 = 5.625$ [6], a more complicated but still
ordered Poincaré map is formed.   As discussed before,
the Poincaré map may be used to answer the question,

394

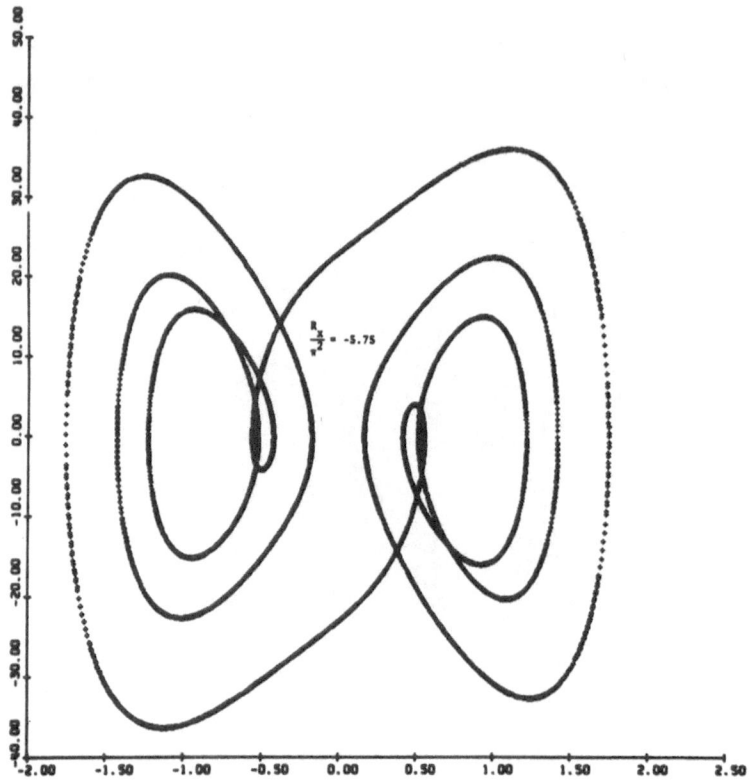

Fig. 2e. Phase plane diagram, velocity vs. displacement.

given some event (the vanishing of the plate velocity
at some point on the plate) and the value of one system coordinate, what is the likely value of another
system coordinate?   Clearly, as seen from Fig. 3, the
range of likely values is rather limited.

The order apparent in the Poincaré maps may be
called "large-scale structure" by analogy to the term
sometimes employed in hydrodynamic turbulence.

Some modest diffusion of the Poincaré map does
occur as $-R_x/\pi^2$ increases from 6 to 7, say.   This, of
course, corresponds to an increase in the range of
possible coordinate values.   The diffusion is more

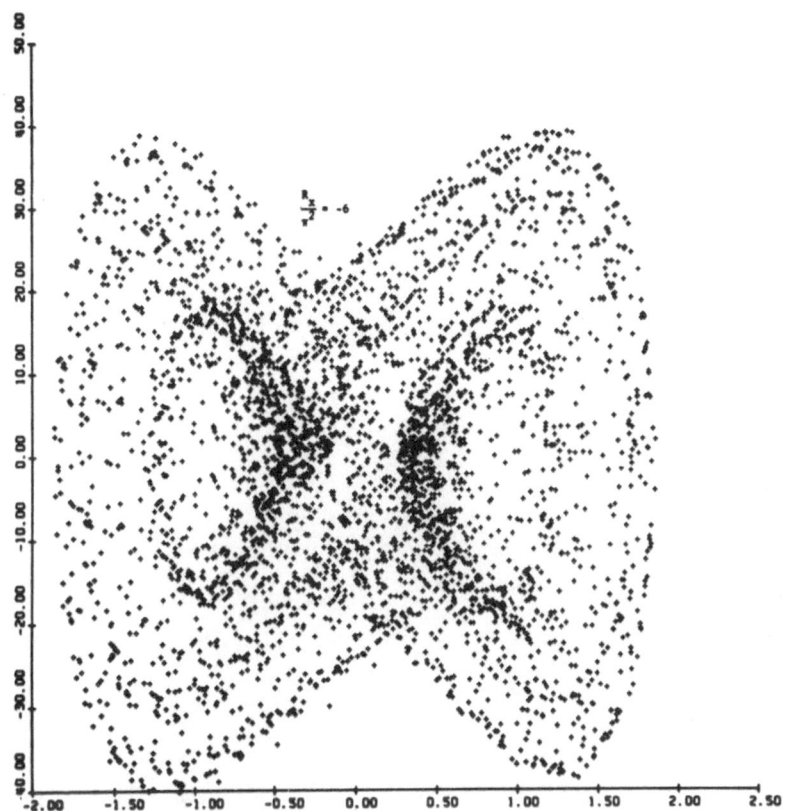

Fig. 2f. Phase plane diagram, velocity vs. displacement.

pronounced, however, if one considers a higher order
representation of the system dynamics. For example,
consider now N = 6 in equation (1).   All previous
results were for N = 2. Results (not shown) were also
obtained for N = 4.

    In Fig. 4 Poincaré maps are presented for $-R_x/\pi^2$
= 6, 9, 12, and N = 6.   Note the systematic increase
in the diffusion of the map as 6 → 9 → 12. By examining higher dimensional Poincaré maps, further order
might be perceived even for these relatively diffuse
Poincaré maps. However no such attempt is made here.

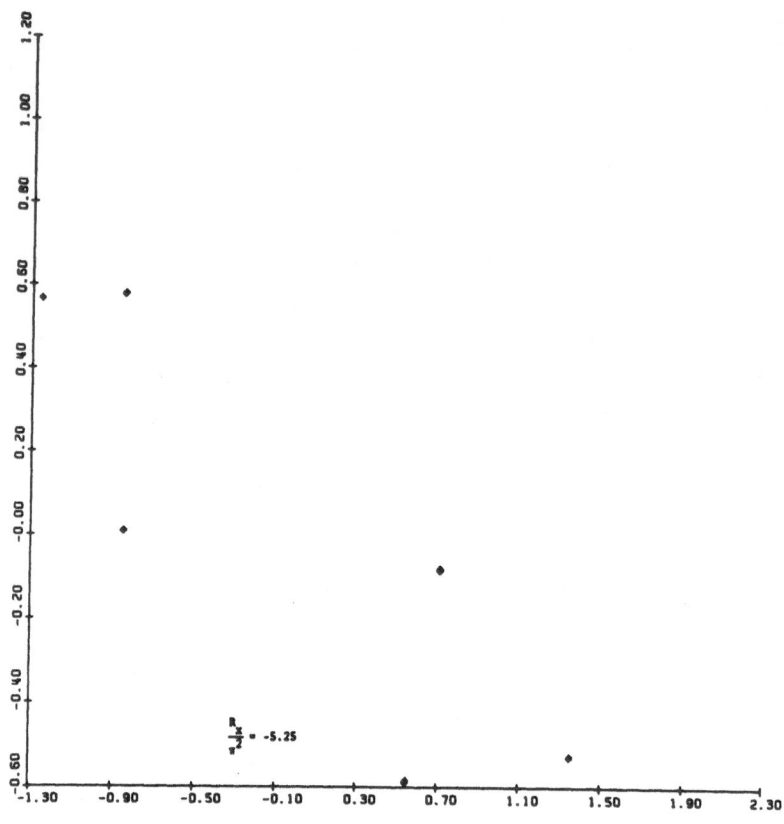

Fig. 3a. Poincaré map, second vs. first Fourier coef-
          ficient, N = 2.

     Note also that chaos occurs for $-R_x/\pi^2$ = 12 and N
= 6 while for N = 2 no chaos appears. Examining the
results for N = 2, 4, 6, it is concluded that an
increase in N leads to a larger range of R_x for which
chaos appears. It is expected that as $-R_x/\pi^2$ increases N must increase to obtain convergence of the series
in equation (1). For the range of parameters examined
N = 6 is sufficient to give good convergence, as can
be established by examining the magnitudes of the a_n
that appear in equation (1). The results for N = 2
are not well converged quantitatively, but are qualitatively comparable to those for N = 6.

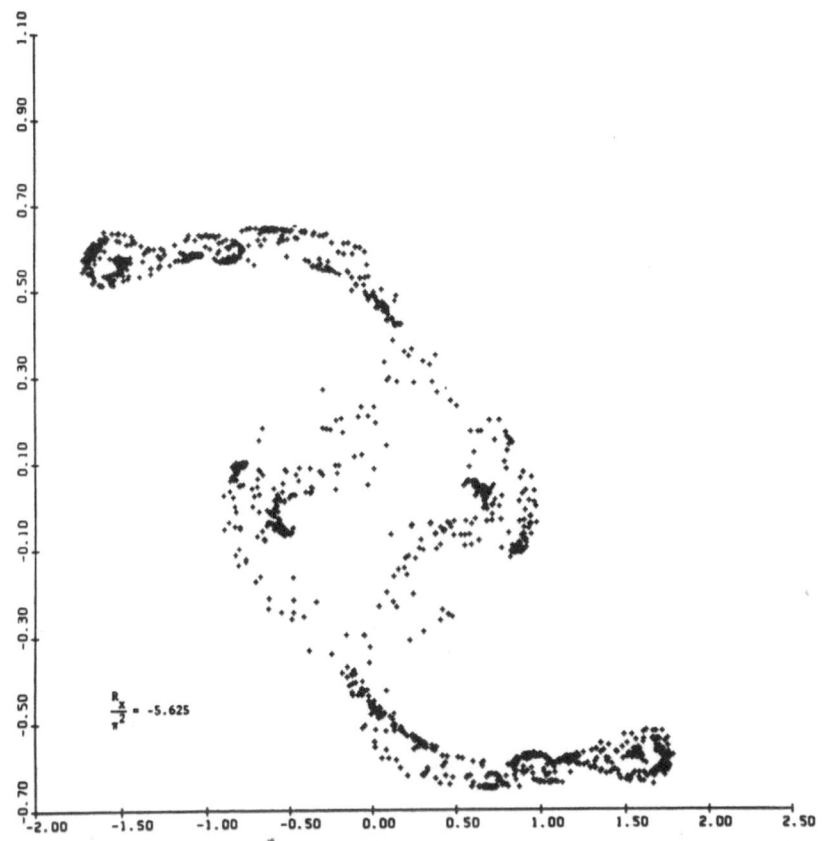

Fig. 3b. Poincare map, second vs. first Fourier coefficient.

Temporal Statistics Of An Event

When chaotic motion occurs, it is of interest to examine the temporal statistics of an event. For example, consider a certain (long) time interval. Within this interval a (large) number of events will occur. One can measure the (small) time between two successive events, and plot the numbers of events that occur versus the (small) times between two successive events. Then this graph can be normalized by dividing by the total number of events. The result is a probability density function that allows one to determine the likelihood of an event occurring within a certain (small) interval of time given the occurrence of a previous event.

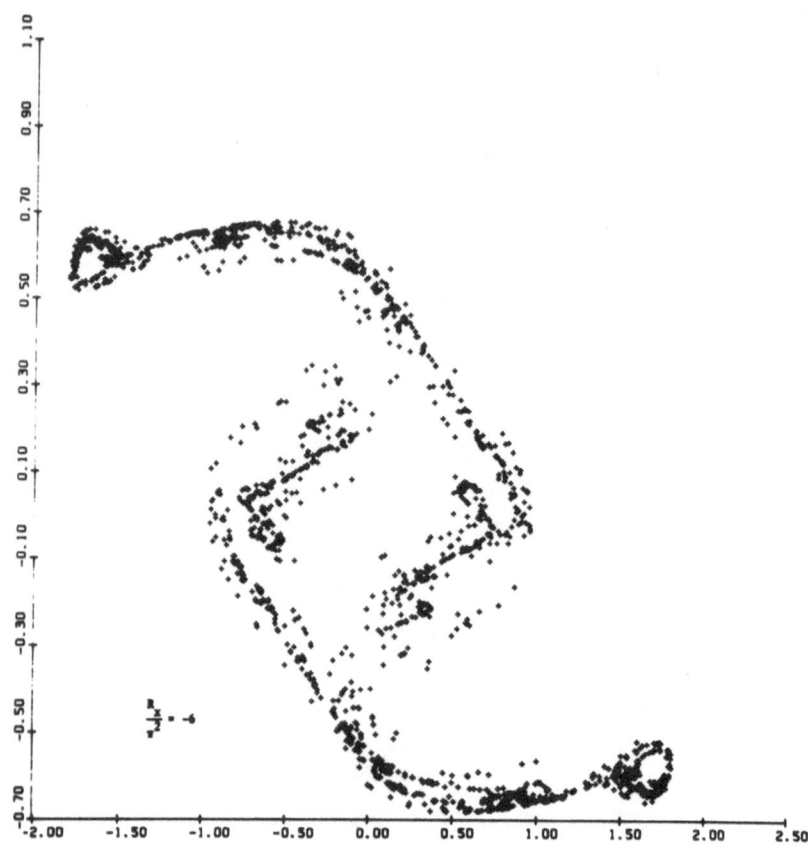

Fig. 3c. Poincare map, second vs. first Fourier coefficient, N = 2.

In Fig. 5 such a probability density function is
displayed for the case R_x/π^2 = -6 and N = 2. The
total (long) time interval considered was τ = 29.4131
→ 66.3981 and the total number of events occurring was
403. To obtain a discrete representation of the
probability density function, the total time interval
was divided into small time intervals of duration,
0.01. The numbers of successive events occurring in
selected time intervals, 0 → 0.01, 0.01 → 0.02, 0.02 →
0.03, etc. was plotted.

Using Fig. 5 one can calculate the probability of
an event occurring within a certain time interval
given the occurrence of a previous event. For exam-

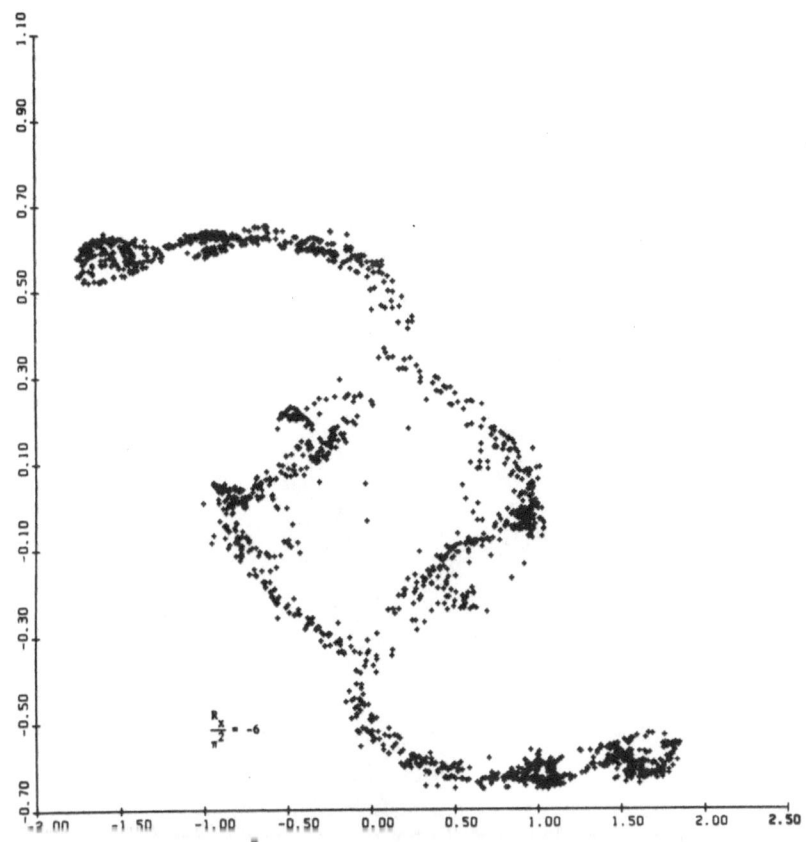

Fig. 4a. Poincare map, second vs. first Fourier coefficient, N = 6.

ple, the probability of an event occurring between τ_1
= 0.10 and τ_2 = 0.12 is 0.097 + 0.092 = 0.189 while
the probability of an event occurring between τ_3 =
0.18 and τ_4 = 0.20 is 0.0025 + 0.0025 = 0.005.   Also
shown in Fig. 5 is the average time between successive
events, $\Delta\tau$, and the variance, $\sigma_{\Delta\tau}$.

It should be emphasized that Fig. 5 is a discrete
representation (delta function) of the probability
density.   To obtain an approximate continuous representation, one divides the ordinates of Fig. 5 by the
selected (small) time interval, in this case, 0.01.
In the limit as the total (long) time interval goes to
infinity and the selected (small) time interval goes
to zero a continuous probability density representa-

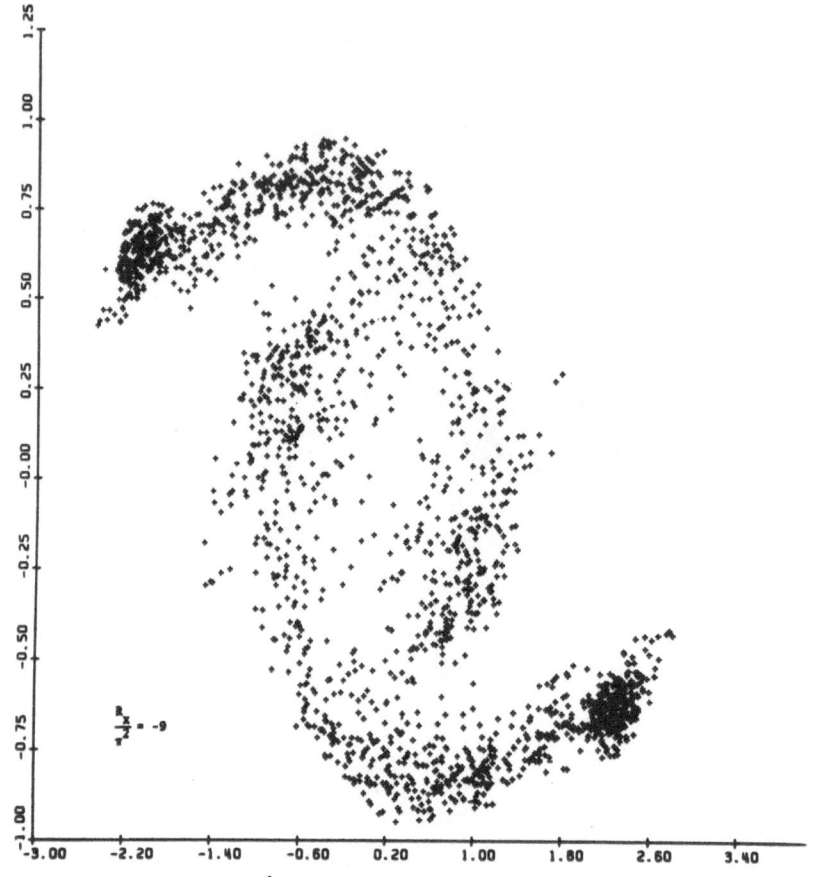

Fig. 4b. Poincaré map, second vs. first Fourier coefficient, N = 6.

tion is obtained. An approximation to this continuous
probability density function based on a selected
(small) time interval of 0.01 is thus obtained by connecting the ordinate points of Fig. 5 and dividing the
vertical scale by 0.01. Integrating the resulting
curve between two times one computes the probability
of an event occurring between these two times given
the occurrence of a previous event.

Other statistics can be obtained, of course, by
returning to the raw event data. For example, given
the occurrence of two events within one time interval,
one could deduce the probability of the occurrence of
a third event during another subsequent time interval.

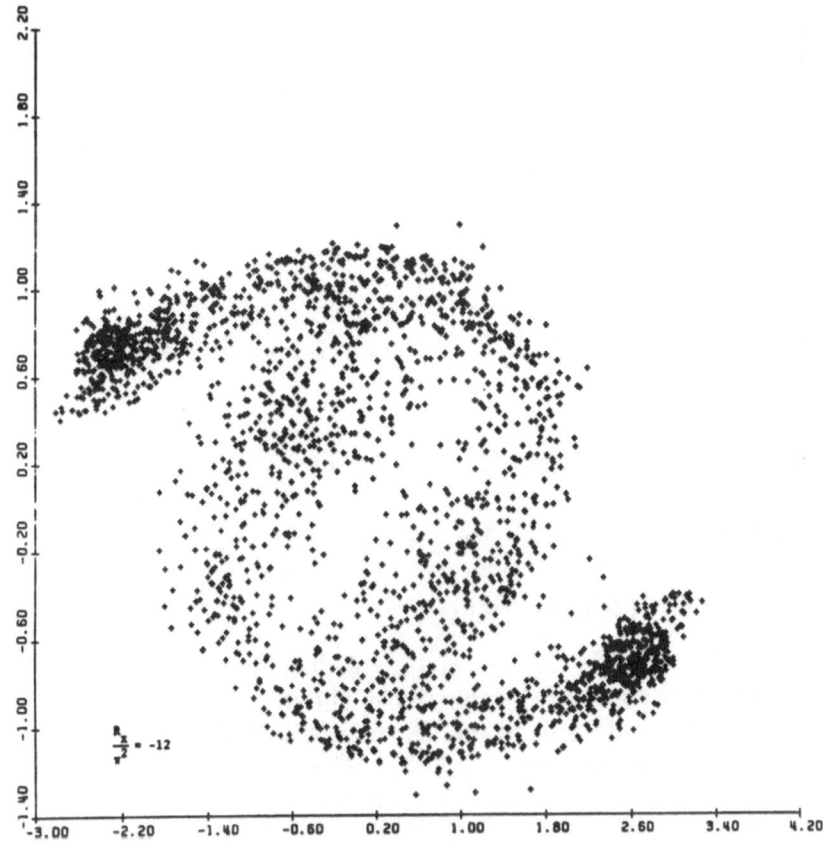

Fig. 4c. Poincaré map, second vs. first Fourier coefficient, N = 6.

By examining the temporal statistics of an event,
further insight may be gained concerning the essence
of the chaotic motion.   For example, Fig. 5 displays
the relatively narrow range of probable time elapse
between two successive events.

Concluding remarks

By using several descriptive techniques, a rich
variety of dynamical behavior can be observed for systems that undergo chaos. As the transition from
simple harmonic, to periodic, to chaotic motion
occurs, the descriptors of most value change from time
histories, to phase plane portraits, to power spectra

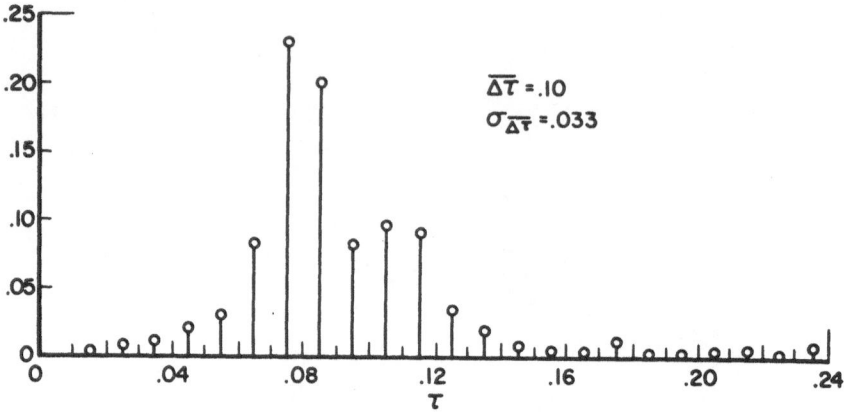

Fig. 5. Probability density of an event.

and Poincaré maps. The temporal statistics of an event which defines the Poincaré map proves to be a useful descriptor as well.

What remains as an open challenge is to translate the information obtained from these descriptors into an improved understanding of why apparently relatively simple systems behave in the complex manner that they do. As Simon [8] has emphasized, "the central task of a natural science is to make the wonderful commonplace: to show that complexity, correctly viewed, is only a mask for simplicity; to find pattern hidden in apparent chaos." Nevertheless, in a sense, it is encouraging that relatively simple mathematical models can give rise to complex behavior including "intermittency" and "large scale (chaotic) structures." Perhaps other physical phenomena exhibiting similar complex behavior may prove to be governed by relatively simple models as well. See, for example, reference [9].

Finally, a few comments are appropriate to relate the work on the present set of equations (see Chapter V for their explicit form) to that which has been carried out on the well-known Lorenz equations [10,11]. In the Lorenz model, above a certain parameter value (analogous to R_x for our equations) the trivial static equilibrium becomes unstable and two new static equilibria exist. These are stable until for a yet higher parameter value they also become unstable. Thus, for a sufficiently high parameter value, three static equilibria exist all of which are unstable (to small perturbations). Above this parameter value, chaos occurs. Chaos appears to be the result of all equilibria solutions being unstable to infinitesimal perturbation; nevertheless the global solutions are still bounded (through understandably chaotic). For parameter values slightly below that at which chaos occurs, long chaotic transients occur [11]. These are similar to those that have been found here for the present equation.

Consider now the present equations and their behavior. As a parameter $(-R_x)$ increases, the trivial static equilibrium becomes unstable. Either two new static equilibria exist or a dynamic limit cycle equilibrium exists. As the parameter increases further, chaos or a long chaotic transient exists and, for further parameter increases, chaos is fully established. What remains to be done for the present equations is to determine whether the appearance of chaos directly follows on the instability (to infinitesimal perturbations) of all possible static and dynamic equilibria. This is a far more formidable task for the present equations than the corresponding one for the Lorenz equations, because of the more complicated equilibria which occur prior to the onset of chaos. For the Lorenz equation there are three _static_ equilibria at most prior to chaos. For the present equations, these equilibria are dynamic periodic motions with multiple frequency content. Indeed it is not clear whether the number of equilibria prior to chaos is finite. However, although tedious, it would appear desirable to systematically find the equilibria and evaluate their stability in the hope that the number of equilibria prior to chaos is finite, as is the case for the Lorenz equations. Even if the number of equilibria is infinite, it may be possible to assess their stability by asymptotic methods. If it should prove to be the case that chaos begins when all equilibria are unstable with respect to infinitesimal disturbance, then a strong circumstantial case would exist for providing an explanation of the creation of chaos. For further discussion, see Chapter VII.

404

REFERENCES

1. Holmes, P. J., "Bifurcations to Divergence and Flutter in Flow-Induced Oscillations: A Finite Dimensional Analysis," J. Sound Vibration, Vol. 53, No. 4, 1977, pp. 471-503.

2. Holmes, P. J., and Marsden, J., "Bifurcations to Divergence and Flutter in Flow-Induced Oscillations: An Infinite Dimensional Analysis," Automatica, Vol. 14, No. 4, 1978, pp. 367-384.

3. Thompson, J. M. T., Instabilities and Catastrophies in Science and Engineering, Wiley, New York, 1982.

4. Dowell, E. H., "Flutter of a Buckled Plate as an Example of Chaotic Motion of a Deterministic Autonomous System," J. Sound Vibration, Vol. 85, No. 3, 1982, pp. 333-344.

5. Marsden, J., and McCracken, M., "The Hopf Bifurcation and Its Applications," Springer Applied Mathematics Series, No. 19, Springer, Berlin, 1976.

6. Guckeheimer, J., and Holmes, P., Nonlinear Oscillations, Dynamical Systems and Bifurcations of Vector Fields, Springer-Verlag, New York, 1983.

7. Dowell, E. H., Aeroelasticity of Plates and Shells, Martinus Nijhoff, Leyden and Boston, 1975.

8. Simon, H. A., The Sciences of the Artificial, M.I.T. Press, Cambridge, Mass., 1969, p. 1.

9. Fenstermacher, P. R., Swinney, J. L., and Collub, J. P., "Dynamic Instabilities and the Transition to Chaotic Taylor Vortex Flow," J. Fluid Mechanics, Vol. 94, Part 1, 1979, pp. 103-128.

10. Lorenz, E. N., "Deterministic Nonperiodic Flow," J. Atmospheric Science, Vol. 20, No. 3, 1963, pp. 130-141.

11. Yorke, J. A., and Yorke, E. D., "Metastable Chaos: The Transition to Sustained Chaotic Behavior in the Lorenz Model," J. Statistical Physics, Vol. 21, No. 3, 1979, pp. 263-277.

APPENDIX B

AN APPROXIMATE METHOD FOR CALCULATING THE VORTEX-INDUCED OSCILLATION OF BLUFF BODIES IN AIR AND WATER*

Summary

In this Appendix an approximate method for calculating vortex-induced oscillation of a bluff body in air and water is given. The structural response of the oscillatory cylinder in the synchronous region and the so-called "lock-in" phenomenon are discussed in detail. Numerical calculations and comparisons with experiment are also given. The calculated results show that if the virtual mass and viscous damping effects are included in the calculations, the broad frequency (or velocity) response of a bluff body oscillating in water (as opposed to air) can be reproduced very well.

Nomenclature

A,B,C	parameters in Eq. (25)
A_1	fluid coefficient
a_1, a_2, a_3, a_4	coefficients in Eq. (28)
C_i	coefficients in Eq. (26)
C_L	aerodynamic lift coefficient
C_{L_0}	lift coefficient for the stationary cylinder

---

*This appendix is based on work by E. J. Cui and E. H. Dowell which was published in Advances in Aerospace Structures and Materials, ASME Winter Annual Meeting, 1981.

D	diameter of the cylinder
DI	discriminant $(= q^2 + p^3)$
F	resultant fluid force per unit length (see Eq. (1))
f_o	natural frequency (Hz)
G,H,h,F	empirical parameters defined by Eq. (24)
H	function used in Eq. (1)
Im	imaginary part
K_o	non-dimensional structural frequency $(= \dfrac{\omega_o D}{\nu})$
K_{o_t}	non-dimensional total frequency $(= \dfrac{\overline{m_s}}{m_t} K_o)$
K_s	non-dimensional Strouhal frequency
$\tilde{L}$	amplitude of lift coefficient $\tilde{C}_L$ [see Eq. (8)]
M	mass of displaced fluid $(= \dfrac{\pi \rho D^2}{4})$
M_v	virtual mass
$m/\rho D^2$	mass ratio
m_s	mass of cylinder per unit length
m_t	total mass (the sum of the structural and virtual mass)
p,q	parameters used in Eq. (25)

Re   real part

Re_ω Reynolds number of oscillating cylinder $(= \frac{\omega D^2}{\nu})$

S_t Strouhal number

s   frequency parameters $(= \frac{\omega_0 D^2}{4\nu})$

s_G reduced damping $(= \zeta/\mu)$

V   flow velocity

V_R reduced velocity $(= V/f_0 D)$

$X = y/D$ non-dimensional transverse deflection

$\bar{x}$ amplitude of x (see Eq. (8))

y   transverse deflection of structure

y_i i = 1,2,3,4, see Eq. (26)

$\dot{y}_i$ derivatives of y_i with respect to time t

α, β, γ parameters used in Eq. (1)

Δ, δ detuning parameters defined in Eq. (13)

ε small parameter in Eq. (7)

ζ damping ratio

ζ_0 damping ratio in in-vacuuo structure

ζ_f viscous damping ratio of fluid given by Eq. (5)

ζ_n equivalent damping factor given by Eq. (4)

ζ_t	total damping ratio (the sum of structural and added fluid damping)
η	parameter in Eq. (25)
λ	parameter in Eq. (26) and (28)
μ	ratio of the displaced mass of fluid to structural mass
μ_t	ratio of the displaced mass of fluid to total mass
ν	kinematic viscosity of the fluid
ρ	flow density
ρ_s	structure density
τ	non-dimensional time $(= t\,\frac{V}{D})$
ϕ	phase of the fluctuating lift coefficient relative to cylinder displacement
ω	oscillating frequency of the cylinder (rad/sec)
ω_0	natural structural frequency (rad/sec)
ω_s	Strouhal frequency (rad/sec)

INTRODUCTION

Vortex-induced oscillation is a very important problem for many practical industrial structures, such as towers, stacks, bridges, heat exchanger tubes, underwater towing and mooring cables, as well as for a space vehicle standing on a launch pad. There have been numerous investigations dealing with these prob-lems, particularly for bodies in air, and various mathematical models have been proposed by Bishop and Hassan [1], Hartlen and Currie [2], Skop and Griffin

[3], Iwan and Blevins [4], Iwan [5], Landl [7], Blevins [6,8], Dowell [9] and Scanlan and Cui [10]. There are also some theoretical and experimental investigations in which the vortex-induced oscillations of bluff bodies in water have been studied [11-16]. Fluid force and pressure distribution measurements have also been made [17-21]. Some of the most recent of these results are summarized in Refs. [16,22,23].

As a typical result, the cross-flow displacement amplitude for a circular cylinder plotted against reduced velocity is given in Fig. 1 [16]. From this figure we can see that the cylinder oscillating in water has a broad frequency (or velocity) response relative to that in air, even though the reduced damping ζ_0/μ is very nearly the same for these two cases. The actual damping ζ_0 differs by approximately a factor of ten, however.

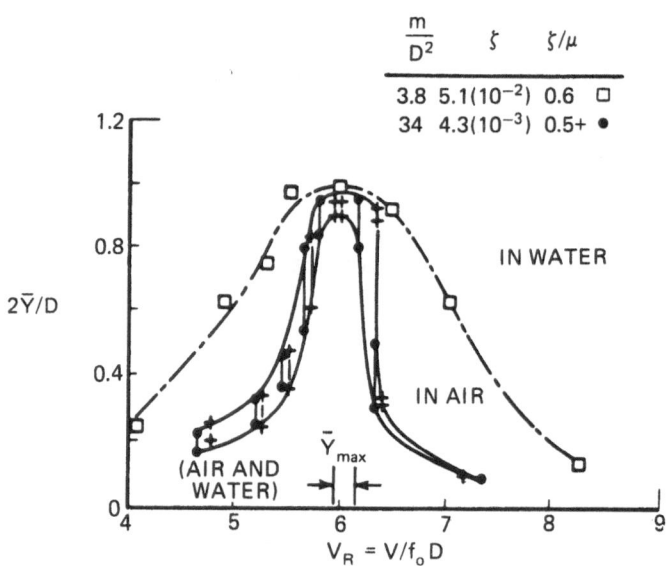

Fig. 1. Cross-flow displacement amplitude 2Y/D for circular cylinder plotted against reduced velocity $V_R = V/f_0 D$. Data points denoted by +, • were measured in wind tunnel using cylinder with natural frequency f_0 = 52Hz and diameter D = 6.1 mm. Data points denoted by □ were measured in water using cylinder with f_0 = 2.2Hz (air), 2.0Hz (water) and D = 25mm.

In the present paper an approximate approach is adopted which is based on the nonlinear oscillator model of bluff bodies given by Dowell in Ref. [9]. By using this approximate method the structural response of an oscillating cylinder in the synchronous region has been studied in some detail. In order to reproduce the measured broad response of the cylinder oscillating in water, it is found that the virtual mass and added viscous damping of the fluid have to be taken into account in the calculations of response. This is done below.

DISCUSSION OF MODEL, SOLUTION AND RESULTS

Calculation of Virtual Mass and Added Viscous Damping

We will use the method given in Ref. [24] to calculate these parameters approximately. Consider an infinitely long cylinder of diameter D oscillating along a diameter with velocity $V\exp(i\omega t)$ in a viscous fluid annulus. The exterior wall at $r = y/2$ is stationary (Fig. 2).

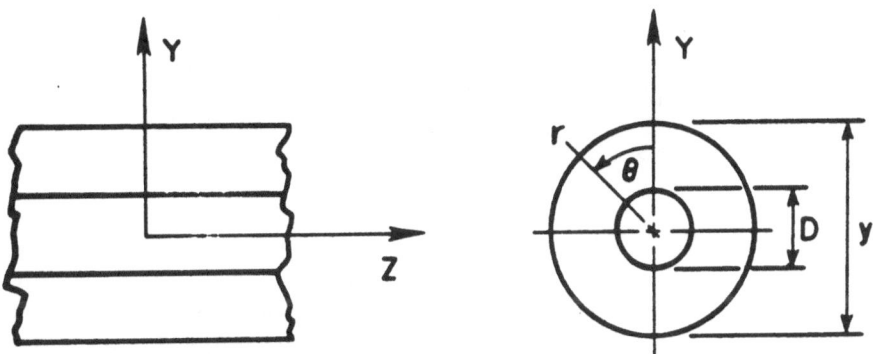

Fig. 2. Schematic and coordinate system.

On the assumption that the amplitude of oscillation is small, the resultant force per unit length is given as [24].

$$F = MV\omega_0[Re(H) \sin\omega t + IM(H) \cos\omega t] \tag{1}$$

where $M = \dfrac{\rho\pi D^2}{4}$, ρ is the density of fluid and H is a very complicated function of the parameters $\alpha = \dfrac{kD}{2}$, $\beta = \dfrac{ky}{2}$ and $\gamma = D/y$. Here $k = (\dfrac{\omega_0}{\nu})^{1/2}$, where ν is the kinematic viscosity of the fluid.

For $y/D > 10$ and $s = \dfrac{\omega_0 D^2}{4\nu} > 2$, by using the asymptotic series expansion for modified Bessel functions of argument $\sqrt{is}$, simplified results which have reasonable accuracy for practical use can be obtained as

$$Re(H) = 1 + K_r \cos(\pi/4) - K_i \sin(\pi/4) \tag{2}$$

$$Im(H) = K_i \cos(\pi/4) - K_r \sin(\pi/4)$$

where

$$K_r = \frac{Ker_1(\sqrt{s}) \, Kei(\sqrt{s}) - Kei_1(\sqrt{s}) \, Ker(\sqrt{s})}{\sqrt{s}\,[Ker^2(\sqrt{s}) + Kei^2(\sqrt{s})]}$$

$$K_i = \frac{Kei_1(\sqrt{s}) \, Kei(\sqrt{s}) + Ker_1(\sqrt{s}) \, Ker(\sqrt{s})}{\sqrt{s}\,[Ker^2(\sqrt{s}) + Kei^2(\sqrt{s})]}$$

$Ker(\sqrt{s})$, $Kei(\sqrt{s})$, $Ker_1(\sqrt{s})$ and $Kei_1(\sqrt{s})$ are Kelvin functions of order 0 and 1, respectively. The values of $Re(H)$ and $Im(H)$ for a specified range of parameters s are plotted in Fig. 3. The virtual mass M_v and the equivalent damping factor ζ_n can be written as

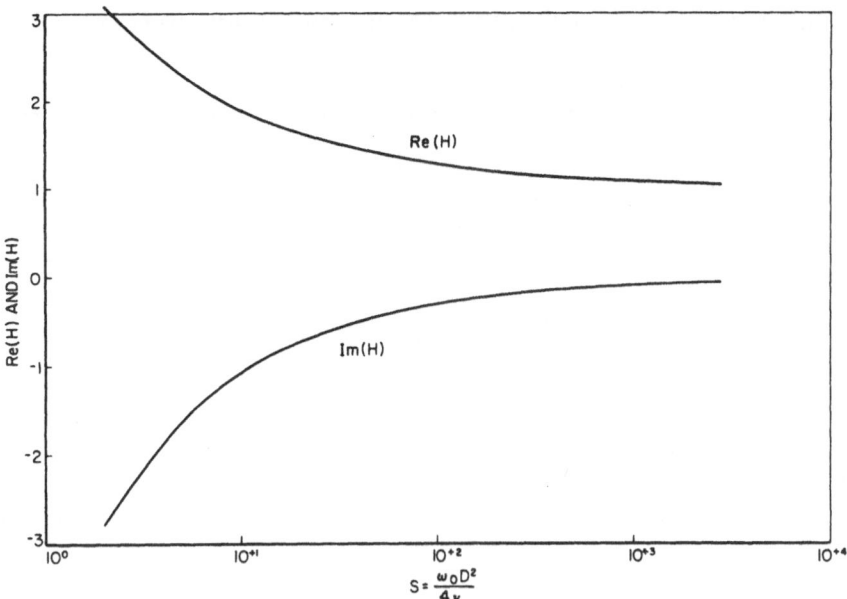

Fig. 3.   Real and imaginary values of H.

$$M_v = Re(H) \, M, \tag{3}$$

$$\zeta_n = -\frac{1}{2} \left(\frac{M}{M_v + m_s}\right) Im(H) \tag{4}$$

where m_s is the mass of cylinder per unit length.

It should be pointed out that the determination of the added damping coefficient of a body oscillating with large amplitude in a flowing fluid accompanied by flow separation and vortex shedding is more complex. The present approximate method is suitable only for small amplitude.

For large amplitudes the estimate given by Eq. (4) is on the low side.   Based on experimental re- sults, Skop, et al. [26] proposed the following empir-

$$\zeta_f = \frac{\rho}{\rho_s} \left[\frac{4.5}{\sqrt{Re}_\omega} + 0.454 \left(\frac{Y}{D} = 0.4\right) H\right] \tag{5}$$

414

ical equation to describe large amplitude effects [*]

Here $Re_\omega = \frac{\omega D^2}{\nu}$ is the Reynolds number of oscillating cylinder, ρ and ρ_S are density of fluid and cylinder respectively. $H = 0$ for $y/D < 0.4$ and $H = 1$ for $y/D > 0.4$.

Approximate solution in synchronous region of nonlinear oscillator model

If the virtual mass and added viscous damping effects of fluid are included in the model, the model equations given in Ref. [9] can be written as (also see Chapter V):

Structural:

$$X'' + 2\zeta_t K_0 X' + K_{0_t}^2 X = \mu_t C_L \tilde{C}_{L_0} \tag{6}$$

Fluid:

$$\tilde{C}_L'' - \epsilon[1-4\tilde{C}_L^2] K_s^2 \tilde{C}_L' + K_s^2 \tilde{C}_L = \frac{K_s^2}{C_{L_0}} A_1 X' \tag{7}$$

---

[*] According to the definition of damping factor

$$\zeta_n = \frac{\text{energy dissipated per cycle}}{4 * \text{(total energy of the structure)}}$$

For a free oscillating elastic circular of density ρ_S, it is approximately given by Eq. (4) which is almost half of the value for small amplitude given by equation (5), i.e., $\zeta_f \cong 2\zeta_n$. Also see: Batchelor: An Introduction to Fluid Dynamics, 1967, p. 357, who gives $\zeta_f = \frac{\rho}{\rho_S} 2 (\frac{2}{s})^{1/2}$, here $s = \frac{\omega D^2}{4\nu}$.

where $X = \dfrac{Y}{D}$, $K_{o_t} = \dfrac{\overline{m_s}}{m_t} K_o$, $K_o = \dfrac{\omega_o D}{V}$, $K_s = \dfrac{\omega_s D}{V}$,

$\tilde{C}_L = \dfrac{C_L}{C_{L_o}}$, $\zeta_t = \zeta_o + \zeta_n$, $\mu_t = \dfrac{\rho D^2}{2m_t}$, $m_t = m_s + M_v$. ε is

a small parameter, A_1 is a fluid coefficient, ζ_o is the natural damping coefficient, ω_o and ω_s are the natural structural and strouhal frequencies, respectively, and C_{L_o} is the peak lift coefficient for the stationary cylinder $Y = 0$.

Suppose a solution to these equations in the synchronous region has the form

$$X = \tilde{X} \sin K\tau,$$

$$(8)$$

$$C_L = \tilde{L} \sin(K\tau + \phi)$$

where $\tau = \dfrac{V}{D}$ is a non-dimensional time.

Substituting Eq. (8) into Eq. (6) and (7), one obtains

$$(K_{o_t}^2 - K^2) \tilde{X} \sin K\tau + 2\zeta_t K_{o_t} K\tilde{X} \cos K\tau$$

$$= \mu_t C_{L_o} \tilde{L} \sin(K\tau + \phi),$$

$$(9)$$

$$(K_s^2 - K^2) \tilde{L} \sin(K\tau + \phi)$$

$$- \varepsilon [1-4\tilde{L}^2 \sin^2(K\tau + \phi)] K_s K \tilde{L} \cos(K\tau + \phi)$$

$$= \frac{K_s^2 A_1}{C_{L_o}} K\tilde{X} \cos K\tau \tag{10}$$

In the synchronous region we have

$$K/K_{o_t} \approx 1,$$

$$K_s/K_{o_t} \approx 1 \tag{11}$$

and

$$(K^2 - K_{o_t}^2) \approx 2K(K - K_{o_t}),$$

$$\tag{12}$$

$$(K_s^2 - K^2) \approx 2[K_s(K_s - K_{o_t}) - K(K - K_{o_t})]$$

Let $\Delta = 2(K_s/K_{o_t} - 1)$,

$$\delta = 2(K/K_{o_t} - 1) \tag{13}$$

Substituting (11), (13) into Eq. (9) and dividing by $K_{o_t}^2$, it can be rewritten as

$$-\delta \bar{X} \sin K\tau + 2\zeta_t \bar{X} \cos K\tau$$

$$= \frac{\mu_t C_{L_o}}{K_s^2} \tilde{L} [\sin K\tau \cos\phi - \cos K\tau \sin\phi] \tag{14}$$

Collecting corresponding terms of $\sin K\tau$ and $\cos K\tau$, through some mathematical operations, one obtains

$$\overline{X} = \frac{\mu_t \, C_{L_o} \, \hat{L}}{K_s^2 \, (\delta^2 + 4\zeta_t^2)^{1/2}} \tag{15}$$

and

$$\phi = \arctan\left(-\frac{\delta}{2\zeta_t}\right) \tag{16}$$

Similarly, Eq. (10) can be written as

$$(\Delta-\delta) \, \hat{L} \, \sin(K\tau + \phi)$$

$$-\epsilon \, [1 - 4 \, \hat{L}^2 \, \sin^2(K\tau + \phi)] \, \hat{L} \, \cos(K\tau + \phi)$$

$$= \frac{A_1 K_s}{C_{L_o}} \, \overline{X} \, \cos K\tau \tag{17}$$

By using a trigonometric identity and neglecting the higher harmonic terms, we have

$$(\Delta-\delta) \, \hat{L} \, [\sin K\tau \, \cos\phi - \cos K\tau \, \sin\phi]$$

$$- \epsilon \, \hat{L}[\cos K\tau \, \cos\phi - \sin K\tau \, \sin\phi]$$

$$+ \epsilon \, \hat{L}^3 \, [\cos K\tau \, \cos\phi - \sin K\tau \, \sin\phi]$$

$$= \frac{A_1 K_s}{C_{L_o}} \, \overline{X} \, \cos K\tau \tag{18}$$

Collecting corresponding terms of $\sin K\tau$ and $\cos K\tau$ and after some mathematical manipulations one obtains [using (15) as well]

$$(\delta - \Delta)\ \delta - \varepsilon\ (L^2 - 1)\ 2\zeta_t = 0 \qquad (19)$$

$$\frac{\mu_t A_1}{K_s} + (\delta - \Delta)\ 2\zeta_t + \varepsilon\ (\tilde{L}^2 - 1)\ \delta = 0 \qquad (20)$$

From Eqs. (19) and (20) one can easily obtain

$$\tilde{L}^2 = 1 - \frac{\mu_t A_1}{\varepsilon\ K_s}\ (\frac{\delta}{\delta^2 + 4\zeta_t^2}) \qquad (21)$$

Substituting Eq. (21) into Eq. (19), it reduces to

$$\delta^3 - \Delta\delta^2 + 4\zeta_t^2\ \delta - (4\zeta_t^2\Delta - \frac{2\zeta_t \mu_t A_1}{K_s}) = 0 \qquad (22)$$

Parametric study

A parametric study has been carried out for configuration I (see Table I) to investigate the separate and combined effects of the parameters included in the approximate solution Eq. (15), (21) and (22).

The parameters under consideration are mass ratio μ_t, damping coefficient ζ_t, fluid mechanical coefficient A_1 and the small parameter of the nonlinear terms ε. The value of $\varepsilon = 0.025$ was selected and the nominal value of the A_1 for configuration I satisfies

$$A_1 = FK_s \qquad (23)$$

in which F is determined from the following empirical relations given by Griffin et al. [3]

$$\log_{10} G = 0.25 - .21 \ s_G,$$

$$H = \zeta_t h, \qquad\qquad (24)$$

$$\log_{10} hs_G^2 = -0.83 + 0.98 \ s_G,$$

$$F = 4G \ s_G/h,$$

$$s_G = \zeta_t/\mu_t$$

The numerical value of A_1 so determined, i.e., $A_1 = 1.1$, is thought to be reasonable even though (24) can be criticized on fundamental grounds [9]. The calculated results are shown in Figs. 2-9, respectively. A_1 was also varied about its nominal value to determine the sensitivity of the results.

From Figs. 4 and 5 we can see that the mass ratio μ_t and damping coefficient ζ_t both have important effects on the maximum amplitude response and the velocity at which it occurs. It is known from Eqs. (15) and (21) that for given parameters A_1, K_E, ε, the amplitude response $\bar{X}$ is independently governed by μ_t and ζ_t. For a relatively highly damped system, the response is somewhat broader. Only for the case, in which δ is smaller as compared with ζ_t, will the amplitude $\bar{X}$ be uniquely determined by the value of $s_G = \zeta_t/\mu_t$. About this point Figs. 6 and 7 provide numerical evidence. Figure 6 shows the maximum amplitude to be only a function of s_G for the parameters chosen; other parameters are specified. The V_R correspond to X_{max}, itself, has a maximum value as a function of s_G.

The effects of A_1 and ε are shown in Figs. 8 and 9. Evidently, the main effect of ε is on the maximum amplitude, but the parameter A_1 also has a strong influence in determining the value of V_R corresponding to the maximum amplitude.

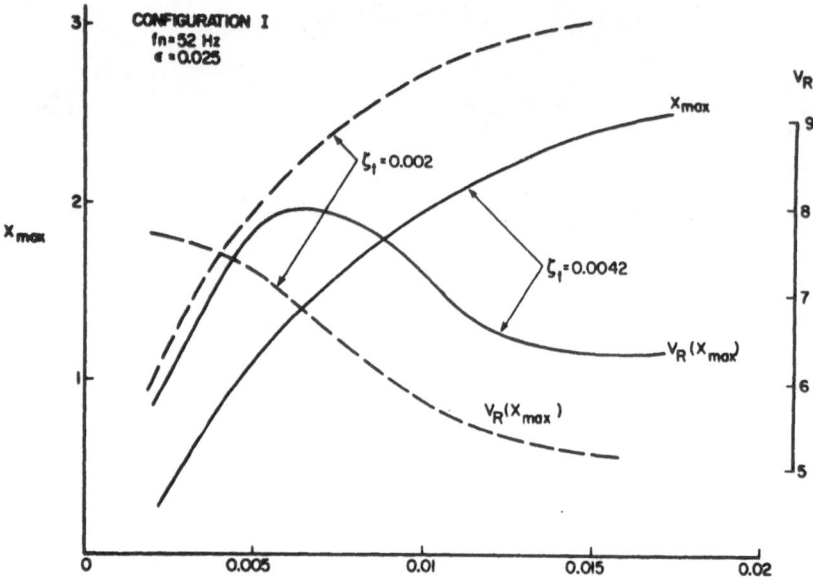

Fig. 4.   Maximum amplitude x_{max} and the corresponding
velocity V_R as a function of mass ratio μ_1.

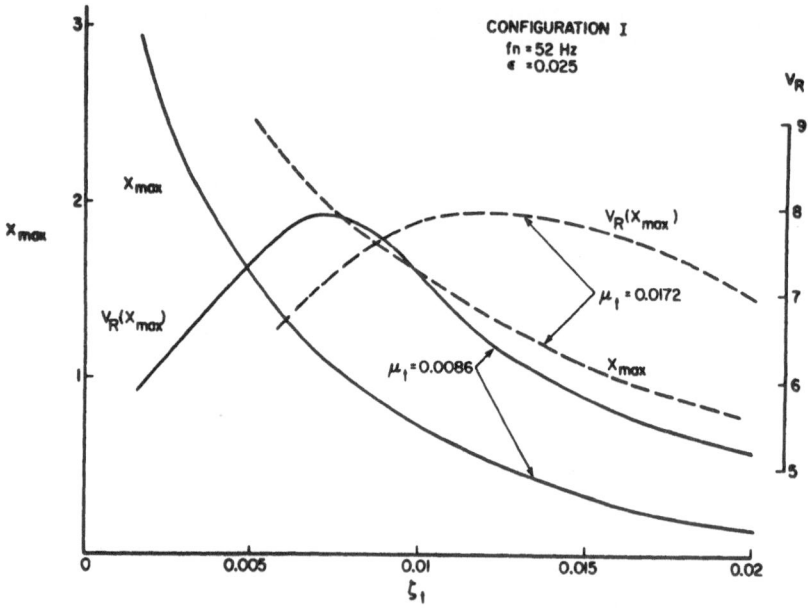

Fig. 5.   Maximum amplitude x_{max} and the corresponding
velocity V_R as a function of damping ratio ζ_1.

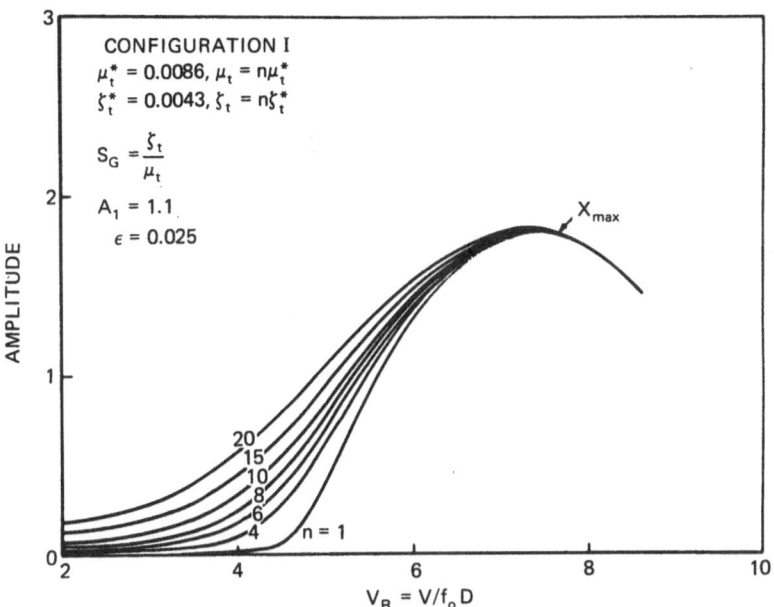

Fig. 6. Amplitude of vortex-induced oscillation for configuration I with different n as a function of V_R.

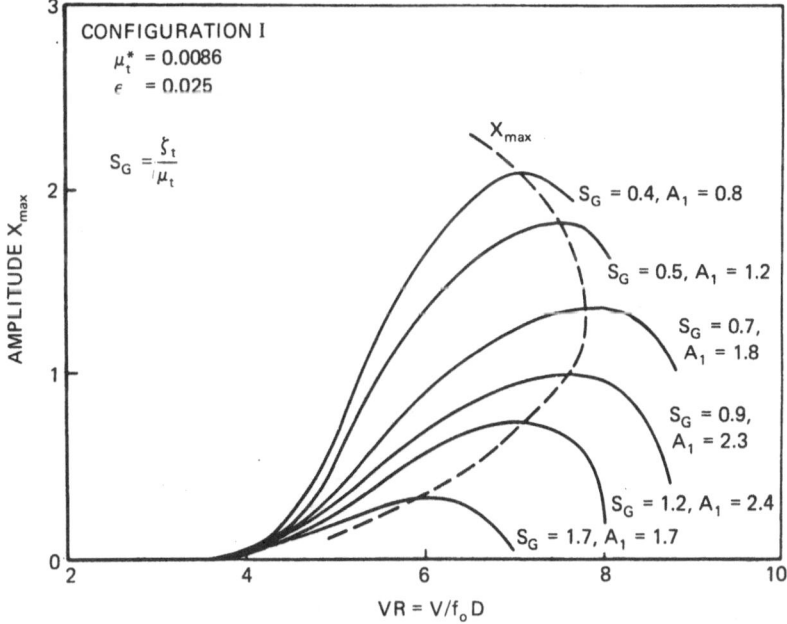

Fig. 7. Amplitude response for configuration I with different S_G but the value of μ_1 is fixed.

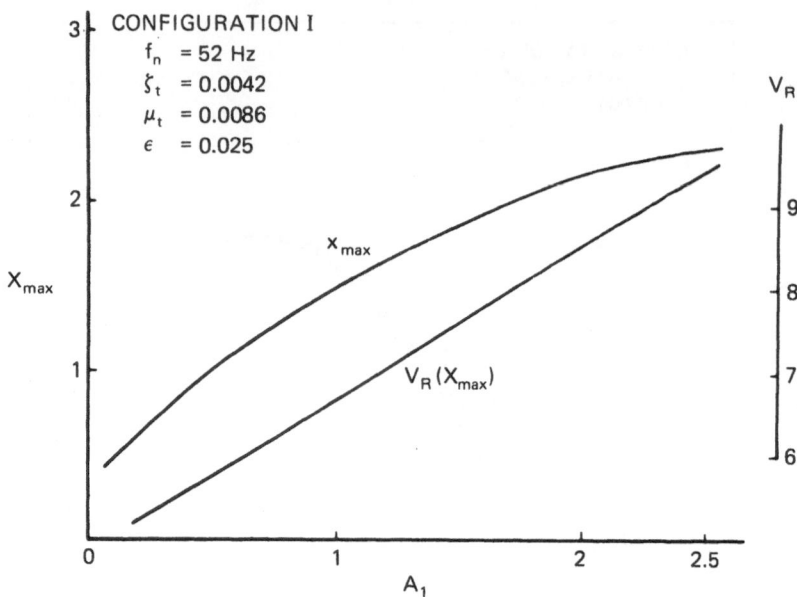

Fig. 8. Maximum amplitude x_{max} and the corresponding velocity V_R as a function of A_1.

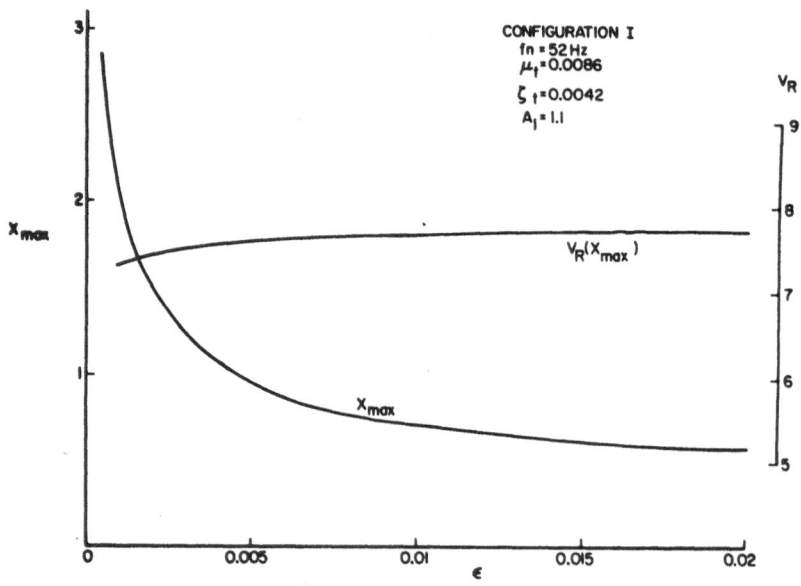

Fig. 9. Maximum amplitude x_{max} and the corresponding velocity V_R as a function of ε.

It is found from many experiments (Refs. [11] and [16]) that the response of bodies oscillating in air and water are somewhat different. In air, the excitation range extends over $5 < V_R < 8$ and the maximum amplitude occurs in the range of $5.5 < V_R < 6.5$, but in water the corresponding values are $4.5 < V_R < 10$ and $6.5 < V_R < 8$, respectively. These differences can also be explained from similar parametric studies. See subsequent discussion.

Numerical results including comparisons with experiment

By using Eqs. (15) and (22) given above, the oscillating response of some physical configurations are calculated. The parameters of configuration I are taken from Refs. [25] and [3] and summarized in Table I.

Table I

Diameter D (mm)	Frequency f_0 (Hz)	Damping Ratio ζ_0	Non-Dimensional Mass μ	Note
6	52	6.8×10^{-4}	8.6×10^{-3}	in vacuo

The calculated parameters for this configuration oscillating in air, based on the data obtained from tests in vacuo and the calculated virtual mass and viscous damping coefficients obtained from Eqs. (2) and (3), are given in Table II.

It is seen from Table II that the calculated parameters are very close to the values obtained from the tests in still air.

The calculated amplitude and frequency variations of vortex-induced oscillation for this configuration as a function of flow speed including comparisons with experiments are given in Figs. 10 and 11.

Table II

D (mm)	f_0 (Hz)	$\frac{2m}{\rho D^2}$	ζ_t	μ_t	Note
6	52	68	4.2×10^{-3}	8.5×10^{-3}	Calculated
			4.3×10^{-3}	8.6×10^{-3}	Experiment in still air

In order to assess the accuracy of the approximate solution method, we have made some numerical simulation computations from the original model equations, Eqs. (6) and (7), by using a fourth-order Runge-Kutta method and also Hamming's predictor-corrector method. Typical results are shown as follows:

Configuration I   Table III

$V_R = 6.0$	Approximate Method	R-K Method	Hamming's Method
Amplitude	1.040	1.054	1.054
Computational Time in IBM-3033 Sec.	0.14	1.73	8.45

The time history of the oscillating amplitude obtained from Runge-Kutta method is given in Figs. 12 and 13 for $V_R = 4.8$ and 6.0.

From this figure we can see that there are "beats" in the total response and the oscillations exhibit a sustained "limit cycle" character for the given specified values of model and flow parameters.

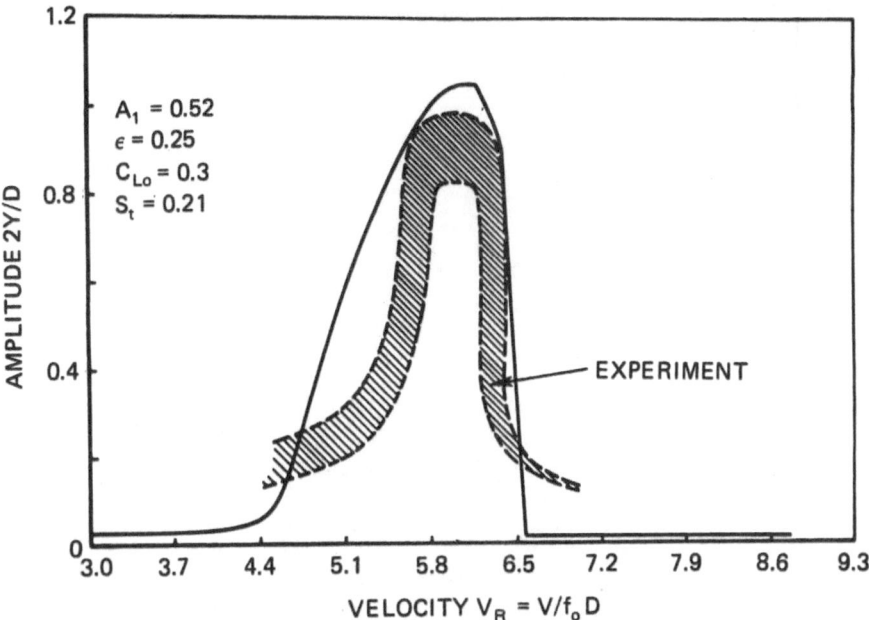

Fig. 10. Amplitude vs. velocity.

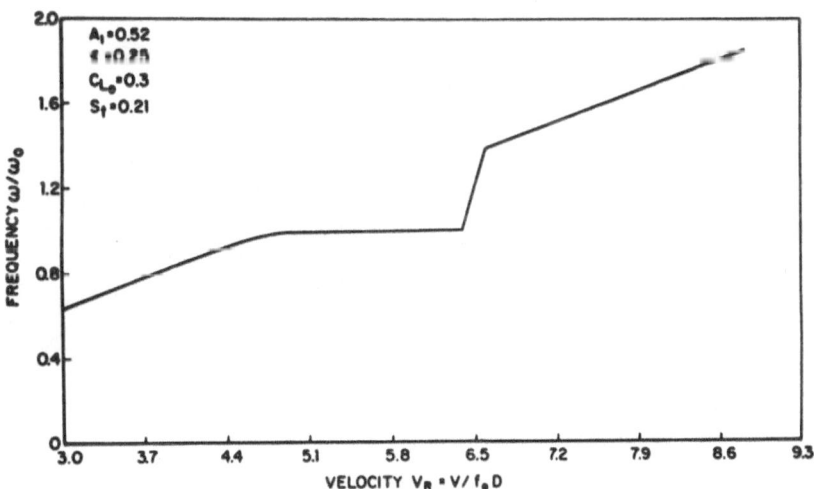

Fig. 11. Frequency vs. velocity.

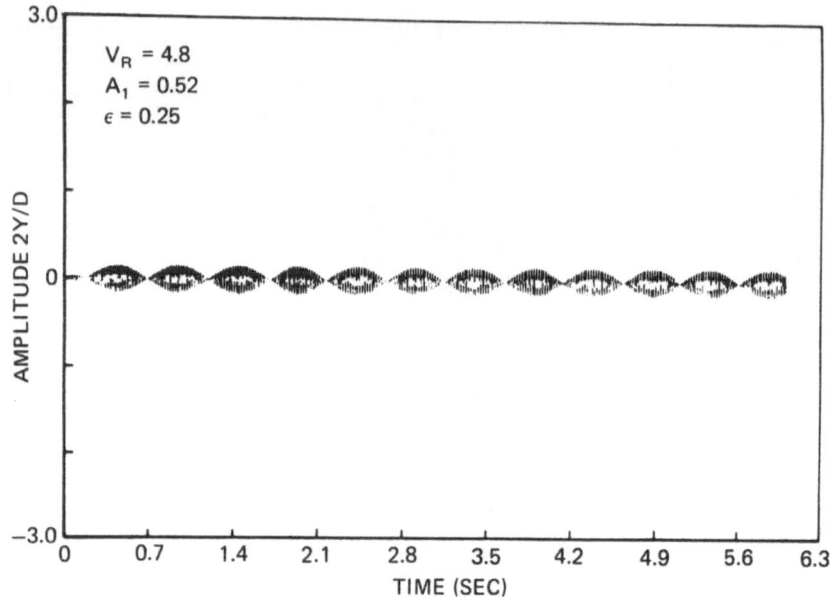

Fig. 12. Time history of amplitude.

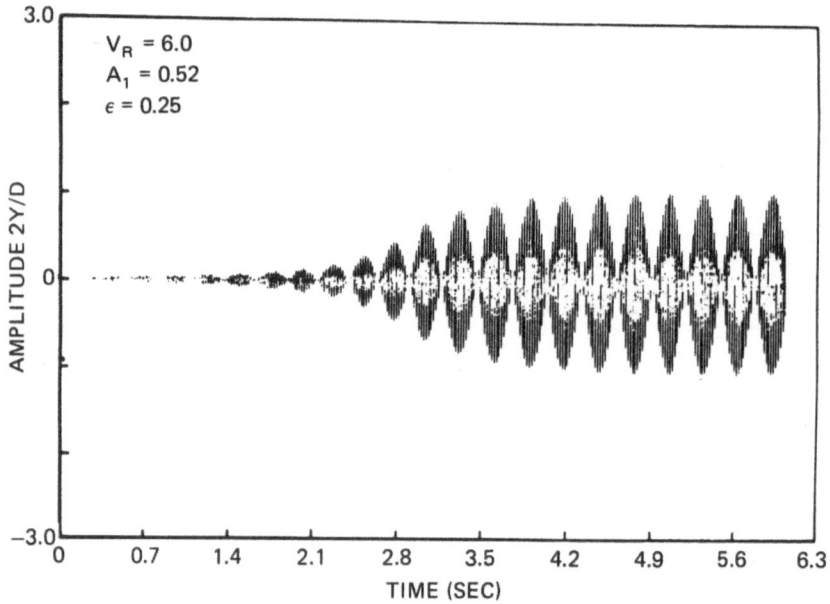

Fig. 13. Time history of amplitude.

For large time, the peak amplitude reaches a fixed value corresponding to the "limit cycle". The beat frequency will change with the flow velocity because the difference between oscillating and Strouhal frequencies changes.

For configuration II oscillating in water, the calculated results are shown in Figs. 14 and 15. The parameters used in the calculations are listed in Table IV.

Table IV

Diameters D (mm)	Natural Frequency f_0 (Hz)	Mass Ratio $\frac{m}{\rho D^2}$	Damping Ratio ζ_t	Non Dimensional Mass μ_t	Note
25	2.0	3.8	5.1×10^{-2}	8.5×10^{-2}	in water

From these figures we can see that the calculated results are in good agreement with experiment.

In order to compare the magnitude of theoretical and experimental lift coefficients, the ratio of the root mean square lift coefficient for the oscillating body to that for the stationary body is shown in Fig. 16. In the calculation A_1 = 6.0, ε = 0.04; other configuration parameters are the same as given in Jones' paper [27]. Jones' experimental data are also shown in this figure. This entirely fluid dynamic result shows that the model is capable of describing the gross features of the flow. Also see Ref. [9].

Bodies oscillating in air vs. water

In order to have a better understanding of the differences between bodies oscillating in air and in

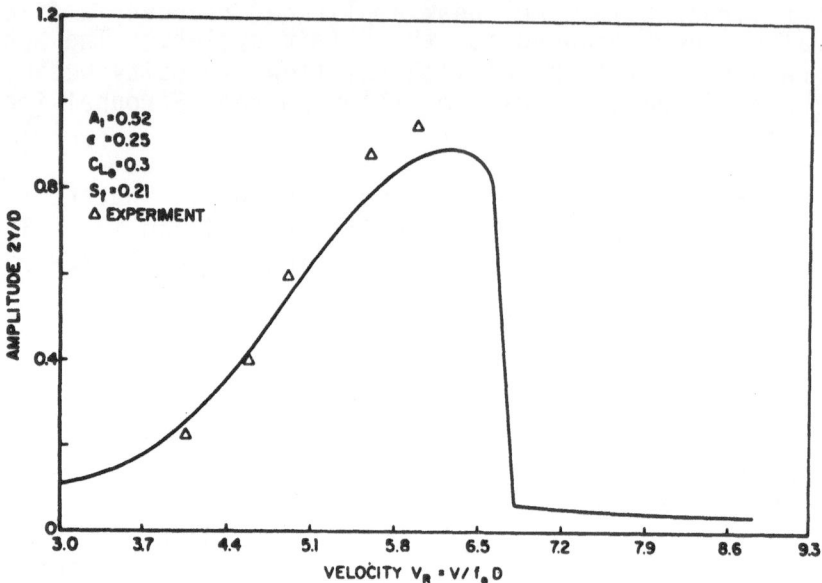

Fig. 14. Amplitude vs. velocity.

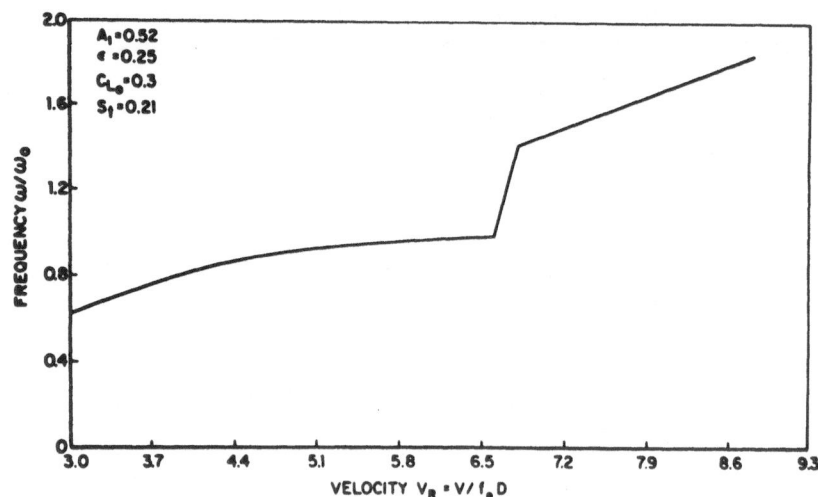

Fig. 15. Frequency vs. velocity.

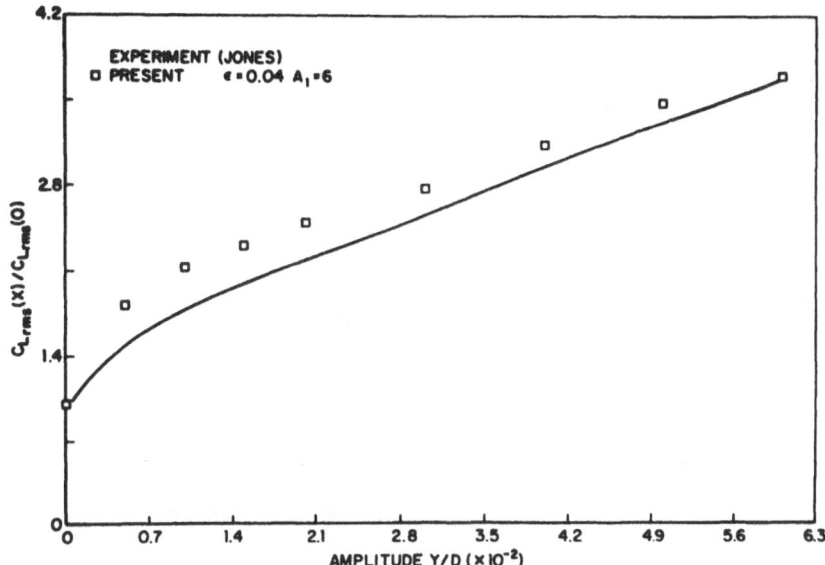

Fig. 16. RMS lift coefficient vs. amplitude.

water, some calculations have been made for a structural model which is made from aluminum tube with O.D. .10 mm and I.D. 9.8 mm. The physical parameters of the model with supporting system are as follows:

model mass $\qquad$ $m_s = 8.54 \times 10^{-3}$ Kg

structural damping $\zeta_0 = 2 \times 10^{-4}$

natural frequency $f_0 = 16$ Hz

By using the method previously described the corresponding parameters of the body oscillating both in air and in water can be obtained as follows:

Table V

Case	Total Mass m_t (Kg)	Non-Dimensional Mass μ_t	Total Damping ζ_t	$s_G = \dfrac{\zeta_t}{\mu_t}$
I in air	8.7×10^{-3}	3.95×10^{-3}	2.0×10^{-3}	0.5
II in water	9.3×10^{-2}	3.1×10^{-1}	3.52×10^{-2}	0.12

It is obvious that the parameter s_G for the model in water is quite different from that in air. In order to obtain a comparable value of s_G, it is necessary to have the structural mass m_s as large as possible and properly adjust the corresponding structural damping of the system. To this end, a model with the same geometric size, but filled with mercury has been assumed. All cases considered in the calculations and the corresponding parameter combinations are summarized in Table VI.

The calculated results corresponding to these cases are shown in Fig. 17 and denoted by I-V respectively.

It can be seen from the curves shown in Fig. 17 that the amplitudes are much higher and the response curves are much sharper for the bodies oscillating in water (case II and III) than those oscillating in air because the effects of the virtual mass for the model in water are larger than those in air. However, if we keep the values s_G as close as possible for these two different cases by properly adjusting the system damping, then nearly the same maximum amplitude can be obtained. Moreover, now the response curve for the model oscillating in water is much broader than that in air (see curves I and IV). Recall Fig. 1. Clearly for complete similarity of behavior between a model oscillating in air and one in water the mass and damp-

Table VI

Case	A_1	m_s	ζ_0	μ_t	ζ_t	s_G
III in water model filled with mercury	Eq. $\begin{Bmatrix} (23) \\ (24) \end{Bmatrix}$ and	1.034	2×10^{-4}	2.42×10^{-2}	3.1×10^{-3}	0.13
IV in water with higher structural damping	Eq. $\begin{pmatrix} (23) \\ (24) \end{pmatrix}$ and	1.034	1×10^{-2}	2.42×10^{-2}	1.29×10^{-2}	0.53
V in air	0.68	8.5×10^{-3}	2×10^{-4}	3.95×10^{-3}	2.0×10^{-3}	0.5

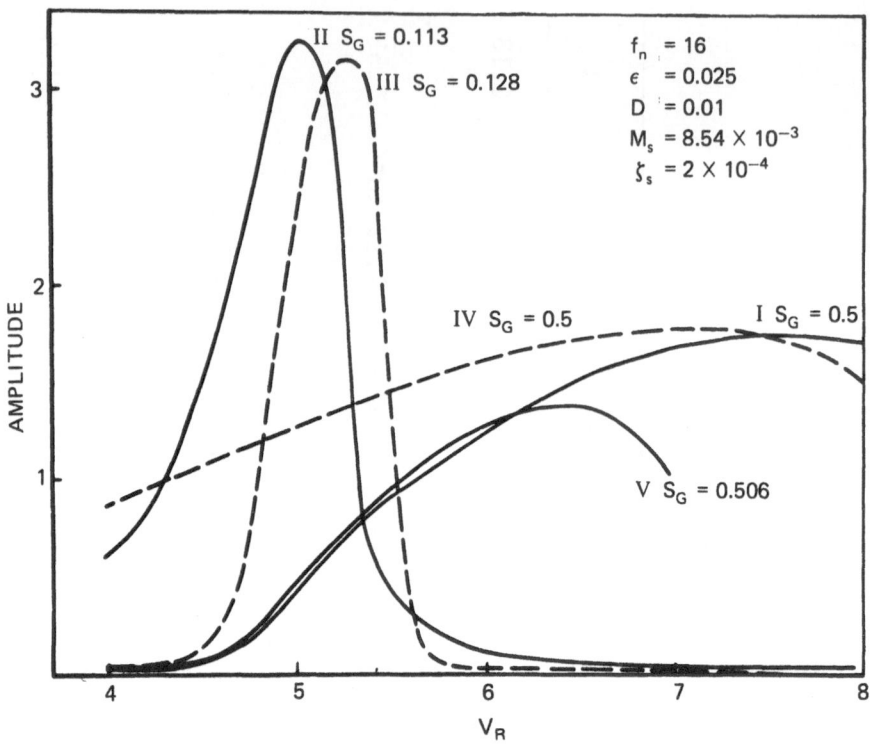

Fig. 17. Amplitude of vortex-induced oscillation for configuration I as a function of V_R.

ing terms should be satisfied separately. Only under nearly resonant condition may these two parameters be lumped together as a combined parameter s_G.

It is apparent that the parameter A_1 has important effects on the maximum amplitude and the velocity V_R at which it occurs (see curves I and V in Fig. 17). It requires careful consideration in the calculations, so that the calculated response can closely match the experimental results.

The prediction of the 'lock-in' range

Physically, when the 'lock-in' phenomenon occurs, the frequencies of vortex shedding and the body oscillation collapse into a single frequency close to the natural frequency of the body.

From the calculated results based on Eqs. (15), (21) and (22) (Figs. 18, 19) we can see that if the shedding frequency is lower or somewhat higher than the natural frequency of the body only one real solution exists corresponding to curves CA and BD in Fig. 18 and the vortex-shedding is governed by the Strouhal relationship. By contrast the 'lock-in' phenomenon occurs over the range AB in which three real solutions exist. One of them has a nearly constant frequency $\omega/\omega_0 \approx 1$ and the highest amplitude (corresponding to curve 1 in Fig. 18). Mathematically, the starting and stopping points of the 'lock-in' range correspond to the two ends of the frequency area in which both solutions of Eq. (22) are all real.

In order to obtain these end points of the 'lock-in' area, we introduce a new parameter $\eta = \delta + \dfrac{A}{3}$ and rewrite Eq. (22) in the form:

$$\eta^3 + 3p\eta + 2q = 0 \qquad\qquad (25)$$

where

$$p = \frac{3B-A^2}{9},$$

$$q = \frac{A^3}{27} - \frac{AB}{6} + \frac{C}{2},$$

$$A = -\Delta,$$

$$B = 4\zeta_t^2,$$

$$C = -(4\zeta_t^2\Delta - \frac{2\zeta_t\mu_t A_1}{K_s})$$

The solution of Eq. (25) depends on the sign of the discriminant $DI = q^2 + p^3$. If $p < 0$ and $D = 0$,

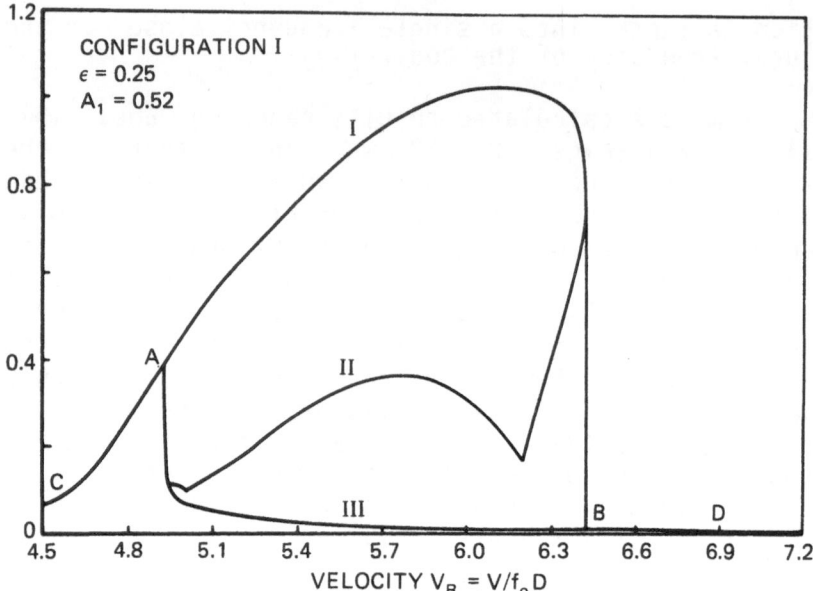

Fig. 18a. Amplitude vs. velocity.

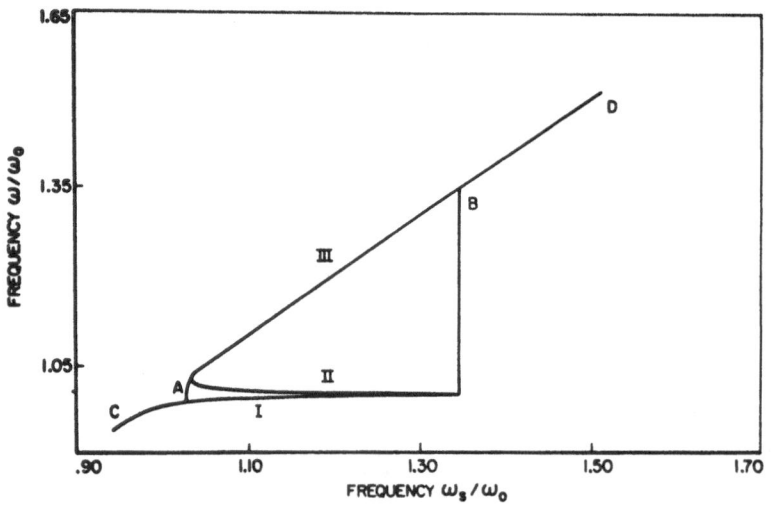

Fig. 18b. Oscillating frequency vs. Strouhal frequency.

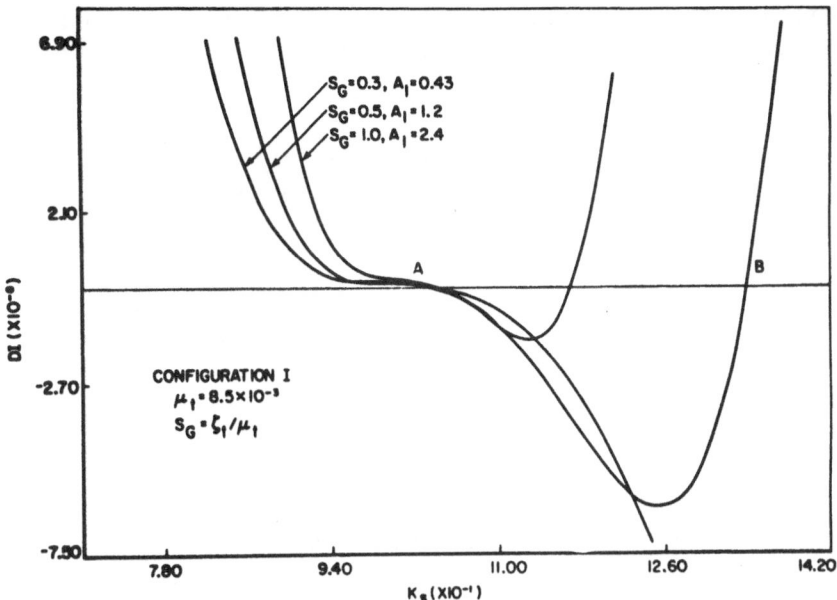

Fig. 19. Variation of discriminant DI = Q2 + P3 with K_S.

then Eq. (25) has triple zero roots which correspond to the starting and stopping points of the 'lock-in' area A and B.   DI is a function of parameters ζ_t, μ_t, Λ_1, K_{u_t} and K_s.   The variation of DI with some of these parameters is shown in Fig. 19.

For configuration I, ζ_t ≃ 0.0042, μ_t ≃ 0.0086, from the calculated results we know that in order to satisfy the conditions p < 0 and DI = 0, the value of K_S must be 1.03 < K_S < 1.36.

Considerations on the stability of the solutions

It has been shown that in the 'lock-in' area, Eq. (22) has three possible solutions. So we want to know which one is most realistic.   In order to answer this question we have to investigate the stability of the solutions.

We will start from the basic equations, Eqs. (6) and (7).   Obviously, since Eq. (7) includes a nonlinear term of $\tilde{C}_L$, the general discussion of stability is difficult. However, we have obtained some specific value of $\tilde{C}_L$ for the given K_S (or V_R), so we can insert this value into Eq. (7) to make the problem simpler.

Let $y_1 = X$, $y_2 = X'$, $y_3 = \tilde{C}_L$ and $y_4 = \tilde{C}_L'$, denote the $\tilde{C}_L$ value at specified K_S as $y_3(K_S)$, and suppose the solutions have the form

$$y_i = c_i\, e^{\lambda t} \tag{26}$$

Then Eqs. (6) and (7) can be transformed into

$$\dot{y}_1 = y_2,$$

$$\dot{y}_2 + \mu_t\, C_{L_o}\, y_3 - 2\zeta_t K_{o_t} y_2 - K_{o_t}^2\, y_1, \tag{27}$$

$$\dot{y}_3 = y_4,$$

$$\dot{y}_4 = \frac{K_s^2 A_1}{C_{L_o}}\, y_2 + \varepsilon[1 - 4y_3^2\,(K_s)]\, K_s^2 y_4 - K_s^2 y_3$$

The characteristic determinant of this system of equations, after expanding, has the form:

$$D(\lambda) = \lambda^4 + a_1\lambda^3 + a_2\lambda^2 + a_3\lambda + a_4 = 0 \tag{28}$$

where

$$a_1 = 2\zeta_t K_{o_t}^2 - \varepsilon[1 - 4y_3^2\,(K_s)]\, K_s,$$

$$a_2 = K_s^2 + K_{o_t}^2 - 2\zeta_t K_{o_t}\, K_s\, \varepsilon[1 - 4y_3^2(K_s)],$$

$$a_3 = 2\alpha_t\, K_{o_t}\, K_s^2 - K_{o_t}^2\, \varepsilon[1 - 4y_3^2(K_s)] - \mu_t K_s^2\, A_1,$$

$$a_4 = K_{o_t}^2\, K_s^2$$

If all of the solutions of Eq. (28) have negative real parts, the oscillation under this condition will be stable; otherwise it is unstable.

As an example, we have solved Eq. (28) numerically for configuration I at a series of given velocities. The results are plotted in Figs. 20a - 20c which correspond to three branches of the solutions shown in Fig. 18, respectively.

It can be seen from these figures that the curve shown in Fig. 20b, corresponding to branch 2 in Fig. 18, has unstable solutions both at the lower and upper end points of the 'lock-in' area A and B. So the oscillation corresponding to this branch is not physically possible. At the lower end point B, only one solution (branch 1), corresponding to the minimum frequency and maximum amplitude in the response curve of Fig. 18, is stable. So the oscillation under this condition is the only one in existence. But at the upper end point, two stable branches (1 and 3) exist. This means that, when the flow velocity decreases from higher velocity, branch 3 in Fig. 18 is also a possible physical state.

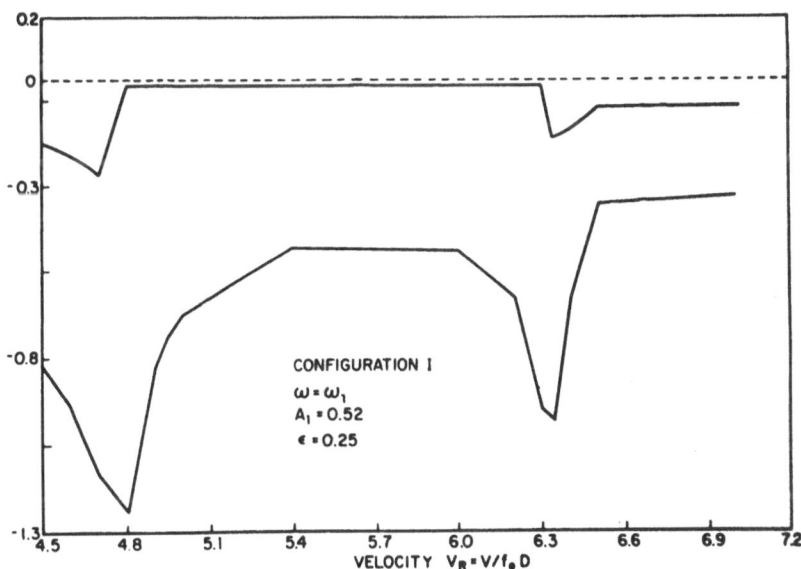

Fig. 20a. Real part of the roots of characteristic determinant.

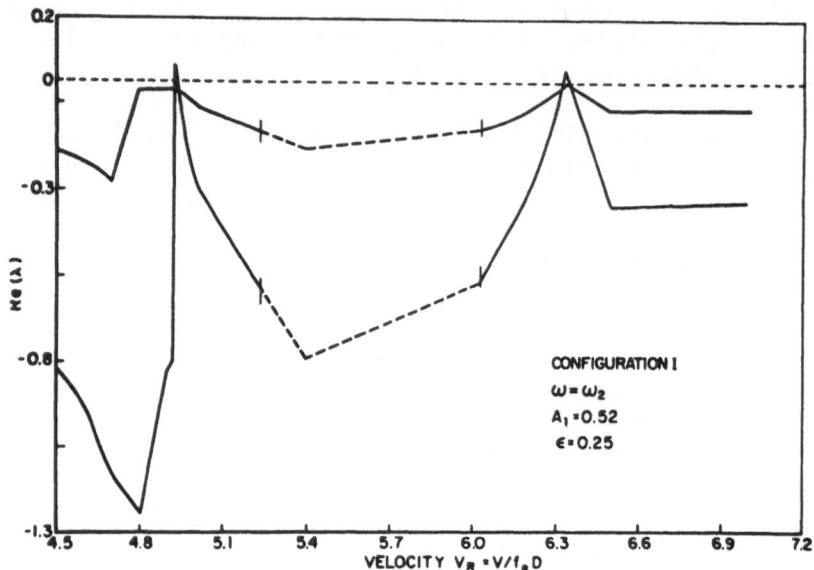

Fig. 20b. Real part of the roots of characteristic determinant.

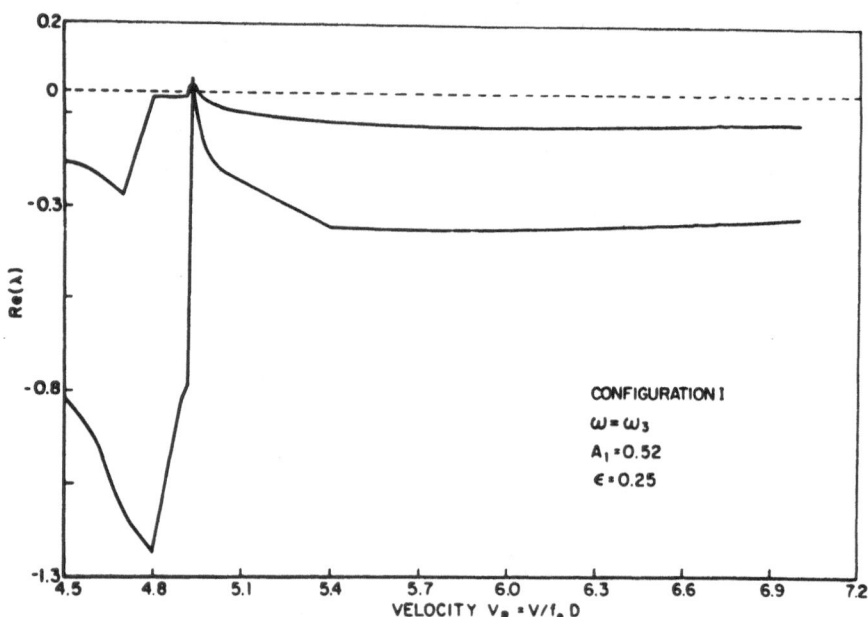

Fig. 20c. Real part of the roots of characteristic determinant.

Conclusions

An approximate method for calculating the vortex-induced oscillation in the synchronous region for the bluff bodies both in air and in water is given, starting from a nonlinear oscillator model. Fluid viscous damping and virtual inertia are included in the model. In order to examine the accuracy of this approximate method some numerical simulation calculations have been made also.

The results presented in this paper show that, by using this oscillator model and the approximate method, the main features of the vortex-induced oscillation of bluff bodies in the synchronous region can be described quantitatively. Because of its simplicity and substantial saving in computational time, this method could be of value in practical engineering problems.

References

1. R. E. D. Bishop and A. Y. Hassan: "The Lift and Drag Forces on a Circular Cylinder Oscillating in a Flowing Fluid," Proc. Royal Society (London) Series A, Vol. 277, 51-74 (1964).

2. R. T. Hartlen and I. G. Currie: "A Lift-Oscillator Model for Vortex-Induced Vibrations," Proc. ASCE, J. Eng. Mech. Vol. 96, 577-591 (1970).

3. O. M. Griffin, R. A. Skop and S. E. Ramberg: "Modeling of the Vortex-Induced Oscillations of Cables and Bluff Structures," Presented at 1976 SESA Spring Meeting, Silver Springs, MD, May 9-14, 1976.

4. W. D. Iwan and R. D. Blevins: "A Model for the Vortex-Induced Oscillations of Structures," Trans. ASME J. Applied Mechanics, Vol. 41, 581-585 (1974).

5. W. D. Iwan: "The Vortex Induced Oscillation of Elastic Structural Elements," Trans. ASME, J. Eng. for Industry, 1378-1832, Nov. (1975).

6. R. D. Blevins and T. E. Burton: "Fluid Forces Induced by Vortex Shedding," ASME J. of Fluids Eng. Vol. 98, 19-26, (1976).

7. R. Landl: "A Mathematical Model for Vortex-Excited Vibrations of Bluff Bodies," J. of Sound and Vibration, Vol. 42, 219-234, (1975).

8. R. D. Blevins and T. E. Burton: "Fluid Forces Induced by Vortex Shedding," ASME J. of Fluids Eng., Vol. 98, 19-26, (1976).

9. E. H. Dowell: "Nonlinear Oscillator Models in Bluff Body Aeroelasticity," Journal of Sound and Vibration, Vol. 75, 251-264 (1981).

10. R. H. Scanlan and E. J. Cui: "A Simplified Model for Vortex-Induced Oscillations of Bluff Bodies," (Unpublished) Princeton University (1980).

11. B. J. Vickery: "Flow-Induced Vibration of Cylindrical Structures," Proc. First Australian Conf. on Hydr. & Fluid Dynamics (1964).

12. R. J. Glass: "A Study of the Hydroelastic Vibrations of Spring-Supported Cylinders in a Steady Fluid Stream Due to Vortex Shedding," ONR Project N0014-69-C-0148, Final Report (1970).

13. F. Angrilli: "Hydroelasticity Study of a Circular Cylinder in a Water Stream," IUTAM-IAHR Symp. on Flow-Induced Structure Vibrations, Karlsruhe, 504-507, (1972).

14. R. King: "On Vortex Excitation of Model Piles in Water," J. of Sound and Vibration, Vol. 29, 169-188, (1973).

15. R. King: "An Investigation of the Criteria Controlling Sustained Self-Excited Oscillations of Cylinder in Flowing Water," Proc. on Turbulence in Liquid. University of Missouri at Rolla, (1977).

16. O. M. Griffin: "Vortex-Excited Vibrations of Marine Cables," J. of the Waterway Port Coastal and Ocean Division, WW2, 183-204 (1980).

17. G. H. Toebes: "Fluidelastic Forces on Circular Cylinder," J. of Eng. Mech. Div., Proc. of ASCE, EM6, 1-20 (1967).

18. A. Protos: "Hydroelastic Forces on Bluff Cylinders," J. of Basic Eng. Trans. of ASME, 378-386, (1968).

19. T. Sarpkaya: "Fluid Forces on Oscillating Cylinders," J. of Waterway, Proc. of ASCE, Vol. 104, No. WW4, 275-290, (1978).

20. P. W. Bearman: "Pressure Fluctuation Measurements on Oscillating Circular Cylinder," J. of Fluid Mechanics, Vol. 91, part 4, 661-677, (1979).

21. C. Dalton: "Pressure Distribution Around Circular Cylinders in Oscillating Flow," J. of Fluid Eng., Trans. ASME, Vol. 102, 191-195, (1980).

22. T. Sarpkaya: "Vortex-Induced Oscillations (A Selective Review)," J. of Applied Mechanics, Trans. of ASME, Vol. 46, 241-258, (1979).

23. P. W. Berman and J. M. R. Graham: "Vortex Shedding from Bluff Bodies in Oscillatory Flow: A Report on Euromech 119," J. of Fluid Mechanics, Vol. 99, part 2, 225-245, (1980).

24. S. S. Chen: "Added Mass and Damping of a Vibrating Rod in Confined Viscous Fluids," J. of Applied Mechanics, Trans. of ASME, 325-328, (1976).

25. O. M. Griffin and G. H. Koopman: "The Vortex-Excited Lift and Reaction Forces on Resonantly Vibrating Cylinders," J. of Sound and Vibration, Vol. 53, 435-448, (1977).

26. R. A. Skop, et al.: "Added Mass and Damping Forces on Circular Cylinders," Naval Research Laboratory Report 7970, (1976).

27. G. W. Jones, Jr.: "Aerodynamic Forces on a Stationary and Oscillating Circular Cylinder at High Reynolds Numbers," NASA TR R-300, (1969).

APPENDIX C

UNSTEADY SEPARATED FLOW MODELS

Introduction

To determine the flow field for unsteady airfoil oscillations in a separated flow is a formidable challenge. Much of our knowledge of these flows comes from experiment. Fortunately some recent work on theoretical phenomenological models gives grounds for optimism. See Refs. 1-18. Ultimately it may prove possible to determine the fluid forces from first principles by a numerical simulation of the Navier-Stokes equations and indeed significant work has been done and undoubtedly will continue along these lines [19]. However, the simpler phenomenological fluid models which have been developed recently have proven successful under some circumstances. Of course it would be desirable to establish a more rigorous fundamental basis for these models.

Among the benefits of validating and further developing these relatively simple fluid mechanical models are

- These models have proven successful in describing unsteady separated flows under some circumstances with sufficient accuracy to predict significant body motion such as flutter.

- These models require computational efforts comparable to those presently employed for attached flows.

- These models, currently developed for two-dimensional, subsonic flow, appear capable of being extended to transonic flow and three-dimensional flow.

- By combining the capabilities of two of the present models, it appears the amount of experimental data required (from either

physical or numerical experiments) can be substantially reduced.

KEY IDEAS

Two phenomenological models are discussed with
respect to their current capabilities and possible
extensions are identified.

The work of Chi

Following a suggestion of Dowell, Chi [1-3] considered a model for unsteady aerodynamics of an airfoil with separated flow. The application of interest
was for flutter of turbomachinery in jet engines.
Fig. 1 displays the nature of the flow model by showing two airfoils of a turbomachinery cascade. Consider the upper surface of the top airfoil. A (relatively thin) region of separated flow is shown. This
separated region occurs without any airfoil oscillation. With airfoil oscillation, if the oscillation
amplitude is sufficiently small or the oscillation
frequency sufficiently high, the region of separated
flow will not change appreciably. Hence the mean,
steady flow about the airfoil will be unchanged by the
oscillation of the airfoil. However, the time dependent, unsteady aerodynamic forces on the airfoil due
to its motion will be influenced by the mean flow and
its separated region. The question is, how to model
its effect?

Chi modeled the mean flow by noting the region of
separation and assuming its effect to be concentrated
in an infinitesimally thin layer over that portion of
the surface where separation occurs. On that portion
of the surface the effect of the separation region on
the unsteady aerodynamics is modeled by stating that
the local (instantaneous) static pressure is equal to
the free-stream static pressure. Everywhere else in
the flow field the potential flow approximation is
made. This modeling is very much in the spirit of the
Kutta condition which accounts for the (essential)
effect of viscosity at the trailing edge of the

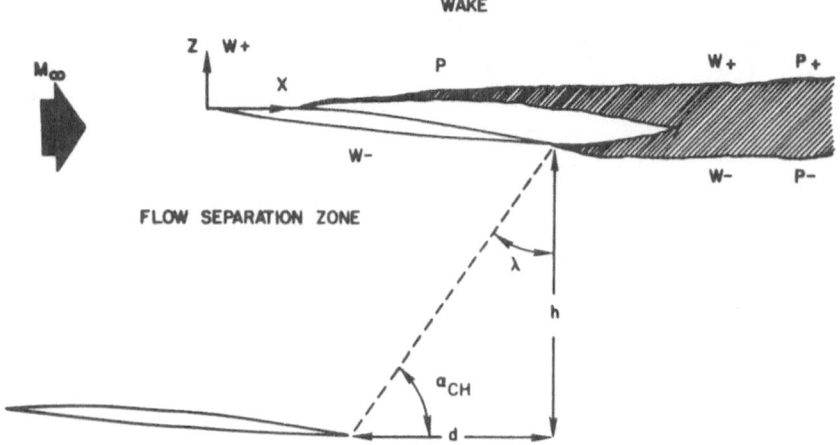

Fig. 1. Cascade geometry.

airfoil in classical aerodynamic potential flow
theory. Perhaps, most importantly, the model has
proven successful in predicting the results of unsteady aerodynamic (physical) experiments involving
cascades in the wind tunnel and also in predicting
stall flutter in jet engines. Although the model is
conceptually simple, it still requires the determination of new Green (Kernel) functions from the governing potential flow field equations and associated
boundary conditions and then the solution of a pair of
singular integral equations.

A representative result is shown in Fig. 2 where
the aerodynamic damping provided to the airfoil by the
flow is shown as a function of reduced frequency,
$\omega c/U$. Results from the model as calculated by Chi are
shown along with results from an experiment by Yashima
and Tanaka [16], an alternative separated flow phenomenological model by Yashima and Tanaka, and the classical attached flow theoretical model which ignores
separation. As can be seen the two phenomenological
models which account for separation are in much better
agreement with experiment than the attached flow
model. Moreover, as Chi has shown, the separated flow
model has proven to be an effective method for stall
flutter prediction in jet engines when attached flow
models fail.

446

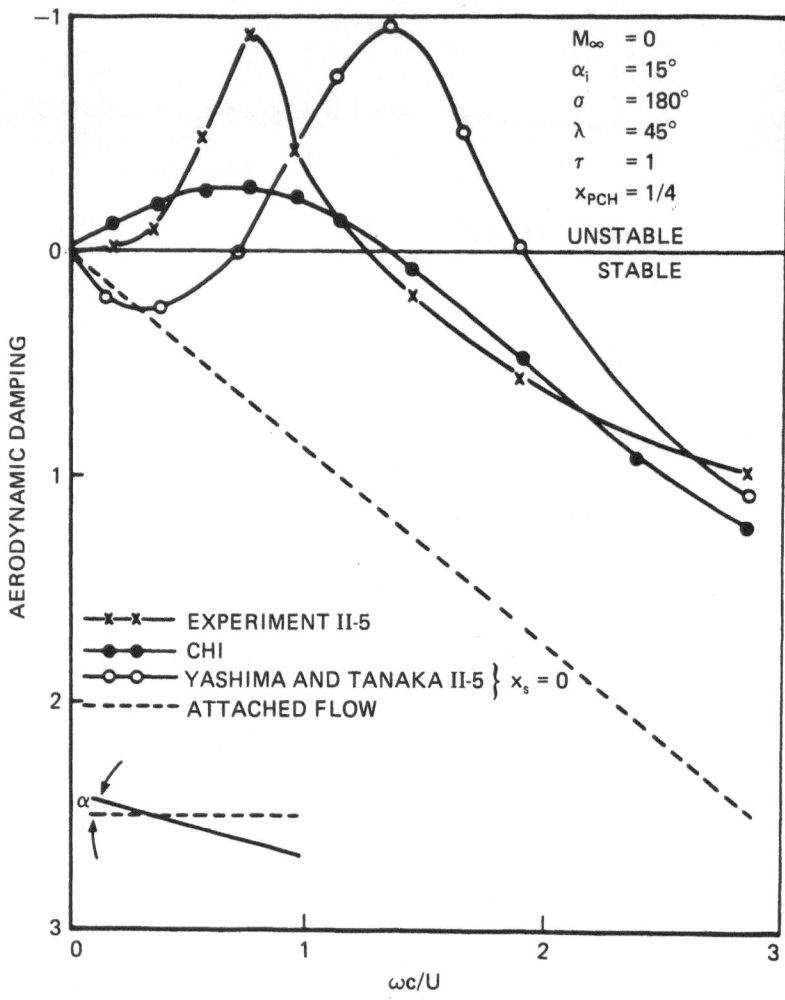

Fig. 2. Aerodynamic damping vs. reduced frequency.

It is clear that the phenomenological model requires as input the specification of the steady flow
separated region.   This might come from a steady flow
experiment, physical or numerical.   Alternatively the
region of separated flow may be treated as a variable
and sensitivity studies conducted.   That is, by varying the region of separated flow in the model, one can
determine the dependence of the aerodynamic forces on
the extent of the separated flow region.   To give an
example, consider Fig. 3, also taken from Chi.   Here

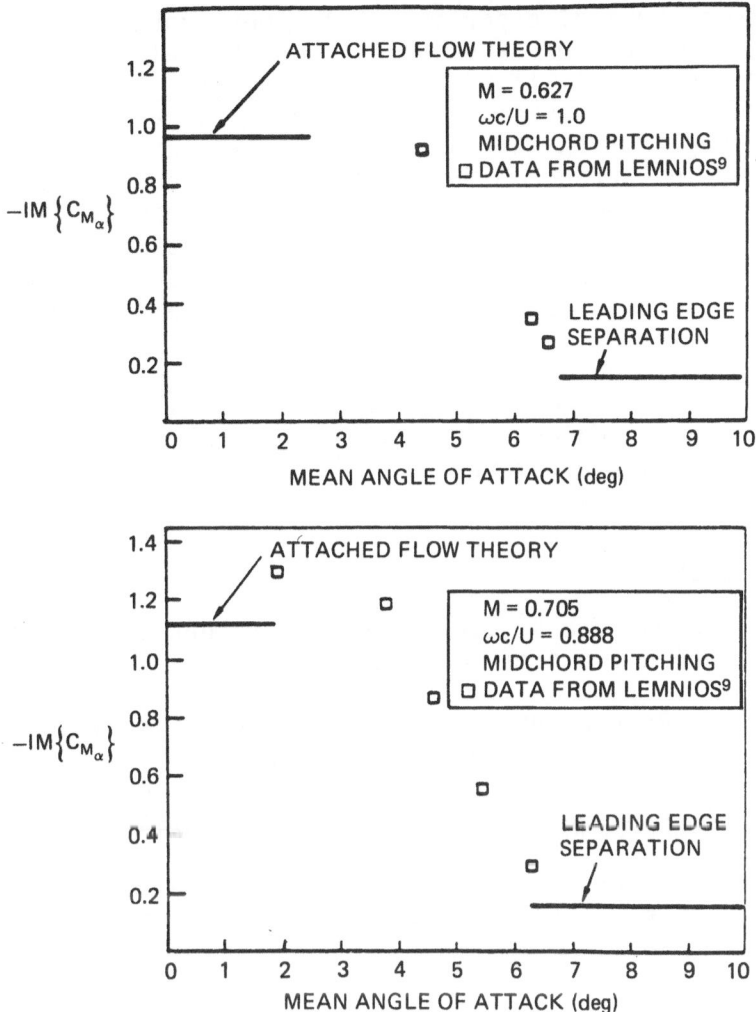

Fig. 3. Torsional aerodynamic damping for section
16(070)(023) of HS 18A20 propeller.

the unsteady aerodynamic pitching moment components
(real and imaginary, or in-phase and out-of-phase with
respect to the airfoil oscillation) are plotted versus
the static mean angle-of-attack of the airfoil, α_0.
In addition to being at a static mean angle-of-attack,
the airfoil also executes a small pitching oscillation
about the static angle of attack.   The symbols are

from a physical experiment. The horizontal lines near $\alpha_0 \equiv 0$ are for attached flow as calculated by Chi. The horizontal lines for large α_0 are for separated flow where separation is assumed to start at the leading edge of the airfoil. These two extreme assumptions of attached flow and leading edge separation provide good agreement with the experimental data at small and large α_0, respectively. By assuming various points of separation along the airfoil intermediate results would be obtained. Unfortunately the steady flow separation points were not measured as a function of static angle-of-attack in the experiments. However this example demonstrates the capability of the model for trend studies where the separation point itself is treated as a parameter.

Extensions

There are several extensions that can be made. We mention two here.

- At present, the model assumes a subsonic free stream and does not incorporate transonic flow effects such as shocks. The extension to transonic flow with shocks is possible.

- At present, the model is two-dimensional and the extension to three-dimensions is desirable.

The work of Tran-Petot

In the work of Chi, although a simplified representation of separated mean flow is used, determination of the unsteady flow field still requires the solution of a boundary value problem through a singular integral equation. In the Tran-Petot [10] model an even more sweeping assumption is made which bypasses the solution of a boundary value problem altogether. In essence (the actual details are somewhat more elaborate than the following description), it is postulated that the lift coefficient of the airfoil, C_L, is related to the motion of the airfoil, say its angle of attack, α, through an ordinary differential equation of the form:

$$\ddot{C}_L + 2\zeta_A \omega_A \dot{C}_L + \omega_A^2 C_L = A_1 \alpha + A_2 \alpha^2 + A_3 \alpha^3 + B_1 \dot{\alpha} \quad (1)$$

where $\cdot$ denotes a time derivative.

It must be emphasized that this equation at present is postulated, not derived from first principles, e.g., from the Navier-Stokes Equations. The form of equation, however, is motivated by physical considerations. On the left hand side it is assumed the fluid lift force has the characteristics of a damped harmonic oscillator because of the fluid's inertia, dissipation, and compliance. On the right hand side, the body motion is assumed to drive the fluid lift oscillator and the driving terms have nonlinear as well as linear components. One could include nonlinear terms in C_L on the left-hand side as well, but Tran-Petot found that an unnecessary complication in their work. It may be noted that in analogous, but independent, work in the field of civil and marine engineering such nonlinear terms in C_L have proven useful, see for example Ref. 11 and Chapter V. Finally, it is emphasized that the $\dot{\alpha}$ and $\dot{C}_L$ terms allow one to model phase shifts between C_L and α.

The challenge in the Tran-Petot or lift oscillator approach is to determine the several coefficients, ζ_A, ω_A, A_1, A_2, A_3, B_1. Once these are known it is a relatively simple matter to solve Equation (1) by analytical or numerical means in combination with the equations of motion of the airfoil. Tran-Petot suggest that ζ_A, ω_A, which can be physically interpreted as a characteristic damping and frequency of the fluid, are in general functions of α. They also presumably depend upon airfoil shape, Mach number and (hopefully weakly) Reynolds number. A_1, A_2, A_3 are assumed to be independent of α. For a given airfoil, Mach number, and Reynolds, ζ_A, etc., are determined from a (physical or numerical) experiment. Briefly the procedure is the following.

First a static experiment is undertaken; hence, Equation (1) simplifies to

$$C_{L_s} = \frac{A_1 \alpha_s + A_2 \alpha_s^2 + A_3 \alpha_s^3}{\omega_A^2} \qquad (2)$$

where the subscript s is added for emphasis to denote static or steady flow conditions. By curve-fitting the experimental data of C_{L_s} vs α_s, the coefficients A_1/ω_A^2, etc., can be determined.

Now the airfoil is oscillated about α_s, i.e.,

$$\alpha(t) = \alpha_s + \hat{\alpha}(t) \qquad (3)$$

To determine ζ_A, ω_A, B_1, it proves sufficient to consider $|\hat{\alpha}| \ll |\alpha_s|$, that is only (infinitesimal) small oscillations about various static angles of attack need be considered to determine all the model parameters.

This has two important consequences:

- One may write, in analogy to Equation (3),

$$C_L = C_{L_s} + \hat{C}_L \qquad (4)$$

  For $|\hat{\alpha}| \ll |\alpha_s|$, then $|\hat{C}_L| \ll |C_{Ls}|$. Moreover, one can derive a linear relationship between $\hat{C}_L$ and $\hat{\alpha}$ under these conditions which significantly simplifies the determination of ζ_A, etc.

- The second consequence is that once ζ_A, etc., are determined, then Equation (1) can be used to determine $C_L(t)$ from a given $\alpha(t)$ even when

the motion is large and transient (not just
small and harmonic in time). The remarkable
result is that when this is done Equation (1)
proves to be reasonably accurate for <u>large</u>
motions. That is, a phenomenological model
whose characteristics have been determined
from a static experiment and a small oscillation experiment is able to predict the behavior of C_L for large oscillations.   Clearly
there must be exceptions to this result.   The
question is when and under   what circumstances?

Extensions

There are several extensions that might be made.
We discuss some of these here.

- Perhaps most important would be the determination of the limits on the ability of the model
  to predict large oscillation results from a
  knowledge of small oscillation behavior. This
  requires further experimental investigations
  in coordination with the evaluation of the
  Tran-Petot model.

- An immediate simplification to the determination of model parameters would be to use step
  changes in α, rather than harmonic oscillations, when determining ζ_A, etc.   This also
  requires experimental verification, however.

- It would be highly desirable to require only a
  static experiment to determine ζ_A, etc.,
  rather than a dynamic one.   This may be done
  (for small oscillations) by using the approach
  of Chi, if in the static experiment one
  measures not only C_{L_s}, but also the separation
  point. Alternatively, the separation point
  could be treated as a parameter variable.

- Although the Tran-Petot approach was developed
  for two-dimensional flow fields, the method is

formally suitable for three-dimensional flows
if the requisite experimental data are available. The same comment applies to compressibility. Of course, the fact that it formally
applies does not mean it will really work in
practice and the effects of three-
dimensionality and compressibility remain open
equations.

Relationship to first principle models of fluid mechanics

Some effort has been made to relate such phenomenological models to first principles. See Ref. 12.
However, so far, progress in this direction has been
modest. Further work is to be encouraged. A place to
begin would be numerical experiments with finite difference Navier-Stokes codes to see if they support the
hypothesis that under some appropriate circumstances
large amplitude motion fluid forces can be predicted
from small amplitude results. It would also be of
interest to relate transient motion fluid forces to
simple harmonic motion forces. The insights gained
from phenomenological models may be used to suggest
important numerical experiments for finite difference
codes which are based upon the first principles of
fluid mechanics.

A candidate Navier-Stokes code is that developed
by Shamroth [19]. To quote from the conclusion of his
report,

"The procedure solves the full time-dependent,
compressible Navier-Stokes equations via an alternat-
ing direction implicit (ADI) method. The calculations
are performed on a highly stretched grid which places
the first grid point off the airfoil in the viscous
sublayer. No slip conditions are applied at solid
boundaries. The calculation has been run extensively
with a mixing length turbulence model. Although not
detailed in the present report, calculations have also
been run with a two equation model for an NACA 0012
airfoil at 6° incidence. The procedure is capable of

453

calculating steady solutions using a matrix preconditioning technique which allows rapid convergence over a Mach number range between virtually incompressible and transonic. Unsteady flows which require transient accuracy, can be made for M > 0.15.

The procedure has been used for several calculations including steady flows about an NACA 0012 airfoil at zero and modest incidence and flow about an NACA 4412 airfoil at modest to high incidence. In all cases excellent agreement between calculated and measured pressure distributions was shown. High incidence unsteady calculations were made for an NACA 0012 airfoil at 19° which showed good agreement with available data. Calculations of an airfoil in pitch below the stall angle also showed good agreement with data. Finally, a calculation was made for an airfoil in deep dynamic stall, 4° < < 20°. This represents a very difficult case which contains dynamic effects, large scale separation and multiple interacting shed vortices. Considering the difficulty of the case, the agreement between calculated and measured pressure distribution was very good."

Although Shamroth's model is sophisticated in its structure and requires an elaborate solution method, it is well to emphasize that even so a number of phenomenological elements are invoked including

- turbulence modeling, via the mixing length or turbulence energy approach

- a judicious use of artificial dissipation to suppress unphysical spatial oscillations

Thus, even at this level of modeling, the approach is fundamentally phenomenological.

REFERENCES

1. R. M. Chi and A. V. Srinivasan, "Some Recent Advances in the Understanding and Prediction of Turbomachine Subsonic Stall Flutter," ASME Paper No. 84-GT-151, 1984.

2. R. M. Chi, "Unsteady Aerodynamics in Stalled Cascade and Stall Flutter Prediction," ASME Paper No. 80/C2/Aero-1, 1980.

3. R. M. Chi, "Separated Flow Unsteady Aerodynamics for Propfan Applications," UTRC, Report R83-356891-1, 1983. Also AIAA Paper No. 84-0874; Journal of Aircraft, Vol. 22, pp. 956-964, 1985.

4. F. Dvorak and B. Maskew, "Prediction of Dynamic Stall Characteristics Using Advanced Non-linear Panel Methods," in AFOSR/FJSRL/U. Colorado Workshop on Unsteady Separated Flows, August 1983, published in 1984.

5. L. W. Carr and T. Cebeci, "Boundary Layers on Oscillating Airfoils," Third Symposium on Numerical and Physical Aspects of Aerodynamic Flows, Long Beach, CA, January 21-24, 1985.

6. A. P. Rothmayer and R. T. Davis, "Massive Separation and Dynamic Stall on a Cusped Trailing Edge Airfoil," Third Symposium on Numerical and Physical Aspects of Aerodynamic Flows, Long Beach, CA, January 21-24, 1985.

7. K. N. Ghia, G. A. Osswald, and U. Ghia, "Analysis of Two-Dimensional Incompressible Flow Past Airfoils Using Unsteady Navier-Stokes Equations," Third Symposium on Numerical and Physical Aspects of Aerodynamic Flows, Long Beach, CA, January 21-24, 1985.

8. F. T. Smith, "Theoretical Aspects of Steady and Unsteady Laminar Separation," AIAA Paper 84-1582, AIAA 17th Fluid Dynamics, Plasma Dynamics, and Lasers Conference, June 25-27, 1984, Snowmass, Colorado.

9. M. H. Williams, R. M. Chi, C. S. Ventres, and E. H. Dowell, "Effects of Inviscid Parallel Shear Flows on Steady and Unsteady Aerodynamics and Flutter," AIAA J., Vol. 15, pp. 1159-1166, 1977.

10. C. T. Tran and D. Petot, "Semi-empirical Model for the Dynamic Stall of Airfoils in View of the Application to the Calculation of Responses of a Helicopter Blade in Forward Flight," Vertica, Vol. 5, pp. 35-53, 1981.

11. E. H. Dowell, "Nonlinear Oscillator Models in Bluff Body Aeroelasticity," J. Sound Vibration, Vol. 75, pp. 251-264, 1981.

12. W. D. Iwan and R. D. Blevins, "A Model for the Vortex-induced Oscillation of Structures," J. Applied Mechanics, Vol. 41, pp. 581-585, 1974.

13. E. H. Dowell, "Generalized Aerodynamic Forces on a Flexible Plate Undergoing Transient Motion in a Shear Flow with an Application to Panel Flutter," AIAA J., Vol. 9, pp. 834-841, 1971.

14. J. P. Rogers, "Application of an Analytic Stall Model to Dynamic Analysis of Rotor Blades," M.S.E. Thesis (D. A. Peters, Thesis Advisor), Washington, U., 1982.

15. R. K. Mehra and J. V. Carroll, "Bifurcation Analysis of Aircraft High Angle-of-Attack Flight Dynamics," in New Approaches to Nonlinear Problems in Dynamics, Edited by P. J. Holmes, SIAM, 1980.

16. S. Yashima and H. Tanaka, "Torsional Flutter in Stalled Cascade," ASME Paper No. 77-GT-72, 1977.

17. W. J. McCroskey, "Unsteady Airfoils," in Annual Review of Fluid Mechanics, Vol. 14, pp. 285-311, 1982.

18. W. J. McCroskey, "The Phenomenon of Dynamic Stall," NASA TM-81264, 1981.

19. S. J. Shamroth, "Calculation of Steady and Unsteady Airfoil Flow Fields Via the Navier-Stokes Equations," NASA CR 3899, August 1985.